THE THEORY OF COSMIC GRAINS

ASTROPHYSICS AND SPACE SCIENCE LIBRARY

A SERIES OF BOOKS ON THE RECENT DEVELOPMENTS
OF SPACE SCIENCE AND OF GENERAL GEOPHYSICS AND ASTROPHYSICS
PUBLISHED IN CONNECTION WITH THE JOURNAL
SPACE SCIENCE REVIEWS

VOLUME 168
CURRENT RESEARCH

THE THEORY OF COSMIC GRAINS

by

F. HOYLE
Bournemouth, United Kingdom

and

N. C. WICKRAMASINGHE
*School of Mathematics, University of Wales,
Cardiff, United Kingdom*

KLUWER ACADEMIC PUBLISHERS
DORDRECHT / BOSTON / LONDON

Library of Congress Cataloging-in-Publication Data

Hoyle, Fred, Sir.
The theory of cosmic grains / authored by F. Hoyle and N.C. Wickramasinghe.
p. cm. -- (Astrophysics and space science library ; v. 168)
ISBN 0-7923-1189-2 (hb : alk. paper)
1. Cosmic grains. 2. Astrophysics. 3. Cosmochemistry.
I. Wickramasinghe, N.C. (Nalin Chandra), 1939- . II. Title.
III. Series.
QB791.2.H69 1991
523.1'125--dc20 91-11105

ISBN 0-7923-1189-2

Published by Kluwer Academic Publishers,
P.O. Box 17, 3300 AA Dordrecht, The Netherlands.

Kluwer Academic Publishers incorporates
the publishing programmes of
D. Reidel, Martinus Nijhoff, Dr W. Junk and MTP Press.

Sold and distributed in the U.S.A. and Canada
by Kluwer Academic Publishers,
101 Philip Drive, Norwell, MA 02061, U.S.A.

In all other countries, sold and distributed
by Kluwer Academic Publishers Group,
P.O. Box 322, 3300 AH Dordrecht, The Netherlands.

Printed on acid-free paper

Printed in the Netherlands

CONTENTS

ACKNOWLEDGEMENT

We are indebted to Mr. G.H. Weston for his unstinting support of this work over many years.

PREFACE

Light scattering and absorption by small homogeneous particles can be worked–out exactly for spheres and infinite cylinders. Homogeneous particles of irregular shapes, when averaged with respect to rotation, have effects that can in general be well–approximated by reference to results for these two idealised cases. Likewise, small inhomogeneous particles have effects similar to homogeneous particles of the same average refractive index. Thus most problems can be solved to a satisfactory approximation by reference to the exact solutions for spheres and cylinders, which are fully stated here in the early part of the book.

The sum of scattering and absorption, the extinction, is too large to be explained by inorganic materials, provided element abundances in the interstellar medium are not appreciably greater than solar, H_2O and NH_3 being essentially excluded in the general medium, otherwise very strong absorptions near 3μm would be observed which they are not. A well–marked extinction maximum in the ultraviolet near 2200Å has also not been explained satisfactorily by inorganic materials. Accurately formed graphite spheres with radii close to 0.02μm could conceivably provide an explanation of this ultraviolet feature but no convincing laboratory preparation of such spheres has ever been achieved.

Certain metal oxides under special conditions show absorptions at 2200Å but only weakly. On the other hand, suites of organic materials such as are actually found in nature have integrated extinction spectra remarkably similar in the ultraviolet to interstellar grains. Indeed on a wide range of counts, organic grains fit data in both visual and ultraviolet wavelengths ranges far better than the inorganic alternatives, data that is both extensive and complex, and which agreement is again confirmed extensively in the infrared.

Organic materials are far more efficiently synthesized biologically than abiologically. The antecedents of modern astronomers existed precariously as minor court officials who were strongly motivated to keep their atronomical studies and discoveries rigorously separated from the world of everyday events. This long history prejudices astronomers today against ideas that might suggest their work could be of practical relevance, or even of great importance in the circumstances that it has to do with the origin of life. Our aim is to take the reader through all the details which in the end support such a view, in the hope that at least some may be able to shake off an intellectually inhibiting attitude from the past.

Life on the Earth is almost as old as the oldest known rocks which are thought to have been too strongly heated for fossil evidence of life to have been retained at the beginning. On this evidence alone, therefore, life could as well be older than the Earth as younger. To the unbiased, the one should be just as much a fit subject of study as the other, taking the exploring mind into the astronomical field considered in this book.

Cardiff, June 1991

F. Hoyle
N.C. Wickramasinghe

1. Introduction

1.1. EARLY IDEAS

Amongst the most startling pictures of the night sky are those that involve interstellar dust clouds. They show up as dark patches and striations against more or less uniform starfields, or as bright nebulosity around individual stars or groups of stars. The Trifid Nebula seen in Fig. 1.1 is an example of a region of the Milky Way where dark lanes are superposed on an irregular–shaped complex of clouds of hot emitting gas and dust. These patches are not recesses through the nebula as they were once thought, but rather are caused by the very effective obscuration of background optical radiation by much cooler clouds containing small solid particles. The term 'interstellar' grains was given to these particles by Lyman Spitzer Jr., one of the great pioneers of interstellar astronomy.

Investigations relating to the nature of interstellar grains have been going on since the early years of the present century. These studies have gathered momentum over the years with approximately 450 research papers appearing annually at the present time that are in some way connected with the properties of grains. Nevertheless, the problem of interstellar grain composition has not yet been finally resolved. It remains one of the most fascinating unsolved problems of modern astronomy.

As a matter of historical interest we note that visual recordings of "dark nebulae" preceded the advent of photography. In 1784 William Herschel catalogued visual sitings of thousands of nebulae – dark nebulae (which we now know to be dust clouds) and bright nebulae, many of which later turned out to be external galaxies. Herschel noted that the latter tended to avoid the plane of the Milky Way. This observation was interpreted by Herschel and most of his successors until about 1910 to imply that the nebulae were a truly galactic population, thus defining a plane of avoidance related to the Galaxy itself. The misconception implied here led to a failure to recognise both the existence of external galaxies and also the presence of interstellar dust in the Galaxy.

Since we are now so accustomed to take the presence of interstellar dust for granted, it is worth reflecting briefly on the early difficulties that were encountered in recognising that any interstellar obscuration could be present at all. Even as late as 1927, after Barnard (1919, 1927) had published a comprehensive atlas of dark clouds, their true nature remained an enigma. It was very much an open question as to whether the markings (seen for instance in Fig. 1.1) were clouds of opaque matter in front of the stars or, as was often suggested, they actually represented holes through the distribution of stars.

Fig. 1.1 *Photograph of the Trifid Nebula showing conspicuous dust lanes*

One of the earliest attempts to resolve this problem was made by F.G.W. Struve as early as 1847. By counting the surface density of stars up to varying limiting values of the apparent magnitude in different regions of the sky, he argued that there may be an extinction of order 1 mag/kpc at the visual wavelength. This conclusion however was considered by later workers to be insecure on account of uncertainties in the stellar density distribution.

1.2. TRUMPLER'S METHOD OF ESTIMATING INTERSTELLAR EXTINCTION

One of the most convincing demonstrations of the existence of a general interstellar extinction was made by Trumpler (1930). In addition to showing that extinction must be present, he was also able to estimate its average amount.

The crucial step in proving the existence of interstellar extinction was to devise a distance scale that was independent of photometric measurements. Trumpler's method was based on measurements of the angular diameters of open galactic clusters. These are groups, each containing of the order of 10^2 to 10^3 individual stars, distributed more or less uniformly in regions close to the galactic plane. The Pleiades and Praesepe are typical examples of such open galactic clusters.

If m and M are the apparent and absolute photographic magnitudes of a particular star in a cluster, the apparent distance r' in parsecs (in the absence of any extinction) is given by

$$5 \log r' = m - M + 5 \tag{1.1}$$

Values of m were determined by measurement, while values of M were obtained from so–called spectroscopic parallaxes. By taking an average value of r′ for several stars of different spectral types within a cluster, a reliable estimate of the apparent cluster distance could be obtained. Further, if α is the angular diameter of the cluster, its apparent linear diameter is given by

$$D' = \alpha r' \tag{1.2}$$

Although there appeared to be little structual similarity among the galactic clusters taken in their entirety, Trumpler managed to arrange them in twelve classes having regard to their degrees of compactness and their richness in stars. Within each category it seemed legitimate (extinction apart) to assume that there was little or no spread in a given physical property, for example, in the linear diameter.

Equations (1.1) and (1.2) may be used to derive D′ from measured values of α. Within a given class of galactic cluster in the absence of extinction there should be no systematic correlation of D′ with r′. What Trumpler in fact found was that the values of D′ increased systematically with increasing r′, indicating that the premise of a perfectly transparent interstellar medium was in error.

Instead of equation (1.1), he therefore took the distance of the cluster r to be given by

$$5 \log r = m - M + 5 - ar \tag{1.3}$$

and the constant a was chosen so as to cancel the systematic increase of linear diameter with distance. The appropriate value of a was found to be

$$a = 0.69 \text{ mag/kpc}. \tag{1.4}$$

A closely similar value was found by Bottlinger and Schnellet (1930) by considering the dispersion of Cepheids perpendicular to the galactic plane, and by van de Kamp (1932) who considered the concentration of extragalactic nebulae towards the pole of our own Galaxy.

More recent estimates of the visual extinction coefficient of dust in the solar neighbourhood have yielded a value higher than that derived by Trumpler. The generally accepted modern value, based on the observations of stars within a few kiloparsecs of the sun in lines of sight along the galactic plane, is ~ 1.8 mag/kpc at $\lambda^{-1} = 1.8\mu m^{-1}$. This is the first crucial datum that relates to the nature and composition of interstellar dust grains.

1.3. THE FIRST COLOUR MEASUREMENTS

One of the most crucial pieces of evidence relating to the dust came with the advent of three–colour photometry. Pioneering observational work in this field was carried out by Stebbins, Huffer, and Whitford (1934, 1939) who made careful measurements of UBV magnitudes of some 1332 B stars. Stars which by their spectral features were expected to have the same intrinsic surface temperatures, and therefore the same colours, were found to possess widely different B–V colours. The existence of very red B stars, stars such as P Cygni and 55 Cygni, whose spectral lines indicated temperatures ~ 20,000° K but whose colours indicated temperatures ~ 6000° K were at first considered difficult to understand.

The colour measurements of Stebbins, Huffer, and Whitford were, however, consistent with the so–far neglected possibility that starlight was reddened owing to interstellar extinction. By comparing heavily reddened and relatively unreddened B stars, and assuming that the stars compared possessed the same intrinsic energy distributions, it was demonstrated that the extinction in magnitudes varied approximately as λ^{-1} over the three wavelengths considered. This is a second crucial datum of relevance to the later chapters of this book.

1.4. THE OORT LIMIT

Not long after the existence of interstellar extinction became established, and its average amount estimated, it also became possible to set an upper limit to the total mass of the interstellar medium. Oort (1932), from an analysis of stellar motions at right–angles to the galactic plane, was able to deduce the normal component of the gravitational field g. The method involved comparing the measured z components of the acceleration of stars with the normal attraction expected for a uniform distribution of gravitating matter occupying an infinite plane parallel slab of thickness 2H. If z is the distance of a point from the mid–plane of such a galaxy model, the gravitational attraction in the z direction is easily shown to be

$$\begin{aligned} F &= 4\pi\, G\rho z\ , & |z| &< H\ , \\ &= 4\pi\, g\rho H\ , & |z| &\geq H\ , \end{aligned} \qquad (1.5)$$

where ρ is the total density of gravitating matter– stars, gas and dust together. The results of Oort's analysis of stellar motions led to the curve in Fig. 1.2.

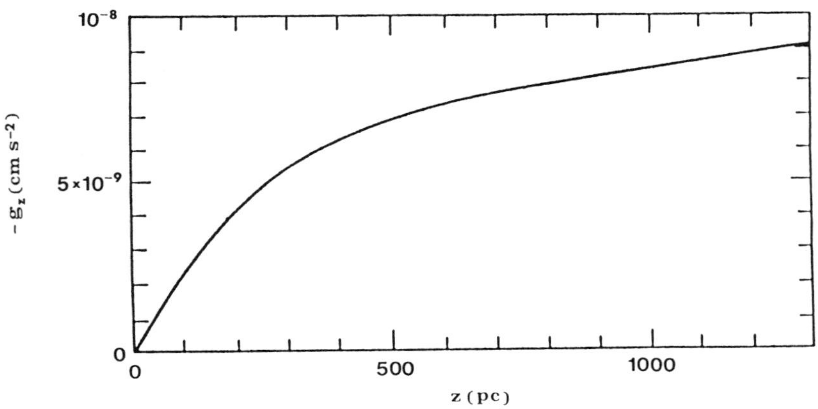

Fig. 1.2 *Normal component of the gravitational field at the height h above the galactic plane. (Adapted from Oort, 1932)*

From this curve we can deduce the effective half–width of the Galaxy to be $H \cong 225$ pc, with a value of the normal acceleration at $z = H$ of about 4×10^{-9} cm s^{-2}. Thus if ρ_s and ρ_i denote the stellar and interstellar components of the density at $z \lesssim H$, we have

$$4\pi\, GH(\rho_s + \rho_i) \cong 4.10^{-9} \ . \tag{1.6}$$

From (1.6), following Oort (1932), we get

$$\rho_s + \rho_i \cong 6.9 \times 10^{-24}\ \text{gm/cm}^3 \ . \tag{1.7}$$

[From a later study, Oort (1952) obtained $\rho_s + \rho_i \cong 6 \times 10^{-24}$ gm/cm^3]. Additionally from star counts Oort (1932) also estimated

$$\rho_s \cong 3 \times 10^{-24}\ \text{gm/cm}^3 \ , \tag{1.8}$$

leaving the balance of $\sim 3 \times 10^{-24}$ gm/cm^3 as an estimate of the density of interstellar matter near the Sun, a third crucial datum.

Several subsequent direct determinations of density for individual components of the interstellar gas have generally been found to be consistent with this early estimate. The neutral atomic hydrogen in the Galaxy has been sampled for different regions using 21 cm measurements and also Lyman–α absorption measurements. Also to be included is the contribution from molecular H_2 for which ultraviolet absorbtion lines have been observed. It was pointed out by P.M. Solomon and one of the present authors that the path length needed to make H_2 self–shielding in the dissociative Lyman band decreases as the square of the hydrogen density, so that the transition from HI clouds to H_2 clouds is exceedingly sharp, occurring whenever the total hydrogen density exceeds ~ 100 cm^{-3} (Solomon and Wickramasinghe, 1969). Taking account of all the observations relating to hydrogen, it is quite remarkable that the total gas density deduced from modern data does not depart from equation (1.8) to any really significant degree.

1.5. DATA RELATING TO INTERSTELLAR CLOUDS

Interstellar clouds have irregular shapes and overall dimensions ranging from 0.01 – 50 pc. The most frequent size of a cloud is ~ 5 pc. Typical extinction through such a cloud is ~ 0.5 mag at the photographic wavelength. The smallest sizes of $\sim .01$ pc are appropriate to the so–called 'Bok globules' which show up against emission line nebulae and also to the compact HII regions and OH/IR sources. They probably have extinctions of $\gtrsim 10$ mag at optical wavelengths and are thought to represent early stages of star formation.

Larger accumulations of interstellar material have recently been recognised in the form of the so–called giant molecular clouds, which showed up first in maps of the Galaxy in radio and millimeter lines of the molecules H_2CO and CO. A typical dimension of such a cloud is ~ 50 pc with several magnitudes of visual extinction through it. The masses of molecular clouds generally range from $\sim 10^4\ M_\odot$ to $10^6\ M_\odot$ with a median near $5 \times 10^5\ M_\odot$.

As we have already noted, the distribution of dust in the Galaxy is highly non–uniform on a variety of distance scales. However, to a zero–order approximation, one might characterise the dust layer by some average density ρ_g which is a function of the height z above or below the galactic plane. If $2\tau_0$ is the optical thickness of this layer at some assigned wavelength, the extinction as seen from the galactic plane at this wavelength of objects (e.g. globular clusters, RR Lyrae stars and external galaxies) situated outside the layer depends on galactic latitude b as

$$\tau = \tau_0 \operatorname{cosec} |b| \tag{1.9}$$

Measurements of the difference of the extinction at standard B and V wavelengths, E_{B-V}, should of course follow the same law,

$$E_{B-V} = (E_{B-V})_0 \operatorname{cosec} |b| \tag{1.10}$$

and it is possible by means of observations to determine $(E_{B-V})_0$, the so–called colour excess of objects towards the poles (at $b = 90°$). The best stellar observations indicate a value close to $(E_{B-V})_0 = 0.06$, while at $b = 0$ the measured value E_{B-V} is 0.6 kpc^{-1}.

To a reasonable approximation we can describe the mean density of dust at height z by an equation of the form

$$\rho_g(z) = \rho_g(o)\, e^{-z/\beta} , \tag{1.11}$$

where the half–width β depends on the velocity dispersion of the dust clouds $\overline{v_z^2}$ perpendicular to the galactic plane and on the gravitational potential. The value of β indicated by observation is about 60 km, significantly less than Oort's value of 112.5 km, which was broadened by the stellar contribution. This smaller value of β was obtained from Doppler measurements of $\overline{v_z^2}$, which give about 7 km s^{-1} (Kaplan and Pikelner, 1970).

1.6. CORRELATION BETWEEN GAS AND DUST CLOUDS

More or less concurrently with the general acceptance of the existence of interstellar dust clouds in the 1930's there followed a series of observations which led to the discovery of interstellar gas clouds. The existence of interstellar gas was proposed as early as 1904, by Hartmann on the basis of his observations of the H and K–lines of Ca II in the spectrum of a spectroscopic binary δ Ori. The circumstance that these line strengths are independent of the phase of occultation led quickly to a recognition of their interstellar origin. Data on the H and K–lines of Ca II and the lines of Na I in a large number of high–temperature stars confirmed this belief. The appearance of multiple Doppler shifts led to the concept of interstellar clouds, and an analysis of the strengths of these bands in relation to amounts of extinction led to the early belief that gas and dust clouds are probably identical objects.

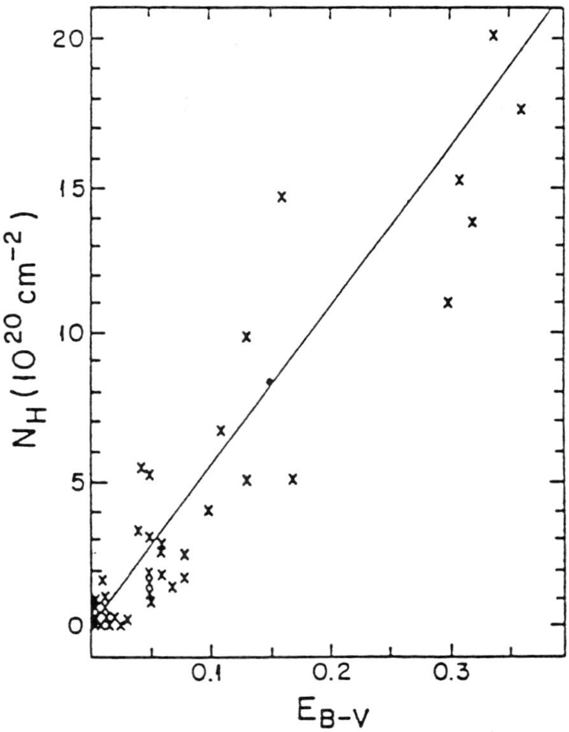

Fig. 1.3 *Plot of visual extinction against column density of neutral hydrogen and molecular hydrogen (from Spitzer, 1978)*

A positive correlation was considered as proved by Lilley (1955) on the basis of comparisons of the optical extinction data and the 21–cm radio data in the directions of the Orion, Taurus and Perseus dust clouds (see Fig. 1.3). McGee et al. (1963) later obtained similar results for the whole sky observable from Australia. However, the situation was not as clear–cut in detail as it first seemed. Wesselius and Sancisi (1971), employing Kendall's rank test, showed that a significant positive correlation is not found in all the latitude belts considered in $|b^{II}| < 40°$, while Heiles (1969) showed that in the interiors of individual dark clouds the HI and dust are negatively correlated, indicating H_2 formation at the surfaces of interstellar grains.

A wealth of new data concerning the distributions of both gas and dust has come more recently from three different sources: (1) studies of far–infrared radiation (IRAS satellite data) at 100μm, (2) studies of millimeter–wave observations of CO, and (3) ultraviolet line observations of interstellar atoms at different ionization levels. We defer a fuller discussion of (3) to a later chapter, but merely state here that the gas phase densities of some heavy elements, e.g. Fe, Si, Al, Ca, appear to be significantly anticorrelated with the total gas density, possibly indicating their condensation on grain surfaces.

The Infrared Astronomy Satellite (IRAS) has mapped the sky through four broad filters with effective wavelengths at 12, 25, 60 and 100μm. An almost complete survey of the celestial sphere has revealed a very large number of point sources of radiation (Mira Stars and OH/IR stars) at the shorter wavelengths, and a distribution of dust clouds of various sizes and densities radiating mainly at 60 and 100μm. A smooth component of the 100μm emission is found to follow the same cosec $|b|$ law as for E_{B-V} given in equation (1.10) (Hauser et al., 1984). The 100μm radiation arises from thermal emission by 15° K dust grains and is therefore a good tracer of normal HI dust clouds. The far–infrared luminosity of clouds per unit cloud mass is found to be remarkable independent of their mass, with an average value of ~ 10L⊙/M⊙ (Solomon and Rivolo, 1987). Superposed upon this general 100μm background radiation are diffuse bright patches ("cirrus clouds") seen at high galactic latitudes (Low et al., 1984). The "cirrus" clouds range in size from ~ 10′ to ~ 2′, the latter being the resolution limit of the IRAS survey. Fig. 1.4 gives the IRAS sky survey at λ ~ 100μm, showing the cloud structure of the interstellar medium.

Fig. 1.4 IRAS map of the Galaxy at 100μm, showing cloud structure of ISM

The millimeter–wave fundamental rotational transition ($J = 1 \rightarrow 0$) of CO at $\lambda = 2.6$ mm has provided a most valuable tool for mapping interstellar clouds (in H_2) because it is collisionally excited by H_2 molecules, even in clouds of very low kinetic temperature (Solomon and Rivolo, 1987). CO observations have revealed an axisymmetric ring of molecular hydrogen extending over the region $4 < R < 8$ kpc with a density peak occurring at $R = 5$ kpc. The total mass of H_2 associated with this ring is ~ $1 - 3 \times 10^9$ M⊙ , possibly arising from clouds with masses ~ 10^4 M⊙ . Fig. 1.5 shows the axisymmetric distribution of H in this ring.

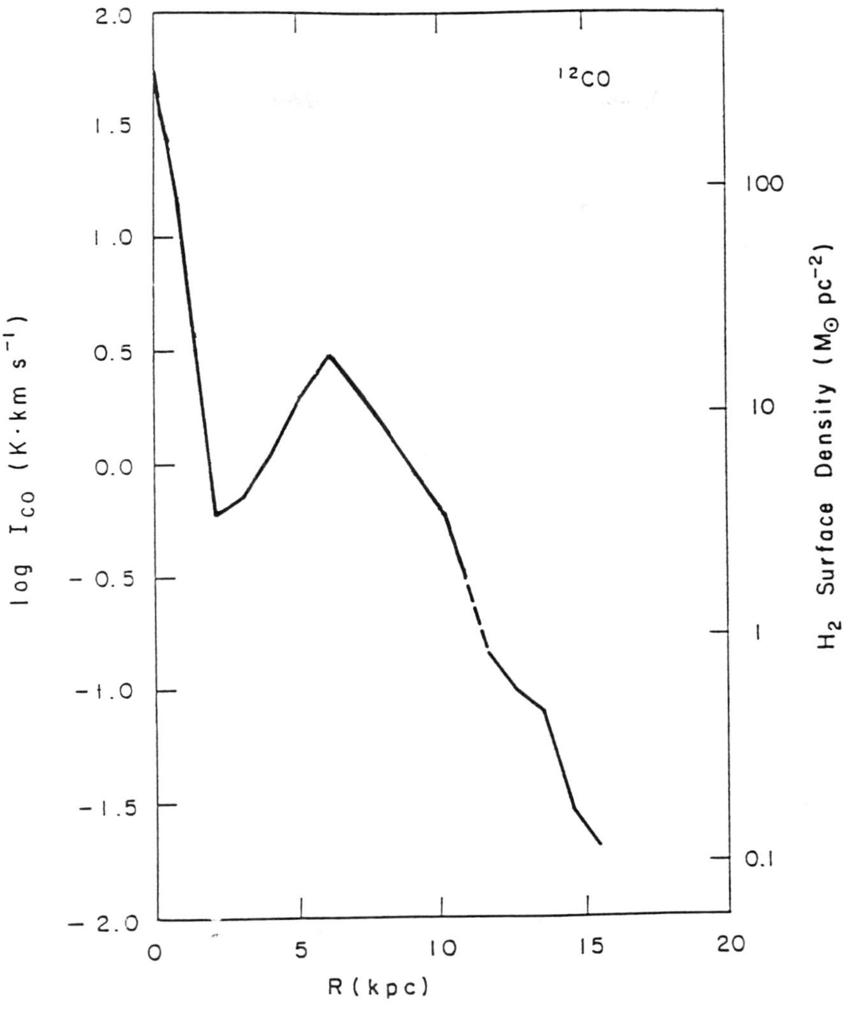

Fig. 1.5 *Axisymmetric distribution of CO and H_2 (after Solomon and Rivole, 1987)*

1.7. COMPOSITION OF GRAINS

A major problem for the early investigators was to isolate a particular absorber/scatterer responsible for interstellar extinction. It was necessary to explain an approximately λ^{-1} variation of the visual extinction in magnitudes, as well as a total extinction at $\lambda = \lambda_V$ amounting to 1–2 mag/kpc in directions close to the galactic plane. A further constraint was that the mass density of the absorbing/scattering agent could not be too high. As an extreme upper limit, the overall density of this material had to be less than $\sim 3 \times 10^{-24}$ g cm^{-3}, the density of the entire interstellar medium set by equation (1.8).

Several possibilities were considered and dismissed quite quickly, either on the grounds that they required excessive mass densities, and/or that they could not readily reproduce an approximately λ^{-1} reddening law. The properties of some of these early candidates for interstellar grains are summarised in Table 1.1 below.

TABLE 1.1

Type of Particle	Extinction Law	Density Required to Produce 1 mag/kpc
Free electrons	λ^{0}	300 cm^{-3}
Atoms, or molecules (Rayleigh scattering)	λ^{-4}	10^{4} cm^{-3}
Nonconducting particles, small compared with wavelength	λ^{-4}	$< 10^{-26}$ gm/cm^{3}
Solid particles, large compared with wavelength	λ^{0}	$> 10^{-25}$ gm/cm^{3}

The exclusions implied in Table 1.1 did not seem to leave many obvious possibilities. Small metallic particles with sizes of the order of 0.01 micron were among the first to be seriously considered. For such particles, Schalen (1939) was able to obtain a reasonable match with the λ^{-1} extinction law. The mass density of particles necessary to produce an extinction of 1 mag/kpc was of the order of 10^{-26} gm/cm^{3}, nearly one percent of the density of the whole interstellar medium given by the Oort limit in equation (1.8). A metallic composition for the particles seemed plausible at the time for other reasons which have since disappeared. The profusion of iron lines in the spectra of dwarf stars made it seem in the 1920's as if iron was as common an element cosmically as hydrogen.

Lindblad (1935) made the first significant attack on the grain problem from the standpoint of their formation. Arguing for a spatial correlation between gas and dust clouds, he put forward the hypothesis of grain growth by condensation of interstellar gas. This line of thought was pursued further by Oort and van de Hulst (1946), and subsequently by many other investigators.

However, there has been no general agreement amongst astronomers on either the composition or the sizes of grains that may result from such a condensation process. Dielectric grains, consisting mainly of H_2O (ice) were proposed by van de Hulst (1946, 1949), particles with sizes of the general order of the wavelength of visible light. Large molecules with unfilled electronic shells have been proposed by Platt (1956) as being a possible end–product of an interstellar condensation process. This latter proposal has been revived in modern times in connection with discussions relating to the presence of aromatic molecules in interstellar space.

As an alternative to an interstellar condensation process, the present authors were the first to introduce the idea of grain formation in the atmospheres of cool giant stars. We discussed the possibility of graphite particles forming in the photospheres of N–type carbon stars (Hoyle and Wickramasinghe, 1962) and later extended the same ideas to include mixtures of refractory particles forming in mass flows from oxygen–rich giant stars. We also discussed grain formation in the ejecta of supernovae and in explosions of supermassive stars (Hoyle and Wickramasinghe, 1968, 1969, 1970). More recently, we introduced the idea of grains composed of complex organic polymers (Wickramasinghe, 1974, 1975) and of biological material (Hoyle and Wickramasinghe, 1977; Hoyle et al., 1982).

The composition of interstellar grains has been a subject of extensive discussion for over 40 years but no final solution has yet emerged. In chronological sequence the following models have been proposed at different times:

(1) Iron grains, 1939–1946
(2) Ice grains, 1946–1962
(3) Graphite grains; composite core mantle grains, 1962–1970
(4) Refractory grain mixtures – silicates, iron, graphite, 1970–
(5) Organic polymers, 1974–
(6) Biological grains, 1979–

The arguments advanced in favour of these various models will occupy a major part of this book.

TABLE 1.2

Solar Abundance Ratios

Element	AW	Rel. Abundance
H	1	$3{\cdot}18 \times 10^{10}$
O	16	$2{\cdot}14 \times 10^{7}$
C	12	$1{\cdot}18 \times 10^{7}$
N	14	$3{\cdot}64 \times 10^{6}$
Mg	24	$1{\cdot}06 \times 10^{6}$
Si	28	$1{\cdot}0 \times 10^{6}$
Fe	56	$8{\cdot}3 \times 10^{5}$
Al	27	$8{\cdot}51 \times 10^{4}$

We can readily obtain upper limits to the mass densities of the various types of grain materials proposed so far by a simple argument. It is generally believed that the nuclear composition of the outer layers of a star is closely similar to that of the interstellar cloud from which it condensed. This is true in particular for the case of the Sun. Table 1.2 shows a recent compilation of the relative abundances in the Sun (Cameron, 1970).

Table 1.2, together with $\rho_H \cong 3 \times 10^{-24}$ g cm^{-3} (equation 1.8) leads to upper limits of mass densities for various types of grain, as set out in Table 1.3. This is further crucial data.

TABLE 1.3

Upper Limits to Grain Densities

Species	Max. density (g cm^{-3})
H_2O (ice)	$3{\cdot}6 \times 10^{-26}$
C	$1{\cdot}33 \times 10^{-26}$
Fe	$0{\cdot}44 \times 10^{-26}$
$MgSiO_3$	$0{\cdot}94 \times 10^{-26}$
Al_2O_3	$0{\cdot}04 \times 10^{-26}$
SiO_2	$0{\cdot}57 \times 10^{-26}$
Organic polymers	$3{\cdot}34 \times 10^{-26}$

References

Barnard, E.E., 1919, *Astrophys. J.*, **49**, 1 and 360.

Barnard, E.E., 1927, *'Atlas of the Selected Regions of the Milky Way'* (Carneg. Inst. Washington).

Cameron, A.G.W., 1970, *Space Science Reviews*, **15**, 121.

Hartmann, J., 1904, *Astrophys. J.*, **19**, 268.

Hauser, M.G., Gillett, F.C., Low, F.J. et al., 1984, *Astrophys. J. Lett.*, **278**, L15.

Heiles, C., 1967, *Astrophys. J.*, **148**, 299.

Heiles, C., 1969, *Astrophys. J.*, **156**, 493.

Hoyle, F. and Wickramasinghe, N.C., 1962, *M.N.R.A.S.*, **124**, 417.

Hoyle, F. and Wickramasinghe, N.C., 1968, *Nature*, **218**, 1126.

Hoyle, F. and Wickramasinghe, N.C., 1969, *Nature*, **223**, 459.

Hoyle, F. and Wickramasinghe, N.C., 1970, *Nature*, **266**, 62.

Hoyle, F. and Wickramasinghe, N.C., 1977, *Nature*, **268**, 610.

Hoyle, F., Wickramasinghe, N.C., Al–Mufti, S., Olavesen, A.H. and Wickramasinghe, D.T., 1982, *Astrophys. Sp. Sci.*, **83**, 405.

Kaplan, S.A. and Pikelner, S.B., 1970, *The Interstellar Medium* (Harvard Univ. Press).

Lilley, A.E., 1955, *Astrophys. J.*, **121**, 599.

Lindblad, B., 1935, *Nature*, **135**, 133.

McGee, R.X., Murray, J.D. and Milton, J.A., 1963, *Australian J. Phys.*, **16**, 136.

Oort, J.H., 1932, *B.A.N.*, No. 238.

Oort, J.H., 1952, *Astrophys. J.*, **116**, 233.

Oort, J.H. and van de Hulst, H.C., 1946, *B.A.N.*, No. 376.

Platt, J.R., 1956, *Astrophys. J.*, **123**, 486.

Stebbins, J., Huffer, C.H., and Whitford, A.E., 1934, *Publ. Washburn Obs.*, **15**, Part V.

Stebbins, J., Huffer, C.H., and Whitford, A.E., 1939, *Astrophys. J.*, **90**, 209.

Solomon, P.M. and Rivolo, A.R., 1978, in G. Gilmore and B. Carswell (eds.) *The Galaxy*, D. Reidel.

Solomon, P.M. and Wickramasinghe, N.C., 1969, *Astrophys. J.*, **158**, 449.

Struve, F.G.W., 1847, *'Etudes d'Astronomie Stellaire'*.

van de Hulst, H.C., 1946 and 1949, *Rech. Astron. Obs. Utrecht*, XI, Parts 1 and 2.

Wesselius, P.R. and Sancisi, R., 1971, *Astron. and Astrophys.*, **11**, 246.

Wickramasinghe, N.C., 1974, *Nature*, **252**, 462.

Wickramasinghe, N.C., 1975, *Mon. Not. Roy. Astr. Soc.*, **170**, P11.

2. Electromagnetic Properties of Small Particles

Because interstellar dust, or at least a major fraction of it, is in the form of solid particles with dimensions comparible to or less than optical wavelengths, an understanding of the detailed electromagnetic interactions of such particles is evidently required. In this chapter we consider the behaviour of particles of regular shapes, spheres, cylinders, as well as core–mantle concentric spheres. Spherical particles are considered to provide a satisfactory representation of randomly oriented irregular–shaped interstellar grains, whereas the properties of infinite cylinders are taken to represent the polarising behaviour of elongated partially aligned grains. Core–mantle spheres provide a model for grains which have acquired mantles of materials other than the cores. If the core is replaced by a vacuum cavity we also have a representation for hollow interstellar grains.

We do not attempt a comprehensive treatment of Light Scattering Theory in this chapter but more simply offer a broad overview of those topics and formulae which are of immediate astronomical interest. Readers are referred for details on various aspects of the subject matter of this chapter to van de Hulst (1957), Wickramasinghe (1973) and Bohren and Huffman (1983). These considerations necessarily provide the basis for understanding the properties of interstellar grains.

2.1. HOMOGENEOUS SPHERICAL PARTICLES

The problem of light scattering by a spherical particle is essentially one in classical electromagnetic theory and involves a solution of Maxwell's equations with the appropriate boundary conditions on the sphere. Formal solutions were worked out independently by Mie (1908) and Debye (1909). The basic problem is as follows: a plane polarized electromagnetic wave is incident on a sphere of radius a, complex refractive index m. A long distance away from the particle the forward beam has lost a certain amount of energy. Part is *scattered* out of the forward beam and part is *absorbed* by the sphere. For particles of radius a let these be set equal to the energy flux multiplying the cross–sections $Q_{sca}\ \pi a^2$, $Q_{abs}\ \pi a^2$. One of objectives of the solution developed by Mie is to calculate Q_{sca}, Q_{abs} as functions of

$$x = 2\pi a/\lambda\ , \tag{2.1}$$

$$m = n - ik = \sqrt{\varepsilon - 2i\sigma\lambda/c}\ , \tag{2.2}$$

where λ is the wavelength of the incident radiation, n = refractive index, k = absorptive index, ε = dielectric constant and σ is the optical conductivity of the grain material. The solution gives

$$Q_{sca} = \frac{2}{x^2} \sum_{n=1}^{\infty} (2n+1)\left[|a_n|^2 + |b_n|^2\right]\ , \tag{2.3}$$

$$Q_{ext} = \frac{2}{x^2} \sum_{n=1}^{\infty} (2n+1)\ \mathrm{Re}\,(a_n + b_n)\ , \tag{2.4}$$

$$Q_{abs} = Q_{ext} - Q_{sca} \tag{2.5}$$

The nth term in the series corresponds to 2^n–multipole components. If we write $y = mx$, the coefficients a_n, b_n are given by

$$a_n = \frac{x\psi_n'(y)\ \psi_n(x) - y\psi_n'(x)\ \psi_n(y)}{x\psi_n'(y)\ \zeta_n(x) - y\zeta_n'(x)\ \psi_n(y)} \tag{2.6}$$

$$b_n = \frac{y\psi_n'(y)\ \psi_n(x) - x\psi_n'(x)\ \psi_n(y)}{y\psi_n'(y)\ \zeta_n(x) - y\zeta_n'(x)\ \psi_n(y)} \tag{2.7}$$

ψ_n, χ_n, ζ_n are Riccati–Bessel functions:

$$\psi_n(z) = \left[\frac{\pi z}{2}\right]^{1/2} J_{n+1/2}(z) \tag{2.8}$$

$$\chi_n(z) = (-1)^n \left[\frac{\pi z}{2}\right]^{1/2} J_{-n-1/2}(z) \tag{2.9}$$

$$\zeta_n(z) = \psi_n(z) + i\chi_n(z) \tag{2.10}$$

Procedure to compute a_n, b_n:

We first define $A_n(y) = \psi_n'(y)/\psi_n(y)$ (2.11)

and using recurrence relations for Bessel functions we obtain from equations (2.6) and (2.7)

$$a_n = \frac{\left[\frac{A_n(y)}{m} + \frac{n}{x}\right] \mathrm{Re}\{\zeta_n(x)\} - \mathrm{Re}\{\zeta_{n-1}(x)\}}{\left[\frac{A_n(y)}{m} + \frac{n}{x}\right] \zeta_n(x) - \zeta_{n-1}(x)} \tag{2.12}$$

$$b_n = \frac{\left[m\ A_n(y) + \frac{n}{x}\right] \mathrm{Re}\{\zeta_n(x)\} - \mathrm{Re}\{\zeta_{n-1}(x)\}}{\left[m\ A_n(y) + \frac{n}{x}\right] \zeta_n(x) - \zeta_{n-1}(x)} \tag{2.13}$$

To generate the functions ζ_n we use as following recurrence relations

$$\zeta_0(x) = \sin x + i \cos x\ , \tag{2.14}$$

$$\zeta_{-1}(x) = \cos x - i \sin x\ , \tag{2.15}$$

$$\zeta_n(x) = \frac{2n-1}{x}\,\zeta_{n-1}(x) - \zeta_{n-2}(x)\ . \tag{2.16}$$

For given y (= mx) we may generate $A_n(y)$ as follows:

$$A_n(y) = \frac{\psi'(y)}{\psi(y)} = \frac{\sqrt{\frac{\pi y}{2}}\,[J_{n-1/2}(y) - \frac{n}{y}\,J_{n+1/2}(y)]}{\sqrt{\frac{\pi y}{2}}\,J_{n+1/2}(y)} \tag{2.17}$$

which leads to

$$A_n(y) = -\frac{n}{y} + \left[\frac{n}{y} - A_{n-1}(y)\right]^{-1} \tag{2.18}$$

$$A_0(y) = \cos y/\sin y\ . \tag{2.19}$$

Equations (2.6 – 2.19) enable us to compute a_n, b_n for n = 1,2,3... and for a given set of values of (m, x). The efficiency factors given by equations (2.3 – 2.5) can then be calculated by adding up the terms in the series. In practice we add up terms in series (2.3) and (2.4), until the next term added is less than .0001 of the partial sum. Convergence is usually quite rapid. The number of terms required, N, increases with x. For example in the cases m = 1.33 – 0.05i, 1.28 – 1.37i we have the following table for N:

TABLE 2.1

Number of Terms for Convergence

x	m = 1·33–0·05i	m = 1·28–1·37i
0·1	3	2
0·5	3	3
1·0	4	4
5·0	7	9
10·0	13	14
15·0	19	20

For most cases of interest the number of terms required for Mie computations is less than 20.

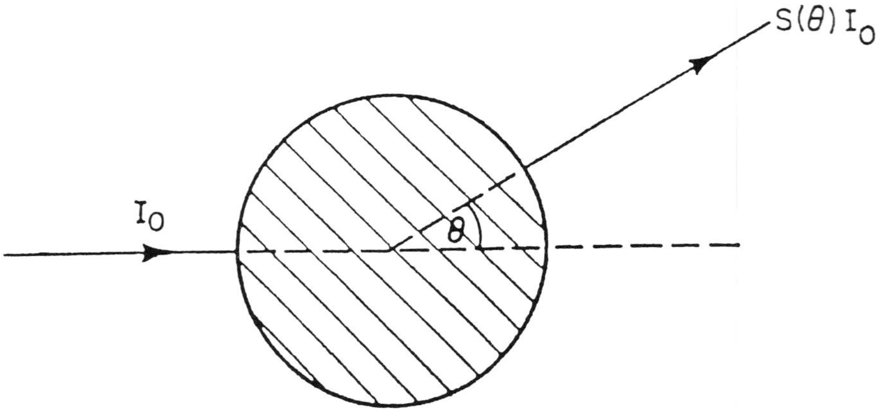

Fig. 2.1 *Definition of scattering angle and phase function.*

Phase Function. In Fig. 2.1 $S(\theta)$ is the fraction of the incident flux scattered into unit solid angle about a direction which makes angle θ with the forward beam and I_0 is the intensity in an incident beam. The Mie formulae give the complex amplitude functions

$$S_1(\theta) = \sum_{n=1}^{\infty} \frac{2n+1}{n(n+1)} \{a_n \pi_n(\cos\theta) + b_n \tau_n(\cos\theta)\} , \tag{2.20}$$

$$S_2(\theta) = \sum_{n=1}^{\infty} \frac{2n+1}{n(n+1)} \{b_n \pi_n(\cos\theta) + a_n \tau_n(\cos\theta)\} , \tag{2.21}$$

and a total scattering function $S(\theta)$ is defined by

$$S(\theta) = \frac{1}{2} \left[\frac{\lambda}{2\pi}\right]^2 \{|S_1(\theta)|^2 + |S_2(\theta)|^2\} . \tag{2.22}$$

In equations (2.20) and (2.21) the functions $\pi_n(\theta)$, $\tau_n(\theta)$ are defined by

$$\pi_n(\theta) = P_n' (\cos\theta) , \tag{2.23}$$

$$\tau_n(\theta) = \cos\theta \, \pi_n(\theta) - \sin^2\theta \frac{d}{d\cos\theta} \pi_n(\theta) , \tag{2.24}$$

where $P_n(\cos\theta)$ is the Legendre polynomial of degree n.

From well-known properties of Legendre polynomials the following recurrence relations may be derived

$$\pi_n(\theta) = \cos\theta \frac{2n-1}{n-1} \pi_{n-1}(\theta) - \frac{n}{n-1} \pi_{n-2}(\theta) \tag{2.25}$$

$$\tau_n(\theta) = \cos\theta[\pi_n(\theta) - \pi_{n-2}(\theta)] - (2n-1)\sin^2\theta\, \pi_{n-1}(\theta) + \tau_{n-2}(\theta)\,, \tag{2.26}$$

$$\begin{aligned} &\pi_0(\theta) = 0\,, && \tau_0(\theta) = 0\,, \\ &\pi_1(\theta) = 1\,, && \tau_1(\theta) = \cos\theta\,, \\ &\pi_2(\theta) = 3\cos\theta\,, && \tau_2(\theta) = 3\cos 2\theta\,. \end{aligned} \tag{2.27}$$

Equations (2.20 – 2.27) permit numerical computations of scattering amplitudes as functions of θ, m, x. We may also compute cos ⊙ defined by the equation

$$\overline{\cos\theta} = \frac{\int_0^\pi S(\theta)\cos(\theta)\sin(\theta)d\theta}{\int_0^\pi S(\theta)\sin(\theta)\,d\theta} \tag{2.28}$$

which gives an index of the forward directivity of the phase function. If the particle is a strong forward scatterer $\overline{\cos\theta} \approx 1$; if it is a nearly isotropic scatterer $\overline{\cos\theta} \approx 0$. The Mie formulae give us:

$$\begin{aligned} \overline{\cos\theta}\, Q_{sca} = \frac{4}{x^2} \sum_{n=1}^{\infty} \{ &\frac{n(n+2)}{n+1} [\mathrm{Re}(a_n)\,\mathrm{Re}(a_{n+1}) + \mathrm{Im}(a_n)\,\mathrm{Im}(a_{n+1}) + \\ &+ \mathrm{Re}(b_n)\,\mathrm{Re}(b_{n+1}) + \mathrm{Im}(b_n)\,\mathrm{Im}(b_{n+1})] + \\ &+ \frac{2n+1}{n(n+1)} [\mathrm{Re}(a_n)\,\mathrm{Re}(b_n) + \mathrm{Im}(a_n)\,\mathrm{Im}(b_n)]\}\,, \end{aligned} \tag{2.29}$$

where a_n, b_n are defined as before.

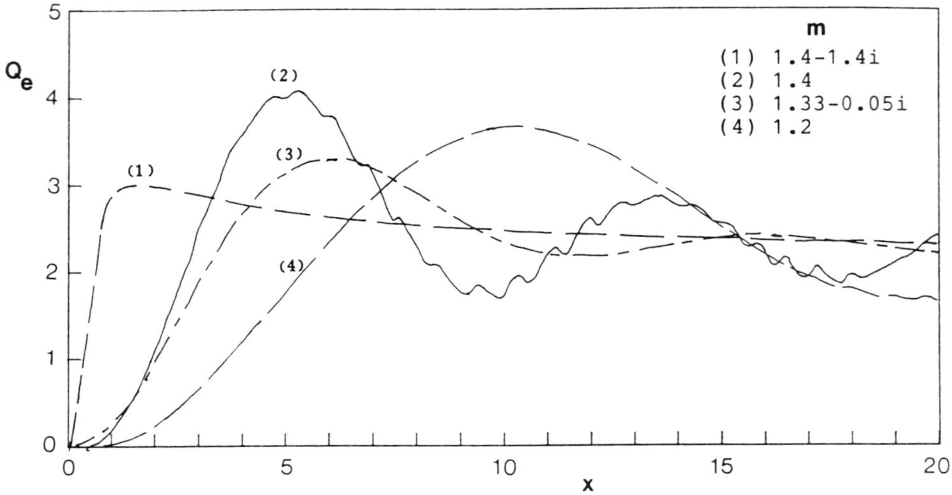

Fig. 2.2 *Extinction efficiencies as functions of x for various values of m.*

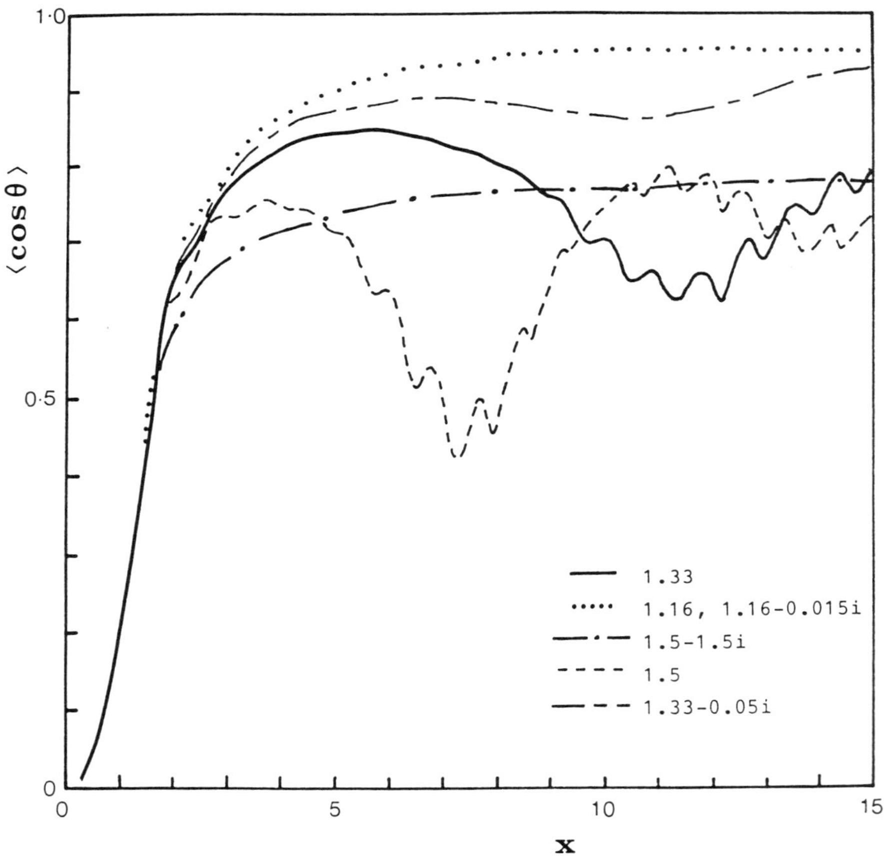

Fig. 2.3 *The phase parameter $\overline{\cos\theta}$ for various values of m.*

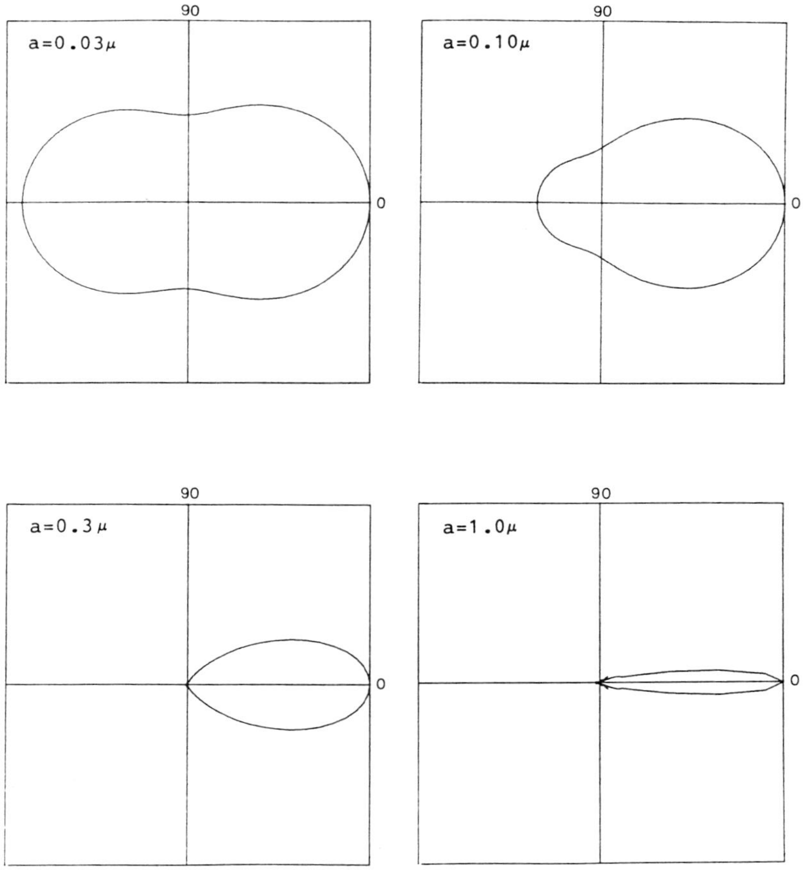

Fig. 2.4 *The total scattering amplitude S(θ) plotted as polar diagrams for the case of spheres with m = 1.4, λ = 0.6μm.*

Fig. 2.2 shows $Q_{ext}(x)$ for various values of m. Fig. 2.3 shows curves for $\overline{\cos\theta}$ in representative cases, and Fig. 2.4 shows the total scattering amplitude $S(\theta)$ plotted as normalised polar diagrams for the case m = 1.4, λ = 0.6μm.

In Fig. 2.5 we show the plots of $|S_1(\theta)|$ and $|S_2(\theta)|$ as functions of θ for m = 1.33 and m = 1.55. From Figs. 2.4 and 2.5 we see that the scattering becomes more and more forward directed as x increases above unity.

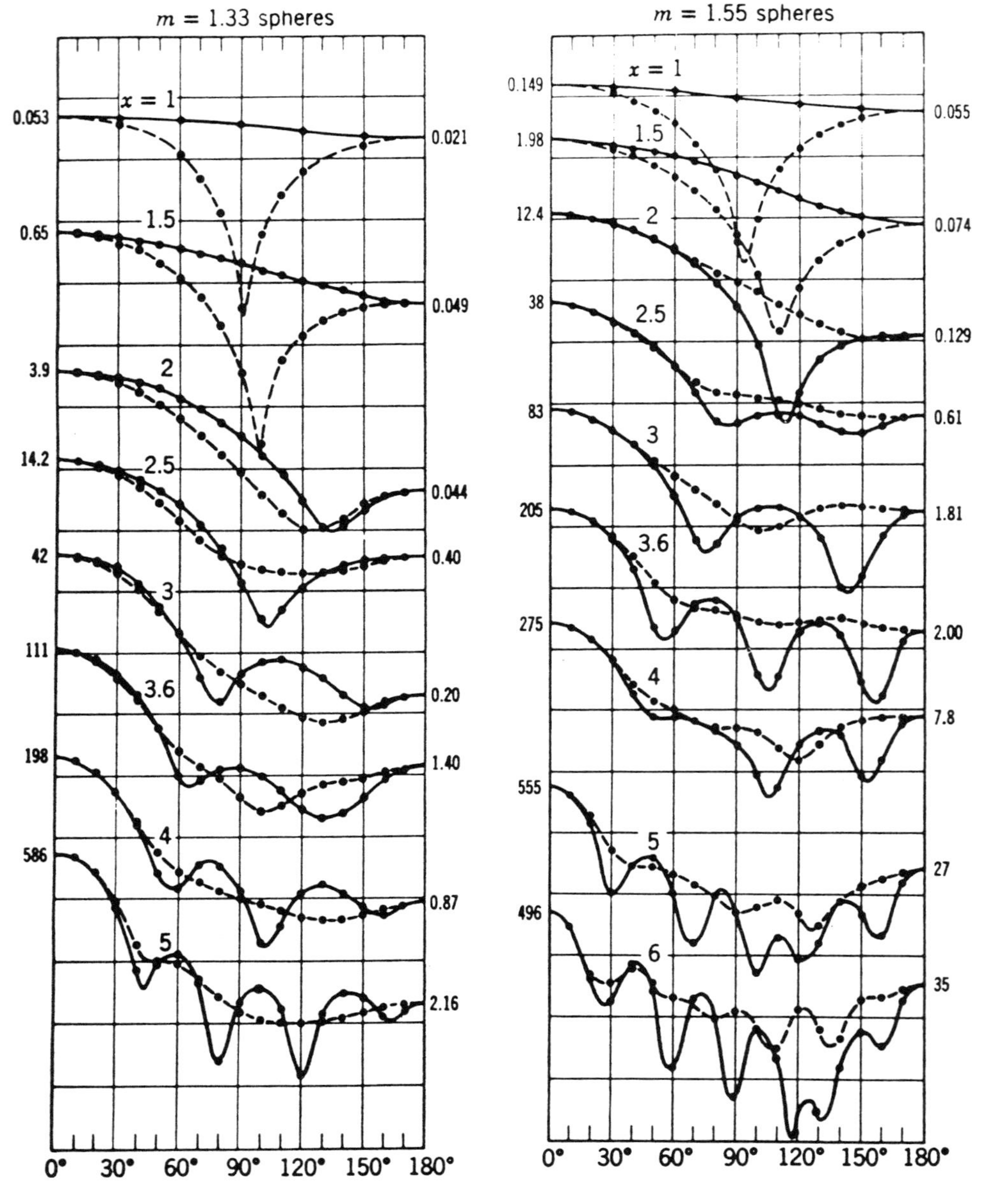

Fig. 2.5 $|S_1(\theta)|$ *and* $|S_2(\theta)|$ *are plotted as solid and dashed curves for the cases* $m = 1.33$ *and* $m = 1.55$ *for several values of* x *(adapted from van de Hulst, 1957).*

Asymptotic Formulae for Homogeneous Spheres. For small values of x << 1 the following approximate formulae are available for complex values of the refractive index m (appropriate to absorbing grains)

$$Q_{abs} = -4x \, \mathrm{Im}\left[\frac{m^2-1}{m^2+2}\right]$$
$$Q_{sca} = \frac{8}{3} x^4 \, \mathrm{Re}\left\{\left[\frac{m^2-1}{m^2+2}\right]^2\right\} \tag{2.30}$$

For non–conducting dielectric particles with m real approximate formulae give

$$Q_{sca} = Q_{ext} = \frac{8}{3} x^4 \left|\frac{m^2-1}{m^2+2}\right|^2$$
$$Q_{abs} = 0 \tag{2.31}$$

Spheres with $|m-1| << 1$
For particles that are so porous that $|m-1| << 1$ a simple expression for Q_{ext} follows by writing (van de Hulst, 1957)

$$m = n - ik$$
$$\frac{k}{n-1} = \tan\beta \tag{2.32}$$
$$\rho = 2x(n-1)$$

The asymptotic expression for Q_{ext} is then

$$Q_{ext} = 2-4e^{-\rho\tan\beta}\left[\frac{\cos\beta}{\rho}\right]\sin(\rho-\beta)$$
$$-4e^{-\rho\tan\beta}\left[\frac{\cos\beta}{\rho}\right]^2\cos(\rho-2\beta)+4\left[\frac{\cos\beta}{\rho}\right]^2\cos 2\beta \tag{2.33}$$

If k = 0 this expression further simplifies to

$$Q_{ext} = 2 - \frac{4}{\rho}\sin\rho + \frac{4}{\rho^2}(1-\cos\rho) \tag{2.34}$$

Table 2 compares the extinction efficiency calculated from (2.33) or (2.34) with that computed from the rigorous Mie formulae for the cases n = 1.1, n = 1.2. For the former case the correspondence is seen to be good, whilst for the latter the differences are more pronounced.

TABLE 2.2

Comparison of Q_{ext} from Mie formulae and asymptotic formula

	n = 1·1		n = 1·2	
x	Q_{ext}(Mie)	Q_{ext}(Asym)	Q_{ext}(Mie)	Q_{ext}(Asym)
0·5	0·0007	0·0005	0·0026	0·0199
1·0	0·0083	0·0199	0·0341	0·0793
1·5	0·0283	0·0448	0·1162	0·1764
2·0	0·0576	0·0793	0·2409	0·3088
2·5	0·1014	0·1233	0·4374	0·4729
3·0	0·1583	0·1764	0·6520	0·6644
3·5	0·2227	0·2384	0·9101	0·8784
4·0	0·2990	0·3088	1·1830	1·1092
4·5	0·3843	0·3871	1·4750	1·3510
5·0	0·4770	0·4729	1·7700	1·5976
6·0	0·6873	0·6644	2·3528	2·0808
7·0	0·9223	0·8784	2·8766	2·5124
8·0	1·1757	1·1091	3·2941	2·8536
9·0	1·4419	1·3510	3·5699	3·0771
10·0	1·7140	1·5976	3·6851	3·1401
11·0	1·9841	1·8428	3·6405	3·1352
12·0	2·2460	2·0808	3·4571	2·9886
13·0	2·4941	2·3057	3·1722	2·7582
14·0	2·7222	2·5124	2·8332	2·4795
15·0	2·9247	2·6963	2·4889	2·1907

2.2. COMPOSITE SHPERES

Formulae similar to the Mie formulae have been derived by Güttler (1952) for a sphere of radius R_0, refractive index m_1 surrounded by a concentric mantle of outer radius R, inner radius R_0 with a refractive index m_2. If C_{ext}, C_{sca} are the total cross–sections for extinction and scattering, we define the corresponding efficiency factors

$$Q_{ext} = C_{ext}/\pi R^2$$
$$Q_{sca} = C_{sca}/\pi R^2 \tag{2.35}$$

These efficiency factors are then given by

$$Q_{ext} = \frac{2}{x^2} \sum_{\ell=1}^{\infty} (2\ell+1)\, \mathrm{Re}(a_\ell+b_\ell) \tag{2.36}$$

$$Q_{sca} = \frac{2}{x^2} \sum_{\ell=1}^{\infty} (2\ell+1) \{|a_\ell|^2+|b_\ell|^2\} \tag{2.37}$$

where $x = 2\pi R/\lambda$ and with a_ℓ, b_ℓ given by

$$a_\ell = \frac{[\psi\psi]'_{\ell,R_0}.[\chi\psi]'_{\ell,R} - [\chi\psi]'_{\ell,R_0}.[\psi\psi]'_{\ell,R}}{[\psi\psi]'_{\ell,R_0}.[\chi\zeta]'_{\ell,R} - [\chi\psi]'_{\ell,R_0}.[\psi\zeta]'_{\ell,R}}$$

$$b_\ell = \frac{[\psi\psi]''_{\ell,R_0}.[\chi\psi]''_{\ell,R} - [\chi\psi]''_{\ell,R_0}.[\psi\psi]''_{\ell,R}}{[\psi\psi]''_{\ell,R_0}.[\chi\zeta]''_{\ell,R} - [\chi\psi]''_{\ell,R_0}.[\psi\zeta]''_{\ell,R}} \tag{2.38}$$

wherein

$$\begin{aligned}
[\chi\psi]'_{\ell,R} &= \chi'_\ell(k_2R).k_3\psi_\ell(k_3R) - k_2\chi_\ell(k_2R).\psi'_\ell(k_3R) \\
[\chi\zeta]'_{\ell,R} &= \chi'_\ell(k_2R).k_3\zeta_\ell(k_3R) - k_2\chi_\ell(k_2R).\zeta'_\ell(k_3R) \\
[\psi\psi]'_{\ell,R} &= \psi'_\ell(k_2R).k_3\psi_\ell(k_3R) - k_2\psi_\ell(k_2R).\psi'_\ell(k_3R) \\
[\psi\zeta]'_{\ell,R} &= \psi'_\ell(k_2R).k_3\zeta_\ell(k_3R) - k_2\psi_\ell(k_2R).\zeta'_\ell(k_3R) \\
[\chi\psi]''_{\ell,R} &= k_2\chi'_\ell(k_2R).\psi_\ell(k_3R) - \chi_\ell(k_2R).k_3\psi'_\ell(k_3R) \\
[\chi\zeta]''_{\ell,R} &= k_2\chi'_\ell(k_2R).\zeta_\ell(k_3R) - \chi_\ell(k_2R).k_3\zeta'_\ell(k_3R) \\
[\psi\psi]''_{\ell,R} &= k_2\psi'_\ell(k_2R).\psi_\ell(k_3R) - \psi_\ell(k_2R).k_3\psi'_\ell(k_3R) \\
[\psi\zeta]''_{\ell,R} &= k_2\psi'_\ell(k_2R).\zeta_\ell(k_3R) - \psi_\ell(k_2R).k_3\zeta'_\ell(k_3R) \\
[\chi\psi]'_{\ell,R_0} &= \chi'_\ell(k_2R_0).k_1\psi_\ell(k_1R_0) - k_2\chi_\ell(k_2R_0).\psi'_\ell(k_1R_0) \\
[\psi\psi]'_{\ell,R_0} &= \psi'_\ell(k_2R_0).k_1\psi_\ell(k_1R_0) - k_2\psi_\ell(k_2R_0).\psi'_\ell(k_1R_0) \\
[\chi\psi]''_{\ell,R_0} &= k_2\chi'_\ell(k_2R_0).\psi_\ell(k_1R_0) - \chi_\ell(k_2R_0).k_1\psi'_\ell(k_1R_0) \\
[\psi\psi]''_{\ell,R_0} &= k_2\psi'_\ell(k_2R_0).\psi_\ell(k_1R_0) - \psi_\ell(k_2R_0).k_1\psi'_\ell(k_1R_0)
\end{aligned} \tag{2.39}$$

Here

$$k_1 = m_1 \frac{2\pi}{\lambda}, \; k_2 = m_2 \frac{2\pi}{\lambda}, \; k_3 = \frac{2\pi}{\lambda} \tag{2.40}$$

and $\psi_\ell(z)$, $\chi_\ell(z)$ and $\zeta_\ell(z)$ are the Riccati–Bessel functions defined in equations (2.8 – 2.10). To compute these Riccati–Bessel functions we use the following recurrence relations

$$\begin{aligned}
\psi_n(z) &= \frac{2n-1}{z} \psi_{n-1}(z) - \psi_{n-2}(z) \\
\psi_n'(z) &= \frac{-n}{z} \psi_n(z) + \psi_{n-1}(z) \\
\chi_n(z) &= \frac{2n-1}{z} \chi_{n-1}(z) - \chi_{n-2}(z) \\
\chi_n'(z) &= \frac{n+1}{z} \chi_n(z) - \chi_{n+1}(z)
\end{aligned} \tag{2.41}$$

together with

$$\begin{aligned}
\psi_0(z) &= \left[\frac{\pi z}{2}\right]^{1/2} J_{1/2}(z) \\
\chi_0(z) &= \left[\frac{\pi z}{2}\right]^{1/2} J_{-1/2}(z) \\
\psi_0'(z) &= \frac{\pi}{4}\left[\frac{\pi z}{2}\right]^{-1/2} J_{1/2}(z) + \left(\frac{\pi z}{2}\right)^{1/2} J_{1/2}'(z) \\
\chi_0'(z) &= \frac{\pi}{4}\left[\frac{\pi z}{2}\right]^{-1/2} J_{-1/2}(z) + \left(\frac{\pi z}{2}\right)^{1/2} J_{-1/2}'(z) \\
J_{1/2}(z) &= \left[\frac{\pi z}{2}\right]^{-1/2} \sin z \\
J_{-1/2}(z) &= \left[\frac{\pi z}{2}\right]^{-1/2} \cos z
\end{aligned} \tag{2.42}$$

We can now compute the efficiency factors Q_{ext}, Q_{sca} for given values of the grain parameters using equations (2.36) and (2.37).

Asymptotic Formulae for Composite spheres. In the approximation $2\pi R/\lambda << 1$ the scattering and absorption cross–sections for the particle discussed above are given by

$$\begin{aligned}
C_{sca} &= \frac{8}{3} \pi k^4 \, |\alpha|^4 \\
C_{abs} &= 4\pi k \, \mathrm{Re}\,(i\alpha)
\end{aligned} \tag{2.43}$$

$$\alpha = R^3 \cdot \frac{R^3(m_2^2-1)(m_1^2+2m_2^2)+R_0^3(2m_2^2+1)(m_1^2-m_2^2)}{R^3(m_2^2+2)(m_1^2+2m_2^2)+R^3(2m_2^2-2)(m_1^2-m_2^2)}$$

where $k = 2\pi/\lambda$.

2.3. INFINITE CYLINDERS

The problem of extinction and scattering of light by homogeneous cylindrical particles has been discussed by van de Hulst (1957), Wait (1959), and Lind and Greenberg (1966). Formulae amenable to numerical computation are available for right circular cylinders of arbitrary cross–sectional radius a and length $l >> a$. Although the theory described here is strictly applicable to such 'infinite cylinders', it is believed that for dielectric particles the results of numerical calculations are valid for finite cylinders, provided l exceeds a by a factor of ~ 3. If $\mathbf{l}$ is the vector denoting particle axis and $\mathbf{k}$ is the propagation vector of plane polarized light, the rigorous formulae give cross–sections C_{eE}, C_{sE} for extinction and scattering respectively associated with electric vector in the plane of $\mathbf{k}$ and $\mathbf{l}$, and C_{eH}, C_{sH} associated with magnetic vector in the plane of $\mathbf{k}$ and $\mathbf{l}$.

Let θ denote the angle $\widehat{\mathbf{kl}}$, m the complex refractive index of the material and μ the magnetic permeability. Following Lind and Greenberg (1966), we make the following further definitions:

$$\begin{aligned}
\alpha &= 90^\circ - \theta \\
\varepsilon &= m^2 \\
k &= 2\pi/\lambda \\
v &= ka \cos \alpha \\
u &= ka(\mu\varepsilon - \sin^2\alpha)^{1/2} \\
s &= 1/u^2 - 1/v^2
\end{aligned} \tag{2.44}$$

The efficiency factors are then given by

$$Q_{eE} = \frac{C_{eE}}{2al} = \frac{2}{ka} \mathrm{Re}\{b_0^E + 2 \sum_{n=1}^{\infty} b_n^E\} \tag{2.45}$$

$$Q_{eH} = \frac{C_{eH}}{2al} = \frac{2}{ka} \mathrm{Re}\{a_0^H + 2 \sum_{n=1}^{\infty} a_n^H\} \tag{2.46}$$

$$Q_{sE} = \frac{C_{sE}}{2a\ell} = \frac{2}{ka}\,[\,|b_0^E|^2 + 2\sum_{n=1}^{\infty}(|b_n^E|^2 + |a_n^E|^2)] \tag{2.47}$$

$$Q_{sH} = \frac{C_{sH}}{2a\ell} = \frac{2}{ka}\,[\,|a_0^H|^2 + 2\sum_{n=1}^{\infty}(|a_n^H|^2 + |b_n^H|^2)] \tag{2.48}$$

wherein

$$a_n^E = in.\sin\alpha.S.R_n\{[B_n(\mu)-A_n(\mu)]/\Delta_n\} \tag{2.49}$$

$$b_n^E = R_n\{[A_n(\mu)B_n(\varepsilon)-n^2S^2\sin^2\alpha]/\Delta_n\} \tag{2.50}$$

$$a_n^H = R_n\{[A_n(\varepsilon)B_n(\mu)-n^2S^2\sin^2\alpha]/\Delta_n\} \tag{2.51}$$

$$b_n^H = -a_n^E \tag{2.52}$$

with

$$A_n(\xi) = [H_n'(v)/vH_n(v)]-\xi[J_n'(u)/uJ_n(u)] \tag{2.53}$$

$$B_n(\xi) = [J_n'(v)/vJ_n(v)]-\xi[J_n'(u)/uJ_n(u)] \tag{2.54}$$

$$R_n = J_n(v)/H_n(v) \tag{2.55}$$

$$\Delta_n = A_n(\varepsilon)\,A_n(\mu)-n^2S^2\sin^2\alpha \tag{2.56}$$

Here $J_n(z)$ is the Bessel function of the first kind of order n and $H_n(z)$ is the Hankel function of the second kind of order n. Writing

$$H_n(z) = J_n(z) - i\,Y_n(z) \tag{2.57}$$

we use the following recurrence relations

$$J_{n+1}(z) = \frac{2n}{z}J_n(z)-J_{n-1}(z) \tag{2.58}$$

$$J_n'(z) = \tfrac{1}{2}[J_{n-1}(z)-J_{n+1}(z)] \tag{2.59}$$

$$Y_{n+1}(z) = \frac{2n}{z}Y_n(z) - Y_{n-1}(z) \tag{2.60}$$

$$Y_n'(z) = \tfrac{1}{2}[Y_{n-1}(z) - Y_{n+1}(z)] \tag{2.61}$$

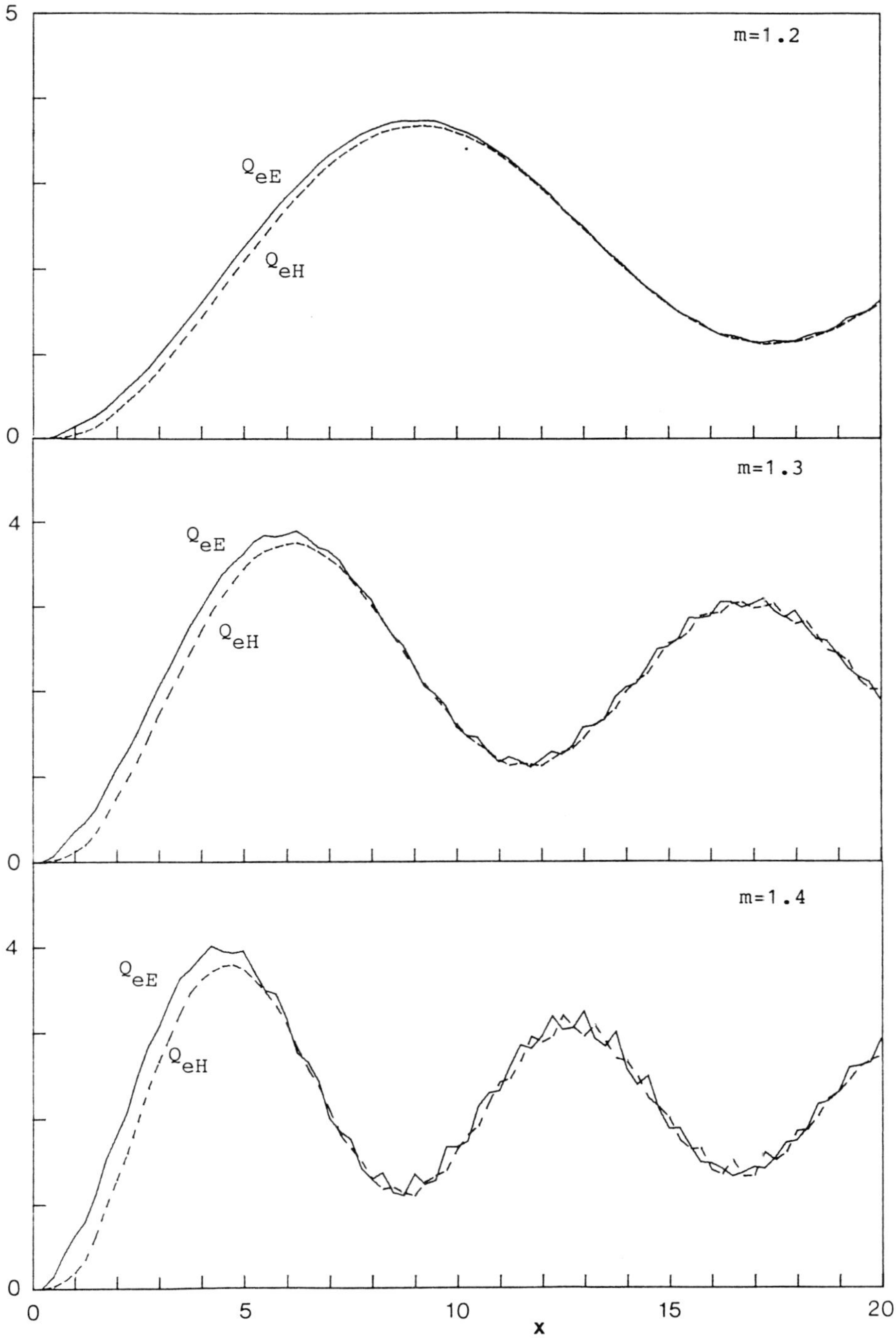

Fig. 2.6 *Efficiency factors for infinite cylinders Q_{eE} and Q_{eH} for $\theta = 90°$ and $m = 1.2, 1.3, 1.4$.*

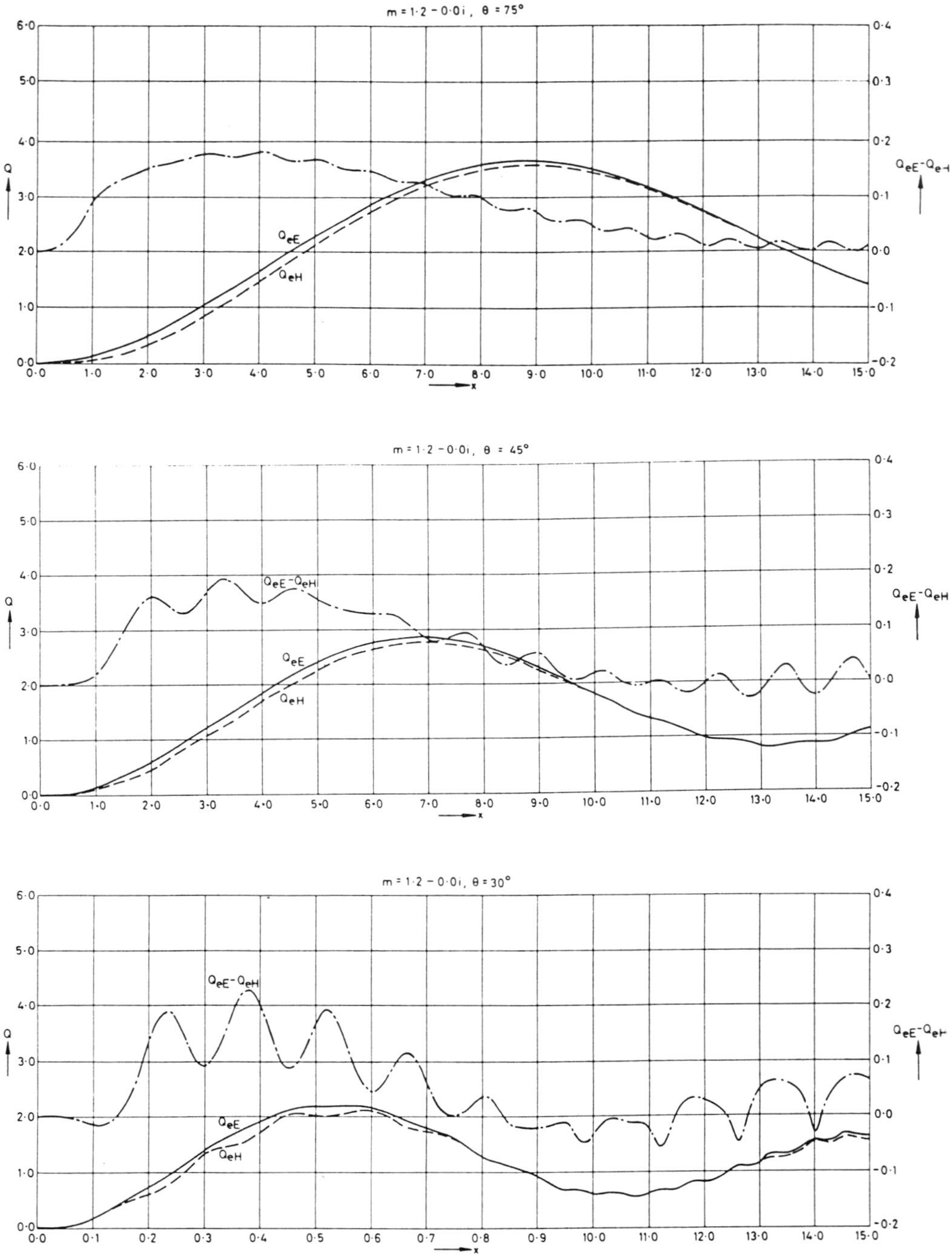

Fig. 2.7 ***Efficiency factors for infinite cylinders with incidence angle $\theta = 75°$, $45°$ and $30°$ and with $m = 1.2$.***

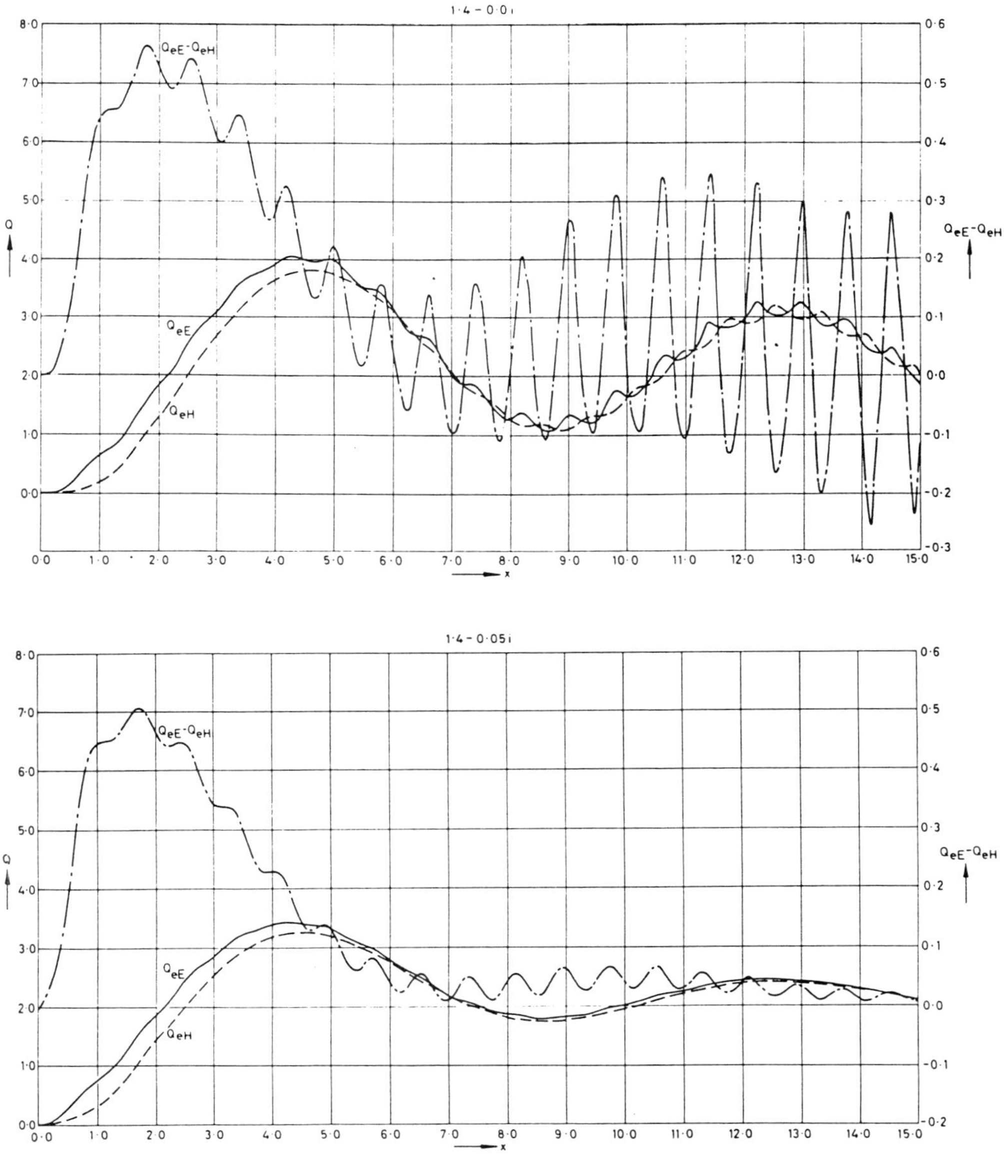

Fig. 2.8 *Efficiency factors for infinite cylinders with normal incidence: m = 1.4−0.00i, m = 1.4−0.05i.*

together with polynomial approximations for $J_{-1}(z)$, $J_0(z)$, $Y_{-1}(z)$, $Y_0(z)$ to generate a_n^E, b_n^E, a_n^H, b_n^H. As for the case of spherical particles, we then proceed to carry out the summations in (2.45 – 2.48). The results of numerical calculations for several representative cases are shown in Figures 2.6 – 2.8.

In addition to Q_{eE}, Q_{eH} we can also compute

$$Q_{cE} = \frac{2}{ka}\,\mathrm{Im}\{b_0^E + 2\sum_{n=1}^{\infty} b_n^E\} \tag{2.62}$$

$$Q_{cH} = \frac{2}{ka}\,\mathrm{Im}\{a_0^H + 2\sum_{n=1}^{\infty} a_n^H\} \tag{2.63}$$

where Q_{cE}, Q_{cH} are the single particle phase lag efficiences in the two directions of polarisation. The degree of circular polarisation caused by a partially aligned set of cylinders is now given by the ratio of the Stokes parameters V/I where

$$\frac{V}{I} \propto (Q_{cE}-Q_{cH}) \cdot (Q_{eE}-Q_{eH}) \tag{2.64}$$

Approximate Formulae for Infinite Cylinders. For infinite cylinders at normal incidence and with $x = 2\pi a/\lambda << 1$ we have

$$Q_{eE} \cong -\frac{\pi x}{2}\,\mathrm{Im}(m^2-1) \tag{2.65}$$

$$Q_{eH} \cong -\pi x\,\mathrm{Im}\left[\frac{m^2-1}{m^2+1}\right] \tag{2.66}$$

$$\begin{aligned} Q_{sE} &\cong \frac{\pi^2}{8} x^3 |m^2-1|^2 \\ Q_{sH} &\cong \frac{\pi^2}{4} x^3 \left|\frac{m^2-1}{m^2+1}\right|^2 \end{aligned} \tag{2.67}$$

Comparison of Cylinders and Spheres. In most astronomical problems the properties of grains are modelled using spherical particles. Cylindrical particles (infinite cylinder models) are used mainly in connection with observations of interstellar polarisation. In this context only a small degree of alignment is required, typically a few percent of the grains effectively lined up. The extinction properties are thus determined by an almost randomly aligned ensemble of cylindrical grains. The average extinction efficiency of such an ensemble is well approximated by considering 3 grains with cylinder axes in orthogonal directions. For an electromagnetic wave incident along one of these directions the average extinction efficiency is

$$\overline{Q}_{cyl} = \frac{Q_{eE} + 2Q_{eH}}{3} \tag{2.68}$$

It is interesting to compare the function $\overline{Q}_{cyl}(x)$ for cylinders from equation (2.68) where $x = 2\pi a/\lambda$, with the corresponding $Q_{ext}(x)$ for a sphere. Such comparisons for $m = 1.2$, and $m = 1.4$ are shown in Fig. 2.9. The 'random cylinder' curves are found to be close to the spherical case if a slightly smaller value is chosen for the cylinder radius, thus justifying the use of the Mie formulae for randomly oriented elongated particles. The factor by which the cylinder radius needs to be scaled down is typically in the range 0.8 – 0.95, varying of course with the value of m.

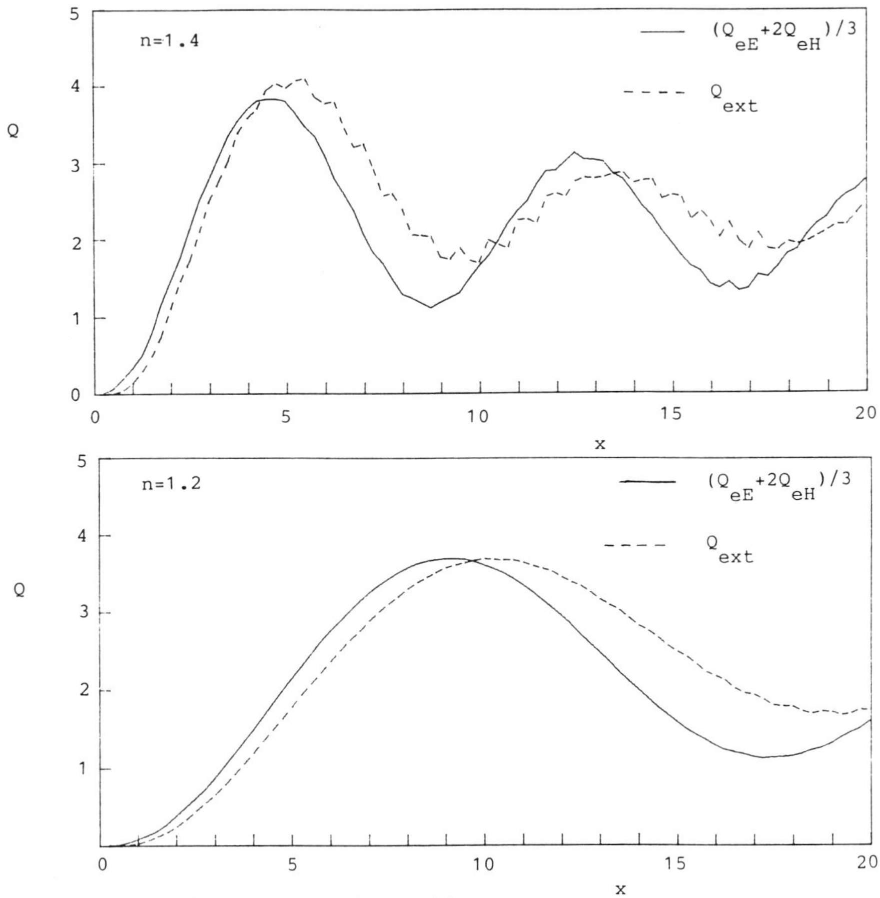

Fig. 2.9 *Comparison of $Q_{ext}(x)$ for a sphere with the average extinction efficiency of cylinders in random orientation, $(Q_{eE} + 2Q_{eH})/3$. Two cases considered are $m = 1.2, 1.4$.*

2.4. RAYLEIGH SCATTERING BY ELLIPSOIDS

Consider a homogeneous ellipsoidal particle with semi–axes of lengths a, b, c and with a total volume V. Let $\varepsilon = m^2$ denote the complex dielectric constant of the material and suppose that the particle dimensions are small compared with the wavelength of incident radiation.

For an applied electric field along one of its principal axes, the polarizability $\alpha_j (j = 1,2,3)$ of the ellipsoid is given by

$$\alpha_j = \frac{V}{4\pi} \frac{\varepsilon - 1}{L_j(\varepsilon - 1) + 1} \tag{2.69}$$

Here the quantities L_j, with $j = 1,2,3$ corresponding to the semi–axes of lengths a,b,c respectively, are given by

$$L_1 = \frac{abc}{2} \int_0^\infty \frac{ds}{(s+a^2)\Delta} \tag{2.70}$$

$$L_2 = \frac{abc}{2} \int_0^\infty \frac{ds}{(s+b^2)\Delta} \tag{2.71}$$

$$L_3 = \frac{abc}{2} \int_0^\infty \frac{ds}{(s+c^2)\Delta} \tag{2.72}$$

wherein

$$s = a + b + c \tag{2.73}$$

and

$$\Delta = \{(a^2+s)(b^2+s)(c^2+s)\}^{1/2} \tag{2.74}$$

From equations (2.70 – 2.74) it can be shown that

$$\sum_{j=1}^{3} L_j = 1 \tag{2.75}$$

In the Rayleigh approximation the cross–sections of the ellipsoid for scattering and extinction of light of wavelength λ are given by

$$C_{sca}^{(j)} = \frac{8}{3} k^4 |\alpha_j|^2 \tag{2.76}$$

$$C_{abs}^{(j)} = 4\pi k \, \mathrm{Re}\{i\alpha_j\} \tag{2.77}$$

where $j = 1,2,3$ correspond to electric vectors parallel to each of the three principal axes and $k = 2\pi/\lambda$. For a set of ellipsoids in random orientation the mean cross–sections are given by the same equations (2.76, 2.77) but with α_j replaced by

$$<\alpha> = \frac{1}{3}\sum_{j=1}^{3} \alpha_j \tag{2.78}$$

These formulae hold good provided

$$\max\{\frac{2\pi a}{\lambda}, \frac{2\pi b}{\lambda}, \frac{2\pi c}{\lambda}\} << 1 \tag{2.79}$$

With this restriction equations (2.69) to (2.77) enable us to compute optical cross–sections of ellipsoids with given values of a, b and c.

We note that in a few important special cases the integrals (2.70) – (2.72) are expressible in simple form:

Oblate spheroid, $b = c$, $a < b$

$$L_1 = \frac{1}{e^2}\left[1 - \frac{\sqrt{(1-e^2)}}{e}\cos^{-1}\sqrt{(1-e^2)}\right]$$

$$L_2 = L_3 = (1-L_1)/2 \tag{2.81}$$

$$e^2 = 1 - \frac{a^2}{b^2}$$

Prolate spheroid, $b = c$, $a > b$

$$L_1 = \frac{1-e^2}{e^2}\left[-1 + \frac{1}{2e}\ln\frac{1+e}{1-e}\right]$$

$$L_2 = L_3 = (1-L_1)/2 \tag{2.82}$$

$$e^2 = 1 - \frac{b^2}{a^2}$$

Sphere, $a = b = c$

$$L_1 = L_2 = L_3 = \frac{1}{3} \tag{2.83}$$

In this case the formulae for optical cross–sections take the simple forms (2.30).

Nearly spherical particle, $a \simeq b$, $b = c$

$$L_1 = \frac{1}{3} + \frac{4}{15}\frac{b-a}{a}$$

$$L_2 = L_3 = (1-L_1)/2 \tag{2.84}$$

Thin circular disk, a << b = c

$$L_1 = 1 \qquad L_2 = L_3 = 0 \tag{2.85}$$

From equations (2.69), (2.76) and (2.77) we then have

$$
\begin{aligned}
C_{abs}^{(1)} &= -Vk\, \mathrm{Im}\left[\frac{m^2-1}{m^2}\right] \\
C_{sca}^{(1)} &= \frac{1}{6}\frac{V^2k^4}{\pi^2}\left|\frac{m^2-1}{m^2}\right|^2 \\
C_{abs}^{(2)} &= C_{abs}^{(3)} = -Vk\, \mathrm{Im}(m^2-1) \\
C_{sca}^{(2)} &= C_{sca}^{(3)} = \frac{1}{6}\frac{V^2k^4}{\pi^2}\left|m^2-1\right|^2
\end{aligned}
\tag{2.86}
$$

Long elliptical cylinder, b, c << a

$$L_1 = 0 \qquad L_2 = \frac{c}{b+c}\,,\ L_3 = \frac{b}{b+c} \tag{2.87}$$

For the case b = c << a this further reduces to

$$L_1 = 0,\ L_2 = L_3 = \frac{1}{2} \tag{2.88}$$

and from equations (2.69), (2.76) and (2.77) we have

$$
\begin{aligned}
C_{abs}^{(1)} &= -Vk\, \mathrm{Im}(m^2-1) \\
C_{sca}^{(1)} &= \frac{1}{6}\frac{V^2k^4}{\pi^2}\left|m^2-1\right|^2 \\
C_{abs}^{(2)} &= C_{abs}^{(3)} = -2Vk\, \mathrm{Im}\left[\frac{m^2-1}{m^2+1}\right] \\
C_{sca}^{(2)} &= C_{sca}^{(3)} = \frac{2}{3}\frac{V^2k^4}{\pi^2}\left|\frac{m^2-1}{m^2+1}\right|^2
\end{aligned}
\tag{2.89}
$$

2.5. HETEROGENEOUS OR POROUS GRAINS

The foregoing discussion presupposed that the complex refractive index m at any given wavelength was a uniform property of a material of which the particle was made. If two materials of refractive indices m_1 and m_2 are assumed to be homogeneously mixed together (the mixture being considered uniform on a molecular scale) an average refractive index m may be calculated by considering volume averages of n–1 and of k separately. If the material of refractive index m_1 occupies a volume fraction f_1 and that of refractive index m_2 occupies a volume fraction f_2 ($f_1+f_2 = 1$) then

$$\overline{m} = 1+(n_1-1)f_1+(n_2-1)f_2+i(f_1k_1+f_2k_2) \tag{2.90}$$

If, on the other hand, very small spheres of material '2' are distributed in a matrix of material '1' an average refractive index $\overline{m}$ is better calculated by combining $\varepsilon_1 = m_1^2$, and $\varepsilon_2 = m_2^2$ according to the Maxwell–Garnet formula as follows (see Bohren and Wickramasinghe, 1977)

$$\overline{m}^2 = m_1^2\left[1 + \frac{3f(\varepsilon_2-\varepsilon_1)/(\varepsilon_2+2\varepsilon_1)}{1-f(\varepsilon_2-\varepsilon_1)/(\varepsilon_2+2\varepsilon_1)}\right] \tag{2.91}$$

where f is the volume fraction of the material 1. However, equations (2.90) and (2.91) give very nearly the same results unless one is dealing with situations where there are strong resonant absorption effects. For instance if the matrix material is a dielectric medium with m = 1.4 and the inclusions are vacuum cavities occupying a total volume fraction f, the two sets of average $\overline{m}$ values are shown in Table 2.3 below.

TABLE 2.3

Mean Refractive Index

f	$\overline{m}$	
	eqn. (2·90)	eqn. (2·91)
0	1·4	1·4
0·1	1·36	1·3592
0·2	1·32	1·3188
0·3	1·28	1·2786
0·4	1·24	1·2387
0·5	1·20	1·19889
0·6	1·16	1·1592
0·7	1·12	1·1195
0·8	1·08	1·0798
0·9	1·04	1·0400
1·0	1·0	1·0

For a hollow sphere (composite sphere with a concentric vacuum cavity) extinction and scattering cross-sections and efficiency factors can be directly calculated from the Güttler formulae. Representing such a hollow sphere by a smoothed out average refractive index given by equation (2.90) and calculating extinction properties from the Mie formulae, we find that a fairly good approximation to the actual optical efficiency factors can be obtained, as is seen in the example of Fig. 2.10. Here we have taken the example of materials with $m = 1.4$ making up a spherical shell with inner and outer radii r_0, r. We consider the particular case of $r_0^3/r^3 = f = 0.6$ giving $\overline{m} = 1.16$ (see Table 2.3). Writing $2\pi r/\lambda = x$ we calculate $Q_{ext}(x)$ from the Mie formulae for $n = 1.16$ and also from the Güttler formulae for $r_0/r = f^{1/3}$. Values are set out in Table 2.4 below. Similar correspondences can be demonstrated rigorously for hollow cylindrical grains by using the appropriate formulae for coaxial infinite cylinders (Jazbi, 1991). In the extinction models discussed in Chapter 8 we accordingly use low refractive index homogeneous spheres and cylinders to represent hollow particles.

TABLE 2.4

$Q_{ext}(x)$ from Mie formulae with $n = 1.16$ compared with the Güttler formulae with $r_0/r = (0.6)^{1/3}$ and $n = 1.4$ for the outer shell

X	MIEAV	GUTT	X	MIEAV	GUTT
0.1000	0.0000	0.0000	3.0000	0.4153	0.3766
0.2000	0.0000	0.0000	3.5000	0.5809	0.5252
0.3000	0.0002	0.0002	4.0000	0.7691	0.7321
0.4000	0.0007	0.0007	4.5000	0.9715	1.0365
0.5000	0.0017	0.0016	5.0000	1.1850	1.1461
0.6000	0.0034	0.0031	5.5000	1.4125	1.3709
0.7000	0.0060	0.0053	6.0000	1.6363	1.6216
0.8000	0.0099	0.0084	6.5000	1.8713	1.8407
0.9000	0.0150	0.0124	7.0000	2.0941	2.0291
1.0000	0.0216	0.0171	7.5000	2.3183	2.3047
1.1000	0.0297	0.0226	8.0000	2.5276	2.4575
1.2000	0.0391	0.0287	8.5000	2.7275	2.6324
1.3000	0.0497	0.0354	9.0000	2.9093	2.8804
1.4000	0.0613	0.0428	9.5000	3.0752	2.9537
1.5000	0.0737	0.0513	10.0000	3.2183	3.1249
1.6000	0.0870	0.0614	10.5000	3.3413	3.3246
1.7000	0.1011	0.0734	11.0000	3.4402	3.3187
1.8000	0.1163	0.0881	11.5000	3.5130	3.4933
1.9000	0.1329	0.1056	12.0000	3.5664	3.6163
2.0000	0.1512	0.1259	12.5000	3.5862	3.5566
2.1000	0.1716	0.1485	13.0000	3.5939	3.7622
2.2000	0.1942	0.1726	13.5000	3.5648	3.7389
2.3000	0.2188	0.1973	14.0000	3.5276	3.6889
2.4000	0.2451	0.2219	14.5000	3.4581	3.9380
2.5000	0.2726	0.2461	15.0000	3.3798	3.7749
			15.5000	3.2805	3.7186

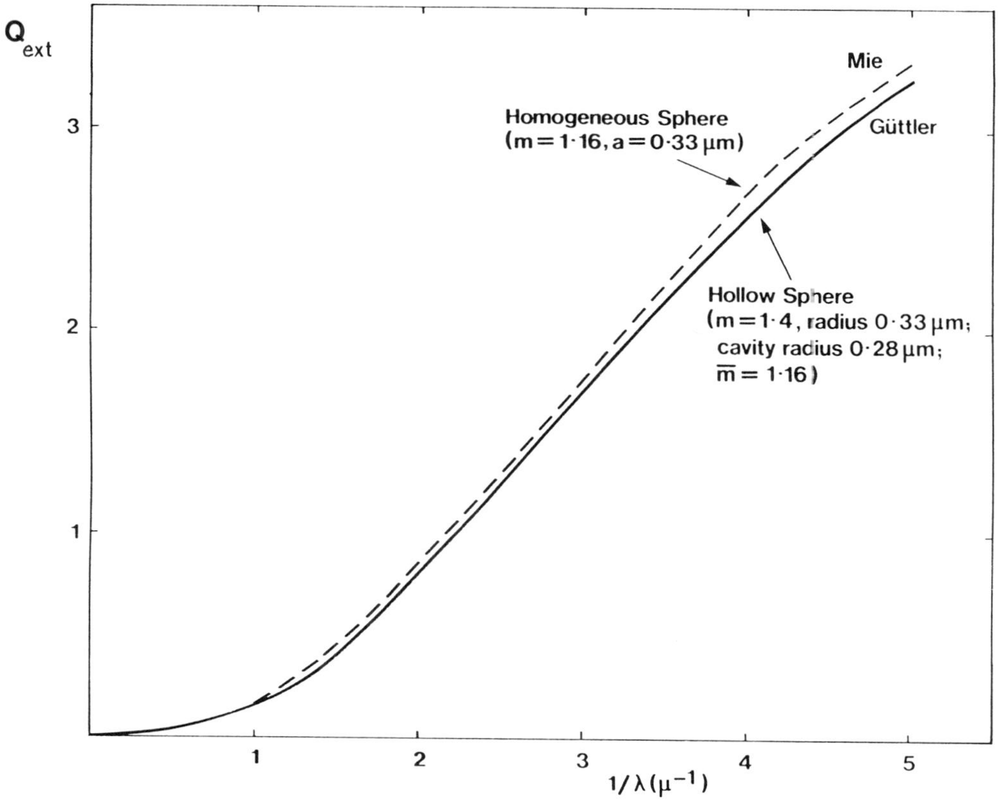

Fig. 2.10 *Extinction efficiency of a hollow sphere of radius $r = 0.33\mu m$ of material with $m = 1.4$ and with a vacuum cavity of radius $r_0 = (0.6)^{1/3}r$. The dashed curve is for a homogeneous sphere of radius $0.33\mu m$ and refractive index 1.16.*

2.6. ABSORPTION CROSS–SECTIONS, BULK ABSORPTION COEFFICIENT AND EMISSIVITY

We have thus far considered the behaviour of individual grains with regard to the scattering and absorption of incident radiation. For an ensemble of particles of any specified type it is often useful to calculate bulk scattering and absorption properties. With spherical grains of radius a and material density s, the bulk absorption coefficient may be defined as

$$\kappa_{abs} = \frac{\pi a^2 Q_{abs}}{\frac{4}{3}\pi a^3 s} = \frac{3Q_{abs}}{4as} \tag{2.92}$$

where Q_{abs} is calculated from the Mie formulae. Similar definitions can be made for κ_{sca} and κ_{ext}. For cylindrical grains equations (2.92) is replaced by

$$\kappa_{abs} = \frac{2a\ell\, Q_{abs}}{\pi a^2 s\, \ell} = \frac{2\, Q_{abs}}{\pi a s} \tag{2.93}$$

Formulae of this type are useful in assessing the behaviour of a given mass of material dispersed in the form of small particles.

The absorption cross–sections calculated from the Mie formulae are also required in discussions concerning emission properties of grains. Since from Kirchhoff's law emission and absorption efficiences must be equal, the emission from a sphere of radius a will be given by an equation of the type

$$F(\lambda) \propto Q_{abs}(\lambda)\, B_\lambda(T) \tag{2.94}$$

where Q_{abs} is calculated from the Mie formulae for a grain of radius a, T is the temperature of the particle and $B_\lambda(T)$ is the Planck function. In most cases of astronomical relevance equation (2.94) needs to be evaluated only at infrared wavelengths, $\lambda > \sim 2\mu m$. At such wavelengths $2\pi a/\lambda < 1$ for particles sizes $\sim 10^{-5}$ cm, so it is usually enough to consider asymptotic formulae in the small particle (strictly small x) limit. Thus for spherical particles we have

$$Q_{abs} = \frac{8\pi a}{\lambda} \operatorname{Im} \frac{1-m^2}{2+m^2} \tag{2.95}$$

to a good approximation in most cases.

Consider now a small spherical particle of radius a comprised of material of refractive index $m = n-ik$ at some infrared wavelength λ. For most solids (outside resonant absorption bands) $n \sim 1$ and $k << 1$. Writing $m = n-ik$ equation (2.95) becomes:

$$Q_{abs} = \frac{48\pi a}{\lambda} \frac{nk}{(2+n^2-k^2)^2+4n^2k^2} \tag{2.96}$$

and from (2.29)

$$\kappa_{abs} = \frac{36\pi nk}{s\lambda} \frac{1}{(2+n^2-k^2)^2+4n^2k^2} \approx \frac{4\pi k}{\lambda} \tag{2.97}$$

using $n \sim 1$, $k << 1$. Remembering that the absorptive index in bulk material is

$k = \kappa s\lambda/4\pi$, where κ is the mass absorption coefficient of a bulk sample, we see from (2.93) and (2.97) that we have the same κ values for small particles as for bulk specimens. This is a useful property in modelling astronomical infrared spectra because bulk properties, which are easily measured in the laboratory, are all that we require.

2.7. TWO SPECIAL CASES

We conclude this chapter by considering computations for two particular cases of astrophysical interest, the importance of which will be explored in later chapters.

Graphite Spheres. Graphite is a strongly anisotropic material. Two distinct complex refractive indices m_c, m_a naturally arise, one corresponding to light polarised with electric vector parallel to the crystal axis, the other at right angles to it. Graphite tends to occur in the form of thin flakes where the crystal axis is normal to the plane of the flake. The available refractive index measurements show that graphite acts as a strongly absorbing material for light with electric vector transverse to the crystal axes, and as a dielectric for light with electric vector parallel to the axis. For small particles resembling minute soot grains, with microscopic crystal domains in random orientation, it has been customary to consider an 'ideal' graphite sphere characterised by an isotropic refractive index given by $m = m_a$. Although this idealisation of graphite may well be challenged we present sample results of Mie calculations for spheres made up of material with complex refractive index data set out in Table 2.5. This data is from measurements of m_a for graphite made by Taft and Phillip (1965).

Fig. 2.11 shows the Q_{ext}, Q_{sca} curves for $a = 0.01\mu m$, $0.02\mu m$, $0.03\mu m$, $0.04\mu m$. We note that the extinction shows a resonant peak at wavelength that varies with particle radius from $\lambda^{-1} = 4.7\mu m^{-1}$ to $4.2\mu m^{-1}$ as a varies from $0.01\mu m$ to $0.04\mu m$. For grains of radius $a = 0.02\mu m$ the extinction peak occurs at a wavelength close to the $\lambda = 2200$ Å.

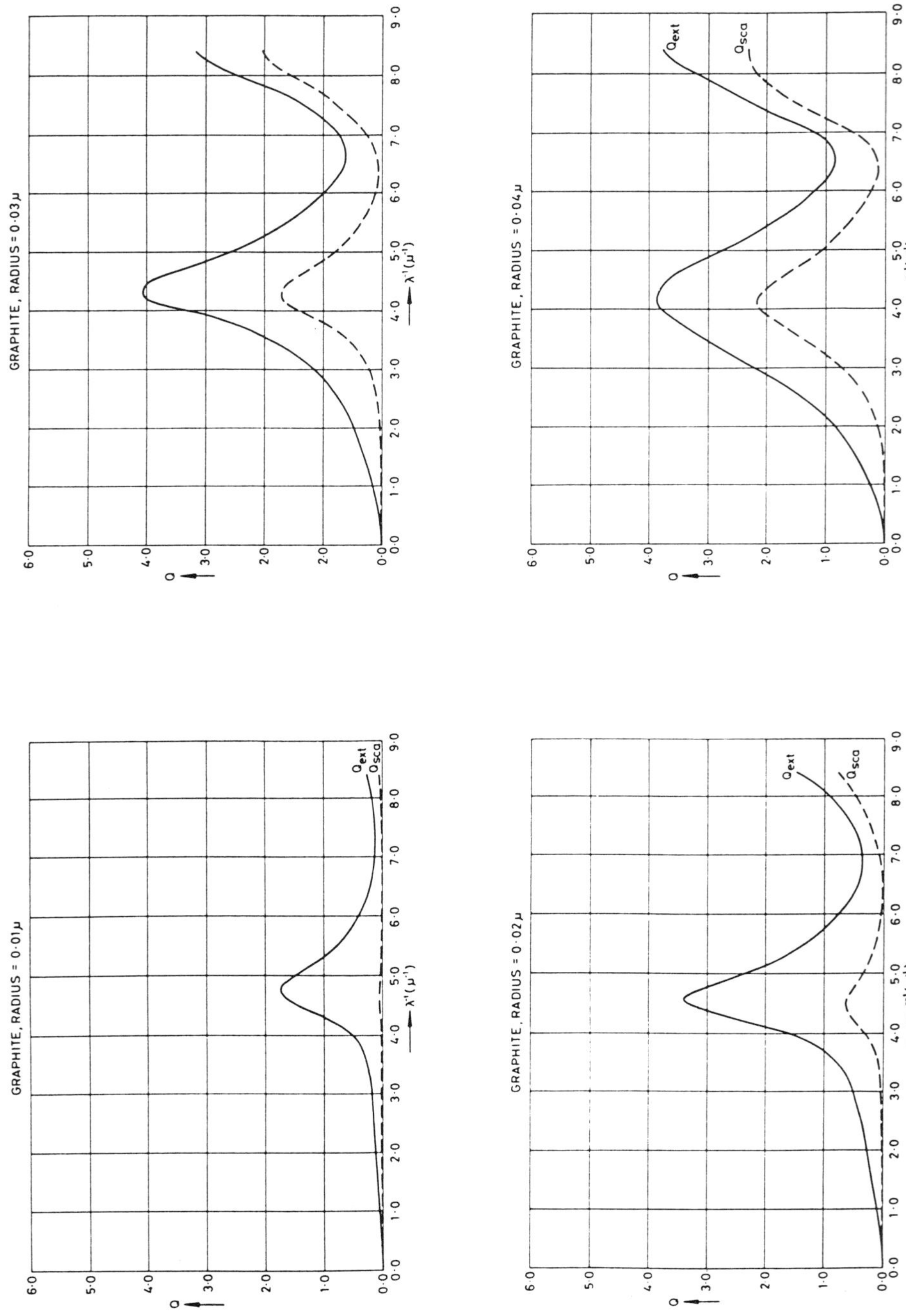

Fig. 2.11 *Extinction efficiencies of graphite spheres of radii 0.01, 0.02, 0.03 and 0.04μm.*

TABLE 2.5

n,k values for graphite (electric vector perpendicular to crystal axis)

λ(μ)	n	k	λ(μ)	n	k
123.96	6.988	58.57	0.288	2.499	2.069
61.98	5.949	28.31	0.270	2.052	2.426
30.99	6.537	14.78	0.258	1.637	2.462
20.66	6.288	10.20	0.248	1.228	2.343
15.50	6.139	7.20	0.238	0.960	2.080
12.40	6.049	6.640	0.225	0.791	1.713
8.26	5.677	5.158	0.206	0.689	1.276
6.20	5.296	4.420	0.191	0.678	0.922
4.13	4.692	3.612	0.182	0.722	0.717
3.10	4.299	3.064	0.177	0.765	0.598
2.48	4.070	2.755	0.172	0.821	0.492
2.07	3.907	2.550	0.167	0.888	0.399
1.77	3.780	2.485	0.163	0.962	0.321
1.55	3.576	2.511	0.159	1.037	0.255
1.38	3.287	2.443	0.157	1.077	0.228
1.24	3.067	2.241	0.154	1.125	0.201
1.03	2.868	1.953	0.147	1.277	0.145
0.885	2.763	1.756	0.141	1.415	0.118
0.688	2.637	1.534	0.135	1.558	0.116
0.620	2.601	1.461	0.130	1.644	0.128
0.546	2.563	1.404	0.124	1.802	0.163
0.495	2.538	1.364	0.118	1.957	0.228
0.413	2.515	1.347	0.108	2.258	0.504
0.354	2.539	1.401	0.099	2.388	1.020
0.310	2.588	1.677			

Metal Whiskers. Laboratory studies have provided a wealth of data relating to the growth of metal particles in the form of long slender needles or 'whiskers' as they are usually called (Nabarro and Jackson, 1958). Iron, aluminium, tin, nickel, zinc, platinum, lead, magnesium, potassium, and mercury have been extensively studied for whisker growth in the liquid and solid phases. Condensation from the vapour phase presents greater experimental difficulties, especially for refractory metals such as iron. Studies of whisker growth from the vapour phase have tended, therefore, to be largely confined to the more volatile metals, particularly mercury and potassium (Nabarro and Jackson, 1958; Sears, 1957); Gomer, 1957, 1958; Dittmar and Neumann, 1958), although more recent work in the Soviet Union has succeeded in obtaining very similar results for aluminium, an example of a refractory metal (Gal'tsov, 1988).

The growth of a condensate begins as a more or less spherical cluster of atoms with a radius that increases to about 0.01μm. At this stage a dramatic change occurs, with the condensate growing linearly with great rapidity up to lengths of ~ 1mm or more, thus giving a ratio of length to diameter of the resulting whisker

TABLE 2.6

λ (μm)	$\sigma = 10^{16}$ s^{-1} Qabs	Qsca	$\sigma = 10^{18}$ s^{-1} Qabs	Qsca	$\sigma = 10^{18}$ s^{-1} Qabs	Qsca	$\sigma = 3 \times 10^{18}$ s^{-1} Qabs	Qsca
2.000E-1	6.243E-1	4.758E-1	3.503E-1	1.048E0	1.249E-1	1.271E0	7.372E-2	1.321E0
2.200E-1	6.282E-1	4.609E-1	3.511E-1	1.060E0	1.251E-1	1.289E0	7.387E-2	1.341E0
2.600E-1	6.545E-1	4.401E-1	3.574E-1	1.094E0	1.273E-1	1.334E0	7.507E-2	1.389E0
2.800E-1	6.732E-1	4.322E-1	3.621E-1	1.113E0	1.289E-1	1.359E0	7.599E-2	1.415E0
3.000E-1	6.941E-1	4.252E-1	3.678E-1	1.133E0	1.307E-1	1.384E0	7.707E-2	1.442E0
3.500E-1	7.516E-1	4.099E-1	3.852E-1	1.184E0	1.363E-1	1.449E0	8.024E-2	1.510E0
4.000E-1	8.117E-1	3.964E-1	4.067E-1	1.234E0	1.428E-1	1.512E0	8.393E-2	1.578E0
4.500E-1	8.711E-1	3.837E-1	4.321E-1	1.282E0	1.499E-1	1.574E0	8.799E-2	1.644E0
5.000E-1	9.284E-1	3.717E-1	4.609E-1	1.329E0	1.576E-1	1.635E0	9.233E-2	1.708E0
5.500E-1	9.828E-1	3.602E-1	4.931E-1	1.374E0	1.656E-1	1.694E0	9.690E-2	1.771E0
6.000E-1	1.034E0	3.492E-1	5.282E-1	1.417E0	1.741E-1	1.751E0	1.017E-1	1.832E0
6.500E-1	1.083E0	3.386E-1	5.661E-1	1.458E0	1.828E-1	1.807E0	1.065E-1	1.891E0
7.000E-1	1.128E0	3.285E-1	6.065E-1	1.498E0	1.918E-1	1.861E0	1.116E-1	1.949E0
8.000E-1	1.210E0	3.096E-1	6.941E-1	1.573E0	2.104E-1	1.966E0	1.221E-1	2.062E0
1.000E0	1.346E0	2.766E-1	8.911E-1	1.706E0	2.498E-1	2.164E0	1.442E-1	2.274E0
2.000E0	1.709E0	1.764E-1	2.116E0	2.157E0	4.687E-1	3.007E0	2.686E-1	3.178E0
5.000E0	1.993E0	8.237E-2	6.171E0	2.547E0	1.368E0	4.969E0	7.137E-1	5.249E0
1.000E1	2.095E0	4.330E-2	1.114E1	2.303E0	3.743E0	7.468E0	1.627E0	7.988E0
3.000E1	2.162E0	1.489E-2	1.791E1	1.234E0	1.987E1	1.364E1	8.161E0	1.630E1
1.000E2	2.184E0	4.514E-3	2.085E1	4.309E-1	8.797E1	1.818E1	5.516E1	3.409E1
5.000E2	2.191E0	9.060E-4	2.174E1	8.987E-2	1.916E2	7.919E0	3.639E2	4.512E1
1.000E3	2.192E0	4.532E-4	2.184E1	4.514E-2	2.071E2	4.280E0	5.120E2	3.175E1
5.000E3	2.193E0	9.066E-5	2.191E1	9.060E-3	2.173E2	8.984E-1	6.367E2	7.897E0
1.000E4	2.193E0	4.533E-5	2.192E1	4.532E-3	2.184E2	4.514E-1	6.484E2	4.021E0

that is of the order 10^5.

Since whiskers in interstellar or extragalactic space will have low temperatures the lattice vibrations that determine the value of conductivity $\sigma \simeq 10^{17}$ s^{-1} of iron at room temperature will be irrelevant to the calculation of κ_{abs} under astronomical conditions. Were it not for lattice defects together with impurities up to a total concentration of about 1% over a distribution of elements from silicon to nickel, very high values of σ would be expected. However, having regard to both impurities and to lattice effects, we prefer not to set σ higher than 3.10^{18} s^{-1}. It may be noted that a value of $\sigma \simeq 10^{18}$ s^{-1} is inconsistent with extrapolations of laboratory data (Domenicali and Otter, 1955) involving iron specimens containing about 0.1% silicon.

We consider a range of σ values varying from 10^{16} s^{-1} to 3×10^{18} s^{-1} for a hypothetical metal whisker. Furthermore, we assume that in the high conductivity limit

$$n \simeq k \simeq \sqrt{\sigma\lambda/c} \qquad (2.98)$$

Setting

$$m = \sqrt{\sigma\lambda/c} - i\sqrt{\sigma\lambda/c} \qquad (2.99)$$

we compute $(Q_{eE} + 2Q_{eH})/3$ for infinite cylinders of radius 0.01μ and for various values of σ. The results are set out in Table 2.6.

References

Bohren, C.F. and Huffman, D.R., 1983, *Absorption and Scattering of Light by Small Particles*, John Wiley, New York.

Bohren, C.F. and Wickramasinghe, N.C., 1977, *Astrophys. Sp. Sc.*, **50**, 461.

Debye, P., 1909, *Ann. Physik.*, **30**, 59.

Dittmar, W. and Neumann, K., 1958, in R.H. Daramus, B.W. Roberts and D. Turnbull (eds.) *Growth and Perfection in Crystals*, J. Wiley, New York.

Domenicali, C.A. and Otter, F.A., 1955, *J. Appl. Phys.*, **26**, 377.

Gal'tsov, D.V., 1987, private communication.

Gomer, R., 1957, *J. Chem. Phys.*, **26**, 1333.

Gomer, R., 1958, *J. Chem. Phys.*, **28**, 457.

Güttler, A., 1952, *Ann. Physik.*, **6**, Folge, Bd. 11, 5.

Jazbi, B., 1991, *Thesis*, in preparation.

Lind, A.C. and Greenberg, J.M., 1966, *J. Appl. Phys.*, **37**, 3195.

Mie, G., 1908, *Ann. Physik.*, **25**, 377.

Nabarro, F.R.N. and Jackson, P.J., 1958, in R.H. Doremus, B.W. Roberts and D. Turnbull (eds.) *Growth and Perfection in Crystals*, J. Wiley, New York.

Sears, G., 1957, *Ann. New York Acad. Sci.*, **65**, 388.

Taft, E.A. and Phillipp, H.R., 1965, *Phys. Rev*, **138A**, 197.

Van de Hulst, H.C., 1957, *Light Scattering by Small Particles*, J. Wiley, New York.

Wickramasinghe, N.C., 1973, *Light Scattering Functions for Small Particles with Applications in Astronomy*, J. Wiley, New York.

3. Interstellar Extinction and Polarisation

In Chapter 1 we referred briefly to the early measurements of interstellar extinction over the UBV wavelength range that gave an approximately $1/\lambda$ extinction coefficient amounting typically to ~ 1–2 mag/kpc. We now discuss both extinction and polarisation observations in more detail with a view to using this data as discriminants for grain models in later chapters.

3.1. EQUATION OF TRANSFER

We first consider the effect of spherical isotropic grains in the line of sight of a star on the intensity of radiation received from it. The intensity of radiation I_ν in a specified direction at a point is defined as the amount of radiant energy crossing a unit area at right–angles to this direction in unit time, through unit solid angle centred about this direction, per unit frequency interval. If $d\sigma$ is an element of area at right–angles to the specified direction, $d\omega$ is an elementary solid angle centred about the same direction, the amount of energy in the frequency range ν, $\nu + d\nu$ crossing this surface through the solid angle $d\omega$ per unit time is

$$I_\nu \, d\sigma \, d\nu \, d\omega \tag{3.1}$$

Consider a right circular cylinder of cross–sectional area dA with length $d\ell$ and axis

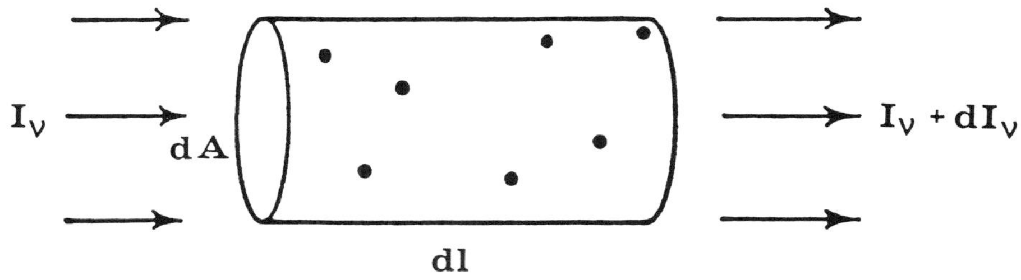

Fig. 3.1 *Elementary cylindrical column in line of sight to a star.*

the line of sight of a star (see Fig. 3.1). The energy in the solid angle $d\omega$ flowing into the cylinder per unit time is

$$I_\nu \, dA \, d\nu \, d\omega \, , \tag{3.2}$$

and the energy in $d\omega$ flowing out per unit time is

$$(I_\nu + dI_\nu)\, dA\, d\nu\, d\omega \tag{3.3}$$

Within the cylinder the energy absorbed and scattered out of the solid angle $d\omega$ is

$$I_\nu\, d\nu\, d\omega\, \pi a^2\, Q_{ext} \text{ x (number of grains)} \tag{3.4}$$

If N_g denotes the number density of grains, the number of grains enclosed in the cylinder is $N_g\, dA\, d\ell$, and (3.4) becomes

$$I_\nu\, d\nu\, d\omega\, \pi a^2\, Q_{ext}\, N_g\, dA\, d\ell\,. \tag{3.5}$$

By conservation of energy, (3.2), (3.3) and (3.4) give

$$I_\nu\, dA\, d\nu\, d\omega = (I_\nu + dI_\nu)\, dA\, d\omega\, d\nu + I_\nu\, d\nu\, d\omega\, \pi a^2\, Q_{ext}\, N_g\, dA\, d\ell \tag{3.6}$$

Or, which is the same,

$$dI_\nu + I_\nu\, N_g\, \pi a^2\, Q_{ext}\, d\ell = 0 \tag{3.7}$$

That is,

$$\frac{dI_\nu}{I_\nu} = -N_g\, \pi a^2\, Q_{ext}\, d\ell \tag{3.8}$$

which integrates to give

$$\ln I_\nu = -N_g\, \pi a^2\, Q_{ext}\, \ell + \text{const.} \tag{3.9}$$

From the above equation it is seen that if N denotes the number of grains (all supposed to be of identical size and composition) in a column of 1 cm^2 cross–section stretching from the star to the observer, the intensity of starlight is diminished due to extinction by the grains by the factor

$$\frac{I_\nu}{I_\nu(0)} = \exp(-N\, \pi a^2\, Q_{ext}) \tag{3.10}$$

Writing $\tau = N\, \pi a^2\, Q_{ext}$, the intensity is diminished by the factor $\exp(-\tau)$ where τ is the optical depth. To express this change of intensity in magnitudes

$$\Delta m = -2.5 \log_{10} \frac{I_\nu}{I_\nu(0)} \tag{3.11}$$

and combining with (3.10), we finally obtain

$$\Delta m = 1.086\ N\ \pi a^2\ Q_{ext} = 1.086\tau \tag{3.12}$$

From (3.12) it is seen that the extinction in magnitudes produced by identical grains in the line of sight of a star differs from the efficiency factor for a single grain only to the extent of a constant factor. As we shall see in the following section, this property is useful in fitting theoretical grain models to the observed interstellar extinction curves.

It should, however, be remembered that (3.12) has been derived on the assumption that grains in the line of sight all have the same radius, a. If instead we have a size distribution of grains and the number of grains with radii in the range a, a+da per unit volume is n(a) da. Then it is easy to show that (3.12) is equivalent to

$$\Delta m = 1.086 \int_0^\infty \pi a^2\ Q_{ext}\ n(a)\ da \Big/ \int_0^\infty n(a)\ da \tag{3.13}$$

Or if we introduce the total grain density n_g and their mass density ρ, (3.13) is equivalent to

$$\rho = \frac{4}{3}\ \pi a^3\ s\ n_g = 3.0 \times 10^{-22}\ as/Q_{ext}\ g\ cm^{-3} \tag{3.14}$$

where s is the density of the material comprising a grain. The average density required for various grain models is shown in Table 3.1, the last column of which gives the maximum grain densities permitted on the assumption of cosmic abundance ratios and a total hydrogen density of 3×10^{-24} g cm^{-3}.

TABLE 3.1

Density Constraints for Grain Models

Grain model (spheres)	ρ(g cm^{-3})	ρ_{max}(g cm^{-3}) (From Table 1·3)
ice (a = 3×10^{-5} cm)	$\sim 10^{-26}$	$3{\cdot}63 \times 10^{-26}$
graphite (a = 3×10^{-6} cm)	6×10^{-27}	$1{\cdot}33 \times 10^{-26}$
iron (a = 2×10^{-6} cm)	$\sim 4{\cdot}5 \times 10^{-26}$	$0{\cdot}44 \times 10^{-26}$
silicate (a = 2×10^{-5} cm)	$1{\cdot}6 \times 10^{-26}$	$0{\cdot}94 \times 10^{-26}$
organics (a = 3×10^{-5} cm)	$\sim 10^{-26}$	$3{\cdot}34 \times 10^{-26}$

3.2. OBSERVATIONS OF INTERSTELLAR EXTINCTION, DEFINITION OF COLOUR INDICES AND COLOUR EXCESSES

For a given set of wavelengths the available techniques determine what are essentially a monochromatic set of apparent magnitudes, $m(\lambda_i)$, $i = ...$, for a star. With D its distance in parsecs apparent magnitudes are related to absolute magnitudes $M(\lambda_i)$ by

$$m(\lambda_i) = M(\lambda_i) + 5 \log D - 5 + A(\lambda_i) , \tag{3.15}$$

where $A(\lambda_i)$ is the interstellar extinction in magnitudes at λ_i. The measured difference of m at wavelengths λ_i, λ_j defines an observed colour index, $(i-j)_{obs}$ say, which is related to the intrinsic colour $(i-j)_{intr}$ at these wavelengths by

$$(i-j)_{obs} = (i-j)_{intr} + A(\lambda_i) - A(\lambda_j) . \tag{3.16}$$

With the intrinsic colour considered known, (3.16) determines $A(\lambda_i) - A(\lambda_j)$ from $(i-j)_{obs}$. The colour excess, E_{i-j}, is defined by

$$E_{i-j} = (i-j)_{obs} - (i-j)_{intr} , \tag{3.17}$$

where the observed colour excess is related to the extinction by

$$E_{i-j} = A(\lambda_i) - A(\lambda_j) . \tag{3.18}$$

Choosing λ_B, λ_V as the standard blue and visual wavelengths, it is easy to see from (3.18) that

$$\frac{E_{\lambda-V}}{E_{B-V}} = \frac{A(\lambda)-A(\lambda_V)}{A(\lambda_B)-A(\lambda_V)} . \tag{3.19}$$

Since $A(\lambda)$ is proportional to $Q_{est}(\lambda)$ the ratio (3.19) can be written equivaliatly as

$$\frac{E_{\lambda-V}}{E_{B-V}} = \frac{Q_{ext}(\lambda)-Q_{ext}(\lambda_V)}{Q_{ext}(\lambda_B)-Q_{ext}(\lambda_V)} \tag{3.20}$$

The left–hand side here being measurable, (3.20) determines $Q_{ext}(\lambda)$ with respect to the extinctions at λ_B, λ_V for any number of filter wavelengths in the set λ. Eight filter wavelengths used by Johnson (1965) are listed in Table 3.2. A large number of extinction curves have been derived using this set of effective wavelengths.

Measurements of Colour Excesses. The basic observational problem involved in interstellar reddening studies is the determination of the wavelength dependence of the extinction in magnitudes $A(\lambda)$ in the above discussion) for a specified group of stars or for a given galactic region. In principle this can be achieved by measuring monochromatic magnitudes or colour index differences between a reddened and unreddened star of the same spectral type and in the same galactic region. In the absence of a suitable unreddened star an alternative procedure involves the

TABLE 3.2

Filter characteristics

	λ_{eff}	λ^{-1}_{eff} (μm^{-1})
U	3450 A	2·90
B	4340 A	2·30
V	5740 A	1·83
R	6420 A	1·56
I	8400 A	1·19
J	1·16 micron	0·86
K	2·14 micron	0·47
L	3·37 micron	0·30

computation of colour excesses using intrinsic fluxes computed from a model of the star being observed. The plot of such magnitude differences or colour excesses against wave–number represents the effect of interstellar extinction and is usually referred to as the extinction law or extinction curve.

The pioneering work on the interstellar extinction law was carried out by Stebbins, Huffer and Whitford (1943, 1945) using the techniques of photographic filter photometry. As an application of their new 3–colour photometry they measured the colours of some 1332 hot B stars. By comparing highly reddened and comparatively unreddened B stars the extinction in magnitudes was found to vary approximately as the reciprocal of the wavelength over the three effective wavelengths 5470 Å(V), 4340 Å(B), 3450 Å(U). Since blue light is more strongly attenuated than red light according to this law, the interstellar extinction curve is sometimes referred to as the 'reddening law'.

Infrared Extinction Observations. The first notable attempt to extend the extinction curve into the infrared was made by Whitford (1948) using an infrared filter with an effective wavelength 2.1μm. The results of this as well as of subsequent investigations indicated that the extinction deviated from a $(1/\lambda)$ law towards longer wavelengths: a dependence on a slightly higher power of $(1/\lambda)$ appeared to be indicated by the data, probably approaching $(1/\lambda)^2$ as $\lambda \to \infty$. H.L. Johnson (1965, 1968) observed a large number of stars at infrared wavelengths. The extinction curve derived by Johnson for the heavily reddened B star VI Cyg No. 12 is shown in Fig. 3.2. This curve is representative of extinction data for early–type stars. Although a variety of other types of infrared extinction curves were also obtained by Johnson (1968), these were mainly for relatively cool giant or supergiant stars which were contaminated by emission from circumstellar dust shells.

Ratio of Total to Selective Extinction. Photometric distances of associations of O and B stars, which are tracers of spiral structure in the galaxy, have to be corrected for total interstellar extinction. A quantity which we require in this connection is

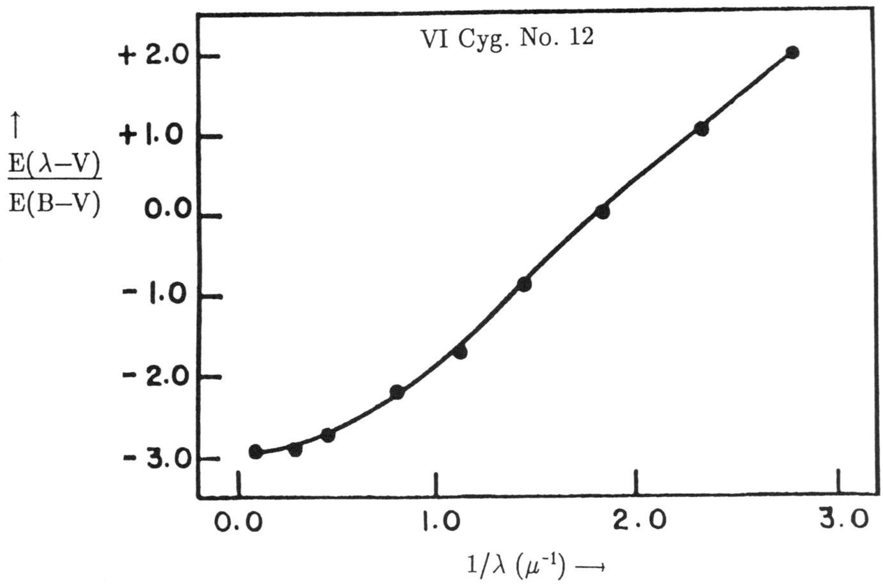

Fig. 3.2 *Johnson's extinction curve for VI Cyg. No. 12.*

the ratio of total to selective extinction, R, defined by

$$R = \frac{Q_{ext}(\lambda_V)}{Q_{ext}(\lambda_B) - Q_{ext}(\lambda_V)} \, . \tag{3.21}$$

By comparing (3.20) and (3.21) it is seen that, provided the grains are such that $Q_{ext}(\lambda) \to 0$ as $\lambda \to \infty$, R is given in terms of measurable colours excesses by

$$R = -\left[E_{\lambda\text{-}V} / E_{B\text{-}V}\right]_{\lambda \to \infty} , \tag{3.22}$$

when the visual extinction is obtained from

$$A_V = R\, E_{B\text{-}V} \tag{3.23}$$

The condition $Q_{ext} \to 0$ as $\lambda \to \infty$ is considered to hold for all grains except metallic whiskers, which are not thought to contribute appreciable to galactic extinction. Taking $\lambda = \lambda_L = 3.37\mu m$, R = 3.03 has been obtained by Nandy (1964 a, b, 1965) for 'normal' stars in the Cygnus direction, while R = 6.7 was obtained for stars associated with the Orion nebula (Johnson, 1968). The best way to estimate R is to extrapolate the extinction data at optical wavelengths to $\lambda^{-1} = 0$ using theoretical models which give the best fit up to the longest wavelength for which reliable measurements are available. Again we end up with values that are closely similar, $R \cong 3$ for the so-called normal stars, and R = 6 for stars associated with dense clouds as in the Orion nebula or in the Ophiuchus region. Higher values of R could be explained on the basis of larger particles – possibly due to mantles condensing on pre–existing grains or the expulsion or destruction of small grains. Large R implies larger amounts of neutral extinction. It is perhaps quite remarkable that values of R that deviate from 3 by significant amounts occur only in exceptional cases, pointing to an overall constancy of the wavelength dependence of visual and infrared extinction.

Optical Extinction Observations. In the wavelength region 9000–3300 Å reliable interstellar extinction observations are available for a large number of stars (Stebbins and Whitford, 1945; Divan, 1954; Johnson, 1968; Nandy, 1964, 1965). In the waveband 9000–4300 Å the extinction curve is remarkably constant for a very large number of stars distributed over widely separated regions of the galaxy. Variations occur mainly at wavelengths shortward of 4300 Å. According to Nandy (1964a,b, 1965) two main types of optical reddening curve occur: one for the Cygnus region and probably representative of the large majority of stars, and another appropriate to stars in the Perseus arm of the galaxy. These extinction curves are reproduced in Fig. 3.3. They may be represented to a high degree of accuracy by two straight lines intersecting at $\lambda^{-1} = 2.3\mu m^{-1}$. The ratio of slopes of the ultraviolet part of the extinction curve to that of the blue–visible part decreases by about 30% in Perseus as compared with the Cygnus region. The change of slope appears to occur over a waveband as little as ~ 100 Å, within the limits of the spectral resolution of Nandy's observations.

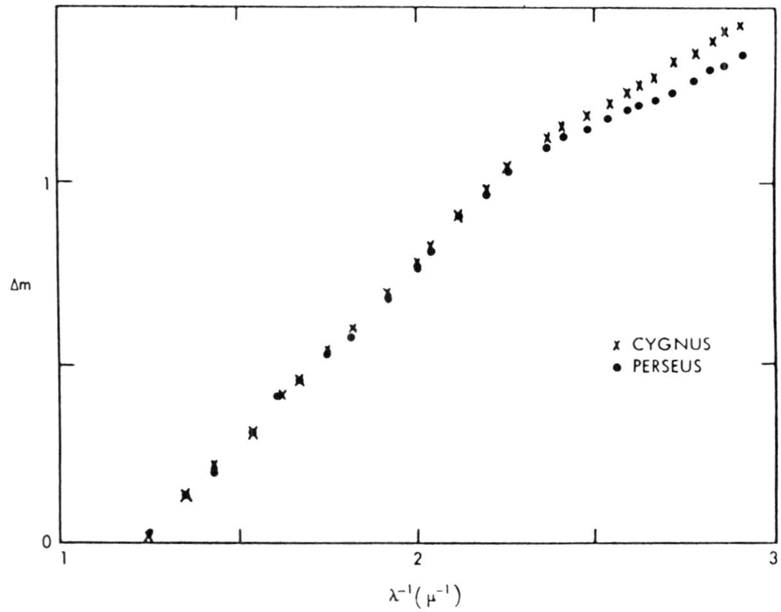

Fig. 3.3 *Nandy's extinction curves in the visual spectral region for stars in the directions of Cygnus and Perseus (Nandy, 1964, 1965).*

The two curves in Fig. 3.3 are derived from interstellar reddening surveys carried out in Edinburgh using several telescopes in conjunction with objective prisms crossed by diffraction gratings. The advantage of using a Schmidt telescope in this way is that a large number of stellar spectra can be obtained simultaneously on a single plate. From a comparison of spectra of reddened and unreddened stars of the same spectral type an extinction curve can be deduced for a given pair of stars; and, averaging over a large number of pairs occurring on the same plate a mean interstellar extinction curve for a particular region of the sky may be obtained. The main source of error in a reddening curve derived from observations of a single pair of stars arises from uncertainties in the determination of spectral class. These errors are reduced by averaging over a large number of stars. Thus in the Edinburgh Survey 28 pairs of stars have been studied in the Cygnus region and 41 pairs in the Perseus region. The details of the observational techniques have been discussed elsewhere (Nandy, 1964, 1965) and will not be repeated here.

It has been recognised for several decades that the extinction curve for stars in the central region of the Orion nebula (which is an ionised region) differs markedly from the mean interstellar extinction curve. Departures from the normal $1/\lambda$ extinction law in the UBV wavelength range were first observed by Baade and Minkowski (1937) and numerous subsequent confirmations followed, notably by Stebbins and Whitford (1945), Borgman (1961), Johnson (1968) and Divan (1971). The extinction curve for the star HD 37061 which lies close to the central region of the Orion nebula is shown in Fig. 3.4 (Nandy and Wickramasinghe, 1971). This curve is also appropriate for stars embedded in the nebulosity and is probably characteristic of dust particles within the nebula (Lee, 1968). The trend in the Orion stars, as we saw earlier, is to find values of total to selective extinction A_V/E_{B-V} considerably higher than the average value of ~ 3.1.

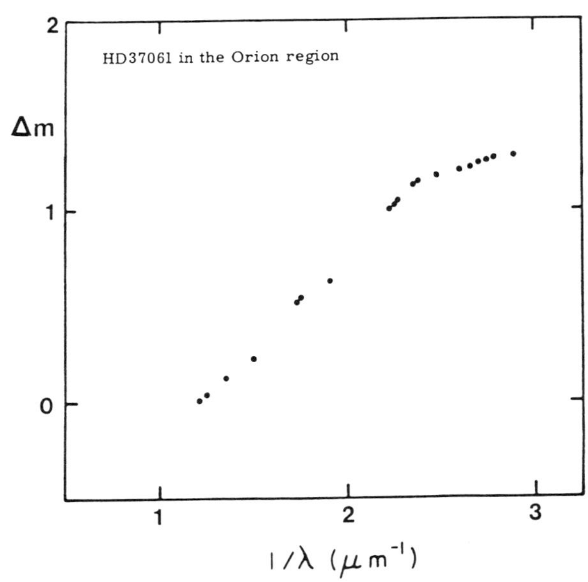

Fig. 3.4 *Normalised extinction data for HD 37061 in the Orion nebula.*

Ultraviolet Extinction Observations. The most important data on the interstellar extinction curve have emerged from its wavelength extension into the ultraviolet, at wavelengths shortward of 3000 Å. Photometry at these wavelengths can be carried out successfully only from above the earth's atmosphere. The first reliable photometric measurements of stars at ultraviolet wavelengths were made by Boggess and Borgman (1964) using rocket–based equipment. From their observations at two effective wavelengths, 2600 Å, 2200 Å together with intrinsic fluxes derived from appropriate model atmosphere calculations, the first mean extinction curve in the ultraviolet spectral region was derived. The ultraviolet extinction (in magnitudes) appeared to rise roughly linearly with $1/\lambda$ up to $\lambda \approx 2200$ Å according to this early data.

Extinction measurements were refined and extended further into the ultraviolet by Stecher (1965, 1969), Bless, Code and Houck (1968) and Bless and Savage (1970). Stecher's original data indicated a conspicuous hump in the extinction centred on $\lambda^{-1} = 4.6 \mu m^{-1}$, and this was confirmed by Bless and Savage (1970, 1972), and Savage (1970) using data obtained from OAO–2. Data now available for a large number of stars in widely separated galactic regions all exhibit the same hump at more or less the same wavelength. The ultraviolet extinction curves for several stars are plotted in Fig. 3.5 in conjunction with the corresponding optical data.

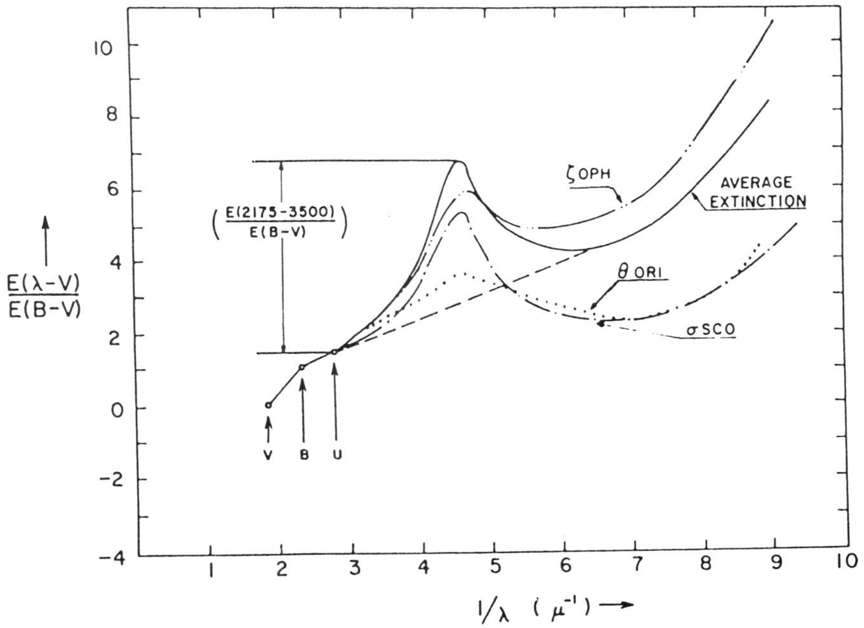

Fig. 3.5 *Normalised ultraviolet extinction data for several stars combined with appropriate optical data (Stecher, 1969; Bless and Savage, 1970).*

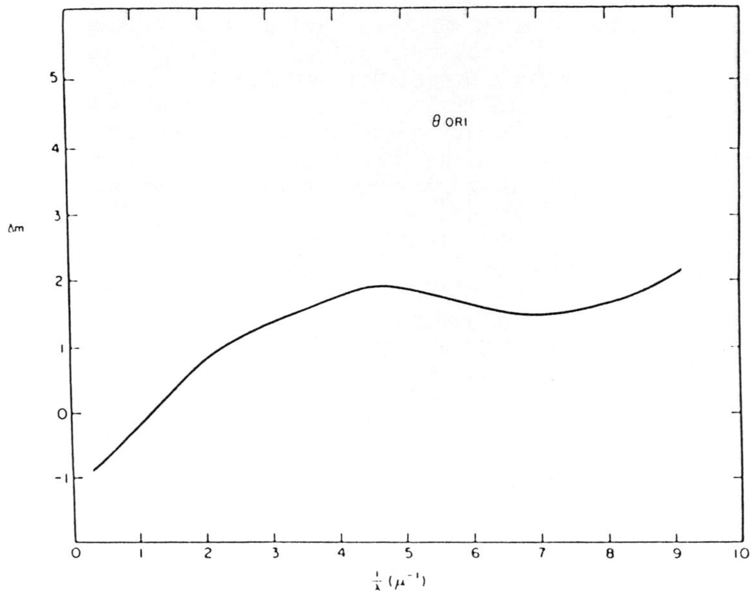

Fig. 3.6 *A typical extinction curve for stars in the Orion region.*

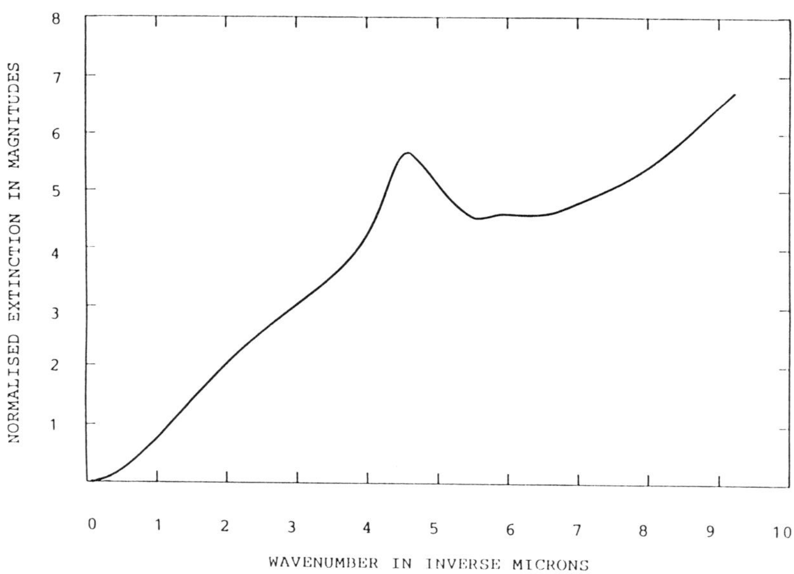

Fig. 3.7 *Normalised mean extinction curve in the Galaxy from the compilation by Sapar and Kuusik (1978). Normalisation is to $\Delta m = 1.8$ at $\lambda^{-1} = 1.8\mu m^{-1}$.*

The normalisation is to $\Delta m = 0$ at $\lambda = \lambda_V$, $\Delta m = 1.0$ at $\lambda = \lambda_B$. The height of the curve near 1000 Å and possibly the width of the $4.4\mu m^{-1}$ hump are found to vary from star to star.

Bless and Savage (1970) have also measured extinction curves appropriate to the Trapezium stars in the Orion nebulae, and Carruthers (1970) has deduced the ultraviolet extinction for θ Orionis from photometric data in the 1030–1180 Å and 1230–1350 Å wavelength ranges obtained in an Aerobee rocket flight. These far ultraviolet extinction data confirm the existence of an extinction law in the Orion nebula which differs significantly from the mean extinction law, pointing to the dominance of effects local to this region. The extinction curves for Orion stars derived by several authors for different spectral regions are presented in Fig. 3.6 (points and dashed curve).

Apart from such notable exceptions as the Orion stars and certain stars in the Scorpio–Ophiuchus region there is a good correlation between strength of the 2200 Å feature and the visual extinction for most stars (Nandy et al., 1975, Savage, 1975). The original OAO–2 observations have now been repeated by several observers using the TD1 and IUE satellites (Nandy et al., Seaton, 1979) and the general shape of the interstellar extinction curve over a wide wavelength range from $0.3 \le \lambda^{-1} \le 9\mu m^{-1}$ appears on the whole to be invariable. The main changes of extinction from star to star occur at the shortwave end of this range. Table 3.3 below shows the average interstellar extinction data from a compilation by Sapar and Kuusik (1978). The normalised mean extinction curve is plotted in Fig. 3.7.

TABLE 3.3

Normalised mean extinction data compiled by Sapar and Kuusik (1978)

$\lambda^{-1}(\mu^{-1})$	$\Delta m(\lambda)$	$\lambda^{-1}(\mu^{-1})$	$\Delta m(\lambda)$
0.30	0.082	4.10	4.497
0.50	0.186	4.20	4.783
1.00	0.792	4.31	5.132
1.20	1.037	4.42	5.487
1.40	1.305	4.50	5.645
1.60	1.561	4.59	5.697
1.80	1.800	4.63	5.691
2.00	2.074	4.72	5.586
2.20	2.313	4.81	5.441
2.40	2.511	4.90	5.266
2.60	2.691	5.00	5.062
2.80	2.866	5.21	4.765
3.00	2.913	5.38	4.619
3.20	3.233	5.56	4.520
3.40	3.443	5.88	4.654
3.50	3.559	6.25	4.579
3.60	3.676	6.67	4.666
3.70	3.810	7.14	4.887
3.79	3.915	7.69	5.190
3.91	4.113	8.33	5.767
4.00	4.282	9.09	6.600

The wavelength at the centre of the ultraviolet extinction feature remains approximately constant even in situations where the band appears weaker as well as broader (Massa and Fitzpatrick, 1986). Fig. 3.8 shows a histogram of the distribution of wavelengths of this extinction peak over a sample of 26 reddened early–type stars accessed from the IUE satellite data bank (Hoyle et al., 1985). A concentration of maxima at 2175 Å together with a relatively small spread of the histogram about the peak is clearly seen. However, individual stars do indeed have peaks that are slightly displaced from the mean, and a couple of stars have recently been found to have peaks displaced to $\lambda < 2130$ Å. Cardelli and Savage (1987) have found the peak to be at 2110 Å for HD 62542 (associated with a dark cloud in the Gum nebula) and 2128 Å for HD 29647 (a star associated with the Taurus cloud). The absorption profiles are also broader and shallower in these two cases.

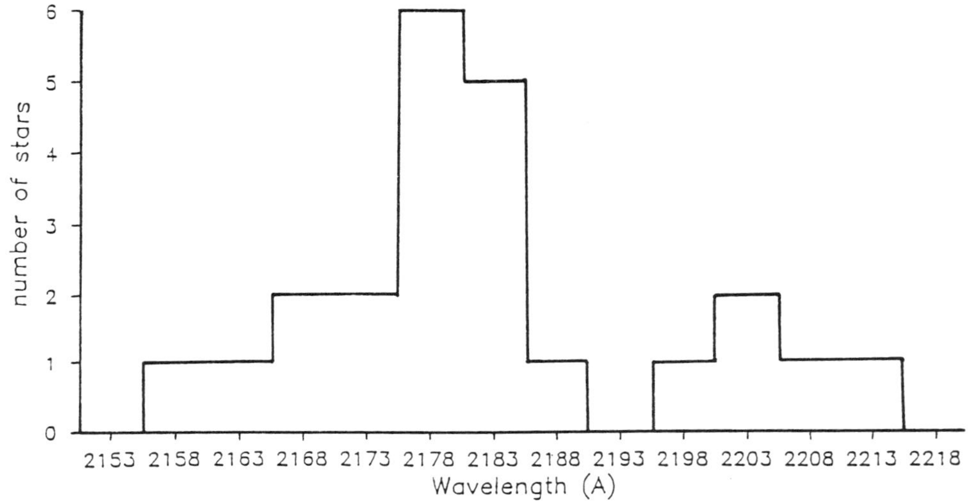

Fig. 3.8 *Histogram of the distribution of ultraviolet extinction peaks for 26 reddened stars from IUE spectra.*

The available interstellar extinction data over the visible and ultraviolet regions of the spectrum may be summarised broadly as follows:

(1) The extinction curves mostly show a definite maximum at $\lambda 2175 \pm 9$ Å, a broad minimum between $\lambda 1350$ Å and $\lambda 1800$ Å and a rise into the further ultraviolet.

(2) The average shape of the extinction curve is generally constant, but the height of the far ultraviolet portion including that of the 2175 Å peak is variable from star to star.

(3) In the visual spectral region a $1/\lambda$ extinction law holds good approximately, with a significant change of slope occurring at $\lambda \simeq 4350$ Å.

For wavelengths $\lambda \gtrsim 0.6 \mu m$ the extinction law is approximately the same for most stars, even those embedded in local dust clouds (Rieke and Lebofsky, 1985). In making such comparisons it is important to note that the extinction curve should be normalised between two reference wavelengths greater than 0.6μm. If this is done the infrared extinction law is found to vary approximately as λ^{-2}.

Extinction Bands in the Infrared. An important aspect of the extension of interstellar extinction curves into the infrared has involved the search for characteristic bands of various types of solid material which were proposed for the grains. Thus, on the basis of extinction by ice grains, a band at $\lambda \simeq 3.1\mu$m was to be expected in the extinction curve of highly reddened stars. Attempts to detect such a band have consistently led to negative results (Danielson, Woolf and Gaustad, 1965; Cudaback and Gaustad, 1969). The stars so far studied are all highly attenuated at visual wavelengths (10 magnitudes of visual extinction in one case), sufficiently bright intrinsically, and/or sufficiently cool, to have measurable fluxes in the infrared. The conclusion to be drawn from this data is that ice grains are unlikely to be the main cause of interstellar extinction. The detection of absorption bands near $\lambda \approx 3\mu$m in infrared sources associated with circumstellar shells does not alter this conclusion for the general interstellar medium.

Another infrared feature which has been actively searched for is the 10μm band characteristic of silicates as well as of organic matter. The occurrence of this band in emission was discovered for several cool stars and in the Trapezium nebula. These results will be discussed in a later chapter. A 10μm feature appears in absorption in the flux curves of several infrared sources, including the well–known BN object. The deep 10μm extinction band also appeared in the spectrum of the galactic centre infrared source GC–IRS 7 (Woolf, 1970).

3.3. OBSERVATIONS OF INTERSTELLAR POLARISATION

The discovery that the light of most stars is partially plane polarized was made by Hiltner (1949) and Hall (1949). They had embarked on a project to search for polarisation effects in early–type eclipsing binaries – a phase dependent feature predicted theoretically by S. Chandrasekhar. Although the effect they were looking for was not found, the project led to the discovery of a phase independent polarisation which was readily attributed to interstellar matter. A similar effect was found by Hiltner (1950, 1956) for a large number of ordinary stars for which interstellar reddening was already established. Subsequently, there has been a considerable accumulation of polarimetric data carried out by a number of investigators.

If I_{max}, I_{min} are maximum and minimum intensities recorded as a polarising device is rotated in the focal plane of a telescope we define p thus:

$$p = \frac{I_{max} - I_{min}}{I_{max} + I_{min}} \tag{3.24}$$

If initially unpolarized light leaves a star, $I_{max} \neq I_{min}$ implies unequal optical depths τ_1, τ_2 for light with electric vectors in two perpendicular directions. Thus we have

$$I_{max} = I_0 \exp(-\tau_1)$$
$$I_{min} = I_0 \exp(-\tau_2) \tag{3.25}$$

and from equation (3.24) we obtain

$$p = \frac{e^{-\tau_1} - e^{-\tau_2}}{e^{-\tau_1} + e^{-\tau_2}} = \frac{1 - \exp(-\Delta\tau)}{1 + \exp(-\Delta\tau)} \tag{3.26}$$

where $\Delta\tau = \tau_2 - \tau_1$. If $|p| < 1$ we get from equation (3.26)

$$\Delta\tau = \ell n(1+p) - \ell n(1-p) =$$
$$= (p - \frac{p^2}{2} + ...) - (-p - \frac{p^2}{2} + ...) \cong 2p \tag{3.27}$$

to sufficient accuracy if $p << 1$. Expressed in magnitudes we have

$$\Delta m_p = 1.086 \, \Delta\tau \approx 2.172 \, p \tag{3.28}$$

where Δm_p denotes the difference in extinction between the two directions of polarisation.

Correlation of p with E_{B-V}. If interstellar polarisation arises from unequal optical depths of the interstellar medium with respect to orthogonal directions of the electric vector it would be reasonable to expect a correlation between p and the total extinction.

From the earliest observations it emerged that Δm_p is at least weakly correlated to E_{B-V}, and hence to Δm (Hall, 1949). Another investigation by Schmidt (1958) expresses this correlation in the form $\Delta m_p/\Delta m \leq 0.065$. Later observations (Serkowski et al., 1975) have suggested lower values for $\Delta m_p/\Delta m$, giving an average of ~ 3% over a path length of ~ 1 kpc in the plane of the Galaxy. These results indicate that total extinction and polarisation are caused by the same or closely related population of grains.

Polarisation and Aligned Grains. A simple explanation for non–zero values of p is that interstellar dust grains have anisotropic shapes (e.g. long whiskers) and are aligned to a small degree in some preferential direction. Consider a model where the individual grains in the line of sight to a star are all in the form of infinite cylinders of radius a, length l, refractive index m and that they are lined up transverse to the propagation vector of starlight in some preferred plane. Let Q_{eE}, Q_{eH} be the efficiency factors for extinction with **E** and **H** respectively in the preferred plane. From an analysis identical to that developed earlier for spherical particles we have

$$\Delta m_E = 1.086(2al \, Q_{eE}) \, N_g d$$
$$\Delta m_H = 1.086(2al \, Q_{eH}) \, N_g d \tag{3.29}$$

where N_g is the number density of grains and d is the path length. Thus we have

$$\Delta m_p = 1.086(2al)\,|Q_{eE} - Q_{eH}|\,N_g d \tag{3.30}$$

and

$$\frac{\Delta m_p}{\Delta m} = \frac{|Q_{eE} - Q_{eH}|}{Q_{eE} + Q_{eH}} \tag{3.31}$$

If only a small fraction f of the cylinders are assumed to be aligned and the rest are in random orientation, then we have

$$\frac{\Delta m_p}{\Delta m} = f\,\frac{|Q_{eE} - Q_{eH}|}{Q_{eE} + Q_{eH}} \tag{3.32}$$

Equations (3.29) – (3.32) enable us to compare observations of interstellar polarisation with the predictions from any particular model of cylindrical grains.

Orientation of the Polarisation. In the following we shall refer to the plane passing through the line of sight and containing the observed maximum of the electric vector as the plane of polarisation of the light. The intersection of this plane with the sky defines a direction which will be called the direction of polarisation, or the polarisation vector.

Even more significant than the correlation between p and A_V, discussed in the previous section, are the observed spatial correlations of the polarisation vector. For stars lying within a few degrees of the galactic equator it is found that the direction of polarisation is generally parallel to the galactic plane over wide ranges of galactic longitude. Recent surveys on this topic include work by Klare and Neckel (1970), Mathewson (1968), Mathewson and Ford (1970), and Martin (1971).

In most regions of the galaxy the polarisation vector is parallel to the direction of the local spiral arm in which the stars are embedded. Since a large–scale galactic magnetic field is present parallel to the spiral arm, it follows that the polarisation vector is also generally parallel to the magnetic field. However, as pointed out by Martin (1971), several local regions do exist where this is clearly not so. The polarisation vector may be highly inclined, even perpendicular, to the mean magnetic field in some cases.

A representative plot of observed polarisation vectors on sky is shown in Fig. 3.9, taken from a diagram given by Mathewson and Ford (1970).

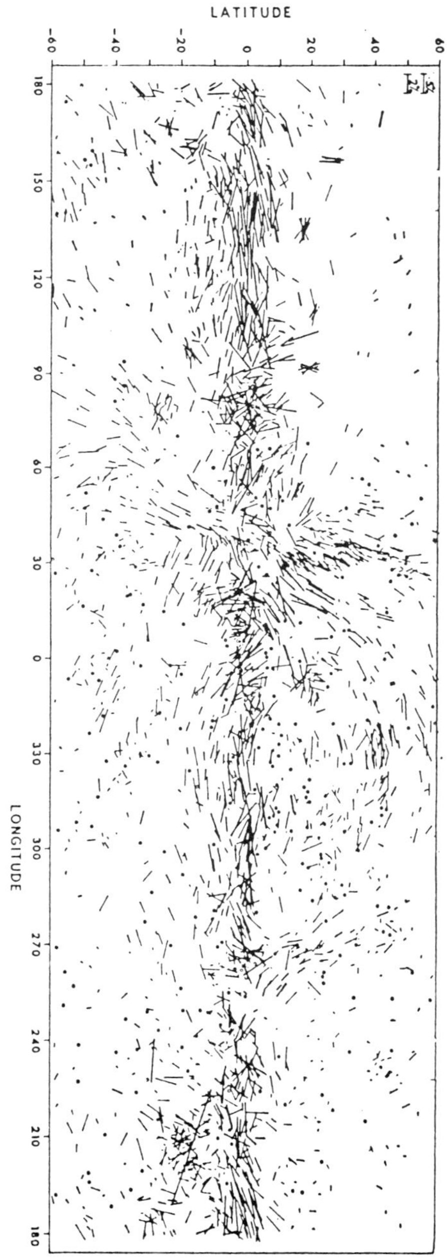

Fig. 3.9 *Plot of polarisation vectors for stars in the Galaxy (from Mathewson and Ford, 1970).*

Wavelength Dependence of Polarisation. The earliest observations indicated that Δm_p is fairly flat over the waveband 4000 – 8000 Å with a weak maximum plateau at $\lambda \approx 5000 - 6000$ Å, but with a fall off at longer and shorter wavelengths.

Unlike the interstellar extinction curve at visual wavelengths, the wavelength dependence of Δm_p is significantly variable from star to star (Coyne and Gehrels, 1966, 1967; Coyne and Wickramasinghe, 1969). In particular, the wavelength of maximum Δm_p is found to vary in the range $\lambda_{max} \simeq 4500 - 6000$ Å, with the most marked variations occurring in cases where the ratio of total to selective extinction is also high. These variations have been interpreted as implying local modifications to the sizes of grains that are effectively aligned. Altough significant variations in λ_{max} occur from star to star, it is found that most of the data is consistent with an empirical formula given by Serkowski (1973),

$$\frac{p(\lambda)}{p_{max}} = \exp\{-K\, \ell n^2(\lambda_{max}/\lambda)\}$$

$$K = 1.15 \tag{3.33}$$

A comparison of observations with this formula is given in Fig. 3.10. The average value of λ_{max} is found to be $\lambda_{max} \cong 5500$ Å.

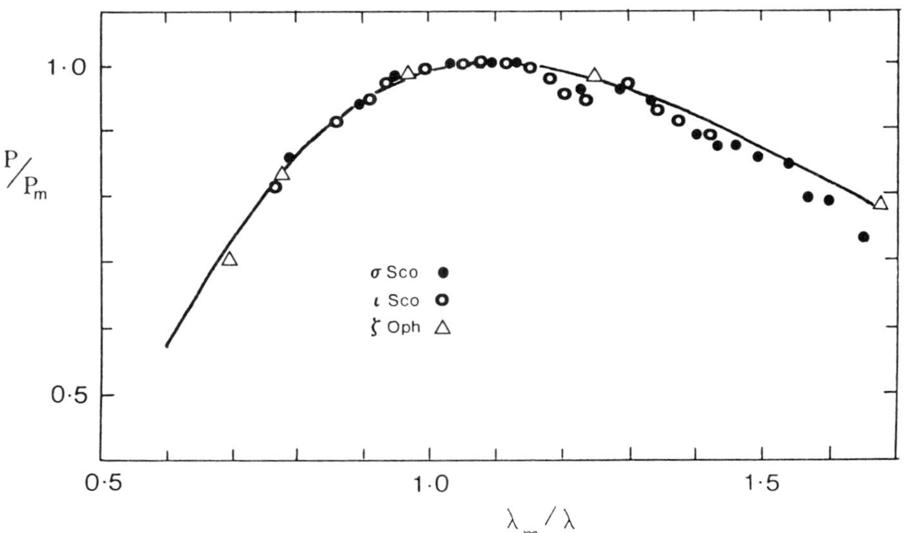

Fig. 3.10 *Wavelength dependence of polarisation for several stars compared with the empirical curve given by Serkowski (1973).*

Circular Polarisation. If interstellar space containing widely separated dust grains is regarded as a scattering medium, the linear polarisation given by equation (3.30) can be regarded as being a consequence of unequal imaginary parts of the complex refractive index of the medium in two orthogonal planes (Purcell, 1969). A corresponding difference between the real parts of the refractive index could lead under suitable conditions to a degree of circular polarisation (Martin, 1973, 1974). The requirement is for light that leaves the source to have a direction of electric vector **E** that is distinct from the directions of each of the principal axes of the medium. Alternatively, directions of grain alignment may be systematically 'twisted' along the line of sight, thus leading to a rotation of the electric vector of the propagating wave.

While the imaginary part of the refractive index of the medium affects the amplitude of a propagated wave, the real part determines the phase. In either of the situations discussed above components of radiation with electric vector in two orthogonal planes will typically show distinct values of the phase velocity, thereby resulting in a relative phase shift which produces a small degree of circular polarisation.

Circular polarisation could be described by an equation of the type (3.24) in which I_{max} and I_{min} are replaced by I_L, I_R, the intensities for left–handed and right–handed circular polarisation. More commonly, however, circular polarisation is measured by the ratio of the Stokes parameters, V/I.

The Crab Nebula is an example of a source where optical emission due to sychrotron radiation is linearly polarized to a high degree, and the direction of the **E** vector is not coincident with either of the principal axes of the interstellar medium.

It is in this object that interstellar circular polarisation was first discovered (Martin et al., 1973). Since then circular polarisation was measured at several wavelengths for a number of stars whose line of sight passes through regions containing grains with systematically different directions of alignment. Circular polarisation measurements for a representative sample of stars are discussed by Martin and Campbell (1976).

Both the linear and circular polarisation effects that result from aligned cylindrical grains can be modelled using calculations based on single particle properties. If $S_1(0)$ and $S_2(0)$ are the complex amplitude functions for forward scattering given by the usual Mie–type series, and $x = 2\pi a/\lambda$ where a is the radius of a single grain, then

$$Q_{eE} = \frac{2}{x}\,\mathrm{Re}[S_1(0)] \tag{3.34}$$

$$Q_{eH} = \frac{2}{x}\,\mathrm{Re}[S_2(0)] \tag{3.35}$$

and similarly we have

$$\begin{aligned} Q_{cE} &= \frac{2}{x}\,\mathrm{Im}[S_1(0)] \\ Q_{cH} &= \frac{2}{x}\,\mathrm{Im}[S_2(0)] \end{aligned} \tag{3.36}$$

where Q_{cE}, Q_{cH} are the single particle phase–lag efficiencies in the two directions of polarisation (van de Hulst, 1957; Shapiro, 1975). The degree of circular polarisation can now be measured by the ratio of the Stokes parameters V/I, where

$$V/I \propto (Q_{cE}-Q_{cH})(Q_{eE}-Q_{eH}) \,. \tag{3.37}$$

Observations of V/I, scaled by an arbitrary factor, are shown by the points with error bars in Fig. 3.11 (Martin, 1974; Shapiro, 1975; Kemp and Wolstencroft, 1972). A consistent trend to be noted is that V/I changes from being positive to negative at a wavelength between 0.5μm and 0.6μm if we take λ_m to be in the range 0.5 – 0.6μm.

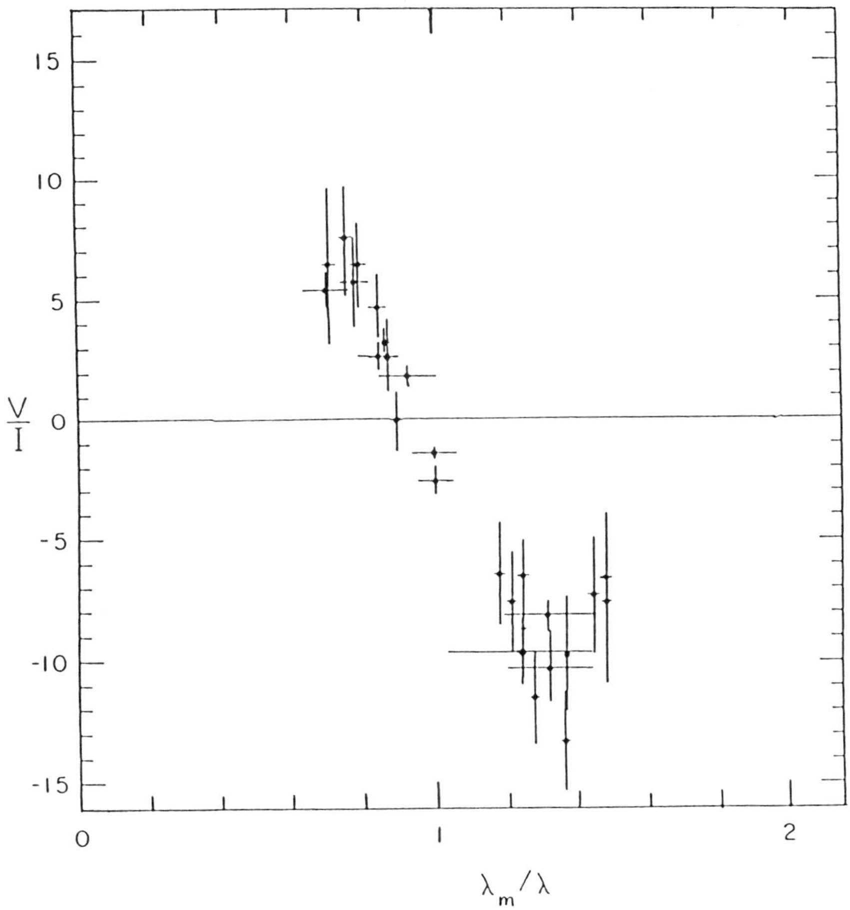

Fig. 3.11 *Measures of circular polarisation V/I for several stars, scaled by arbitrary normalising factors (from Shapiro, 1975; Kemp and Wolstencroft, 1972).*

3.4. DIFFUSE INTERSTELLAR BANDS

We have so far described the observational data relating to the main features of the extinction and polarisation of starlight. The spectral resolution used in these studies, for instance in wide band filter photometry, would not have permitted the detection of any narrow lines or bands in the spectra of reddened stars. There are in fact a number of diffuse lines and bands in stellar spectra which are to date unidentified and which can only be interpreted as fine structure in the wavelength dependence of extinction. They have widths ranging from ~ 40 Å to less than 5 Å – extremely wide compared with known atomic and molecular lines with natural widths $\leq$ 0.01 Å.

TABLE 3.4

Principal Diffuse Bands

Central Wavelength (A)	Width (A)
4428	28
4762	4
4883	30
5778	16
5780	3
5797	3
6175	30
6203	2
6270	2
6284	4
6614	1

Since their original discovery in 1936 by P.W. Merrill, the number of these so–called diffuse interstellar bands has steadily risen. At the present time about 20 such features have been detected and studied. Fig. 3.12 shows the wavelength distribution and widths of the diffuse bands from data supplied by G.H. Herbig. The salient characteristics of the stronger of these diffuse bands are summarized as follows:

(1) Central wavelengths and widths of strongest bands are set out in Table 3.4.

(2) We note that the present list of diffuse bands is restricted to the rather limited wavelength range 4400 – 7000 Å.

(3) Central wavelengths are invariant, but shapes of profiles could vary for a given band from star to star.

(4) For the strongest and most thoroughly investigated band, the λ 4430 Å feature, the excess of extinction at the band centre (relative to the extinction

in the wings of the band) is ~ 0.1 mag per magnitude of continuum extinction at $\lambda = 5000$ Å.

(5) For most stars the colour excess (and hence extinction) and the equivalent width of the λ 4430 Å (or λ 6175 Å) band are positively correlated. This strongly suggests that the carriers of the diffuse bands are contained in or are intimately associated with the grains that cause extinction at visual wavelengths.

(6) A deficiency of interstellar features is often found in stars associated with bright emission line nebulae. In some way the carriers of bands appear to be destroyed or rendered inactive near hot stars.

Extinction and Polarisation Profiles of Diffuse Bands. The main problem involved in obtaining accurate extinction profiles of bands is the determination of the stellar continuum in the vicinity of band. A procedure used by K. Nandy is to fit a set of orthogonal polynomials to the stellar spectrum omitting the band. One has to be careful here to omit all other lines and bands as well. Subtraction of such a fitted continuum from the original spectrum yields the desired band profile. The mean profiles of the λ 4430 Å band for two groups of stars, one covering the spectral types B1 – B3 and another B8 – A0, indicate the presence of an emission wing about 20 Å on the blue side of the band. Such an emission wing is expected if the absorbing agent is dispersed throughout grains of radius ~ 10^{-5} cm and is due essentially to the effect of anomalous dispersion (Wickramasinghe and Nandy, 1970).

If diffuse extinction bands represent fine structure in the interstellar extinction curve, similar fine sructure should also show up in the interstellar polarisation curve at high resolution. This expectation is borne out in an investigatiohn by Nandy and Seddon (1970) for the λ 4430 Å band.

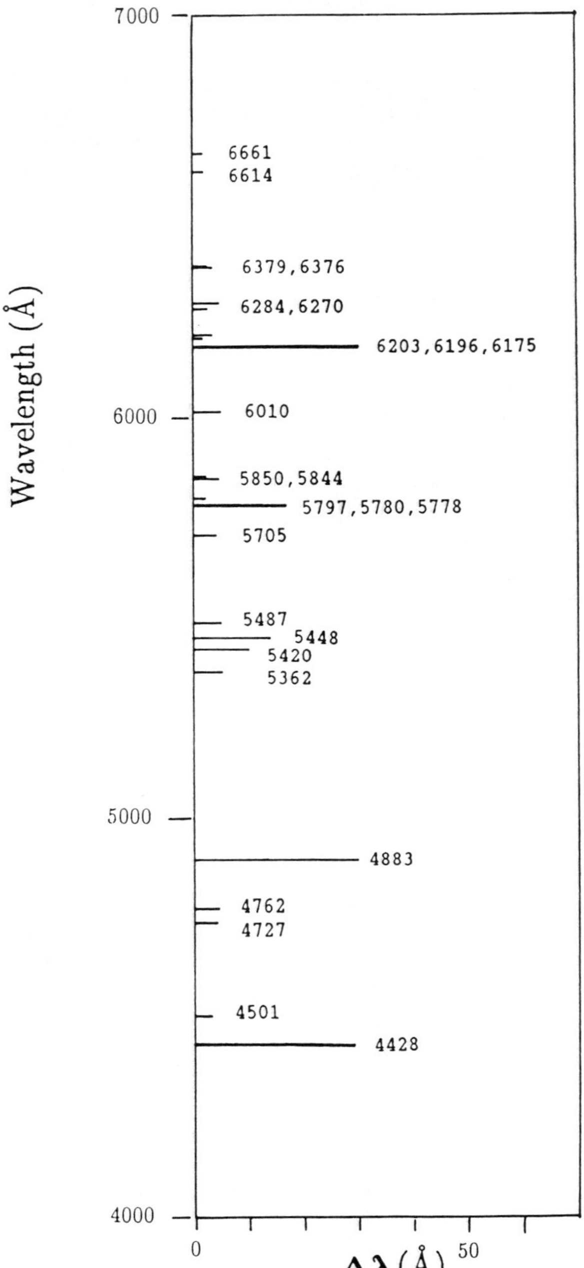

Fig. 3.12 *Distribution of central wavelengths and widths of the diffuse interstellar bands (Courtesy G. Herbig)*

References

Baade, W. and Minkowski, R., 1937, *Astrophys. J.*, **86**, 123.

Bless, R.C., Code, A.D. and Houck, T.E., 1968, *Astrophys. J.*, **153**, 561.

Bless, R.C. and Savage, B.D., 1970, in L. Houziaux and H.E. Butler (eds.) *Ultraviolet Stellar Spectra and Groundbased Observations*, D. Reidel.

Bless, R.C. and Savage, B.D., 1972, *Astrophys. J.*, **171**, 293.

Borgman, J., 1961, *Bull. Astr. Inst. Netherlands*, **16**, 99.

Boggess, A. and Borgman, J., 1964, *Astrophys. J.*, **140**, 1636.

Cardelli, J.A. and Savage, B.D., 1988, *Astrophys. J.*, **325**, 864.

Carruthers, G.R., 1970b, *Astrophys. J. Lett.*, **161**, L81.

Coyne, G.V. and Gehrels, T., 1967, *Astron. J.*, **72**, 887.

Coyne, G.V. and Wickramasinghe, N.C., 1969, *Astron. J.*, **74**, 1179.

Danielson, R.E., Woolf, N.J. and Gaustad, J.E., 1965, *Astrophys. J.*, **141**, 116.

Divan, L., 1954, *Ann. Astrophys.*, **17**, 456.

Divan, L., 1971, *Astr. Astrophys.*, **12**, 76.

Hall, J.S., 1949, *Science*, **109**, 166.

Hiltner, W.A., 1949, *Science*, **109**, 165.

Hiltner, W.A., 1950, *Phys. Rev.*, **78**, 170.

Hiltner, W.A., 1956, *Astrophys. J. Suppl.*, **2**, 389.

Hoyle, F., Wickramasinghe, N.C., Al–Mufti, S. and Karim, L.M., 1985, *Astrophys. Sp. Sc.*, **114**, 303.

Johnson, H.L., 1965, *Astrophys. J.*, **141**, 923.

Johnson, H.L., 1968, *Stars and Stellar Systems*, **7**, 167.

Johnson, H.L. and Borgman, J., 1963, *B.A.N.*, **17**, 115.

Kemp, J.C. and Wolstencroft, R.D., 1972, *Astrophys. J.*, **176**, L115.

Klare, G. and Neckel, T., 1970, *IAU Symp. No. 38*, 449.

Knacke, R.F., Cudaback, D.D. and Gaustad, J.E., 1969, *Astrophys. J.*, **158**, 151.

Lee, T.A., 1968, *Astrophys. J.*, **152**, 913.

Martin, P.G., 1971, *Mon. Not. Roy. Astr. Soc.*, **153**, 279.

Martin, P.G., 1973, in IAU Symp. No. 52, J.M. Greenberg and H.C. van de Hulst (eds.) *Interstellar Dust and Related Topics*, D. Reidel.

Martin, P.G., 1974, *Astrophys. J.*, **187**, 461.

Martin, P.G., Illing, R. and Angel, J.R.P., 1973, in IAU Symp. No. 52, J.M. Greenberg and H.C. van de Hulst (eds.) *Interstellar Dust and Related Topics*, D. Reidel.

Martin, P.G. and Campbell, B., 1976, *Astrophys. J.*, **298**, 727.

Massa, D. and Fitzpatrick, E.L., 1986, *Astrophys. J. Suppl.*, **60**, 305.

Mathewson, D.S., 1968, *Astrophys. J. Lett.*, **153**, L47.

Mathewson, D.S. and Ford, V.L., 1970, *Mem. Roy. Astr. Soc.*, **74**, 139.

Merrill, P.W., 1936, *Astrophys. J.*, **83**, 126.

Nandy, K., 1964a, *Publ. Roy. Obs. Edin.*, **4**, 57.

Nandy, K., 1964b, *Publ. Roy. Obs. Edin.*, **3**, No. 6, 142.

Nandy, K., 1965, *Publ. Roy. Obs. Edin.*, **5**, 13.

Nandy, K. and Seddon, H., 1970, *Nature*, **227**, 264.

Nandy, K., Thompson, G.I., Jamar, C., Monfils, A. and Wilson, R., 1975, *Astro. Ap.*, **44**, 195.

Nandy, K. and Wickramasinghe, N.C., 1971, *Mon. Not. Roy. Astr. Soc.*, **154**, 255.

Purcell, E.M., 1969, *Astrophys. J.*, **158**, 433.

Rieke, G.H. and Savage, B.D., 1987, *Astrophys. J.*, **288**, 618.

Savage, B.D., 1975, *Astrophys. J.*, **188**, 92.

Sapar, A. and Kuusik, I., 1978, *Publ. Tartu Astrophys. Obs.*, **46**, 71.

Schmidt, Th., 1958, *Z. fur Astrophys.*, **46**, 145.

Seaton, M.J., 1979, *Mon. Not. Roy. Astr. Soc.*, **187**, 73p.

Serkowski, K., 1973, in IAU Symp. No. 52, J.M. Greenberg and H.C. van de Hulst (eds.) *Interstellar Dust and Related Topics*, D. Reidel.

Serkowski, K., Mathewson, D.S. and Ford, V.L., 1975, *Astrophys. J.*, **196**, 261.

Shapiro, P.R., 1975, *Astrophys. J.*, **201**, 151.

Stecher, T.P., 1965, *Astrophys. J.*, **142**, 1683.

Stebbins, J. and Whitford, A.E., 1943, *Astrophys. J.*, **98**, 20; 1945, *Astrophys. J.*, **102**, 318.

Whitford, A.E., 1948, *Astrophys. J.*, **107**, 102.

Whitford, A.E., 1958, *Astron. J.*, **63**, 201.

van de Hulst, H.C., 1957, *Light Scattering by Small Particles*, John Wiley.

Wickramasinghe, N.C. and Nandy, K., 1970, *Astrophys. Sp. Sci.*, **6**, 154.

Woolf, N.J., 1970, in J.M. Greenberg and H.C. van de Hulst (eds.) *Interstellar Dust and Related Topics*, D. Reidel, 1970.

4. Reflection Nebulae and the Diffuse Galactic Light

4.1. INTRODUCTORY REMARKS

The gas and dust comprising the interstellar medium shows a marked tendency to exist in the form of discrete clouds. The typical dimension of a cloud is of the order of a few parsecs. Interstellar extinction and polarisation may be regarded as the net effect of several such clouds in the line of sight of a star.

Interstellar clouds show up in the form of dark patches and striations giving rise to the splotchy appearance of photographs of the Milky Way. They may also show up as bright nebulae when they occur in the neighbourhood of early–type stars. Such associations between early–type stars and clouds may arise either by random encounters between stars and clouds, or as a result of stars forming within cloud complexes.

Pioneering observational work on bright nebulae was done by Edwin Hubble as early as 1922. After making a large number of observations of bright nebulae and their associated stars, he classified them into two groups – reflection nebulae and emission nebulae.

Reflection nebulae are those associated with stars of spectral type B or later. The nebular spectrum in such cases is mainly of a continuum nature, showing the absorption features of the star itself. The reflection nebulae are visible in the light diffusely scattered by the dust grains present in them.

Emission nebulae are clouds associated with stars of spectral type O or B0. Their nebular spectra show strong emission lines and a weak background continuum.

In the present chapter we shall deal mainly with reflection nebulae. We shall also consider briefly the phenomenon of the diffuse galactic light – the light scattered diffusely by grains from the general interstellar radiation field.

4.2. THE APPARENT SIZE OF REFLECTION NEBULAE

Hubble's early observations (1922) indicated that there existed a rather well–defined relationship between the apparent magnitude of an illuminating star and the maximum angular distance from the star at which a nebula had sufficient brightness to be photographed. With the sensitivity of plates used by Hubble this limiting brightness was expressible as 23.25 mag/square second of arc. Let a denote the angular separation between the star and the most distant visible point of the nebula in minutes of arc, and m_* the apparent magnitude of the star. The points $(\log a, m_*)$ for a large number of reflection nebulae were plotted by Hubble as in Fig. 4.1. The best straight–line fit to these observations may be expressed by the equation

$$m_* + 4.9 \log a = 11 \tag{4.1}$$

The interpretation of reflection nebulae observations is made difficult owing to uncertainties in the geometry of the nebulae and of the relative positions of star and nebula. Certain rather qualitative deductions are, however, possible if we introduce simplifying assumptions.

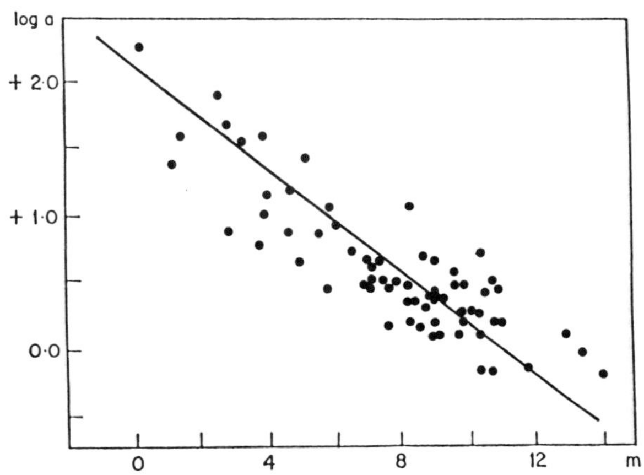

Fig. 4.1 *Hubble's relation between the radius of a reflection nebula and the magnitude of the illuminating star*

Consider a cylindrical element of a nebula with cross–sectional area dA and generators parallel to the line of sight, extending through the whole nebula, as shown in Fig. 4.2. The light scattering by grains within this cylinder would give rise to a patch of luminosity on the projected image of the nebula. Let τ_0 be the optical thickness of the whole length of cylinder, and let $d\tau$ be the optical depth of an element at P of length ds. Let the line joining P to the star make an angle α with the line of sight of the nebula, and let the optical path which lies within the nebula be of optical depth τ_1. Further, let σ be the total cross–section for extinction of all the grains contained in a unit volume of the nebula, i.e.

$$\sigma = \int_0^\infty Q_{ext}\, \pi a^2\, n(a)\, da \tag{4.2}$$

where Q_{ext} is the efficiency factor for extinction, a is the radius of a grain (assumed spherical) and n(a) da is the number density of grains with radii in the range a, a + da. Denote SP by r, and the distance between nebula and Earth by R (see Fig. 4.2).

We make the following simplifying assumptions:

(a) The distribution of grains is uniform throughout the nebula – i.e. σ is constant throughout nebula.

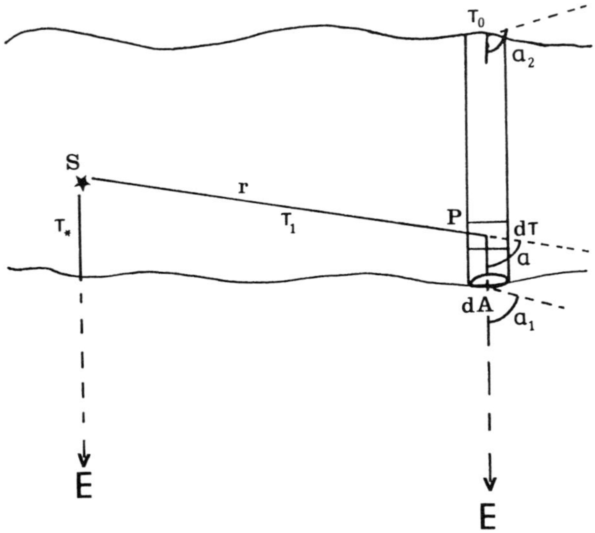

Fig. 4.2 *Section of a star–nebula system for Merope.*

(b) The star is sufficiently distant from the cylinder for r and τ_1 to remain constant as P is taken at different points on the length of cylinder.

(c) The phase function for scattering by grains is isotropic.

The intensity of radiation in erg/cm^2 sec received at P is

$$I = \exp(-\tau_1) \frac{L}{4\pi r^2} \tag{4.3}$$

where L is the luminosity of the star. The total flux removed by the grains in the element of cylinder at P is

$$I\sigma \text{ ds dA} = \exp(-\tau_1) \frac{L}{4\pi r^2} \sigma \text{ ds dA erg/sec} \tag{4.4}$$

Since σ ds $= d\tau$ (σ being extinction cross–section of grains in unit volume) equation (4.4) can be rewritten as

$$\exp(-\tau_1) \frac{L}{4\pi r^2} d\tau \text{ dA erg/sec} \tag{4.5}$$

If γ is the mean albedo of the grains, a fraction γ of this energy is diffusely scattered with an isotropic phase function. The intensity of scattered light from the element of cylinder received at Earth is therefore

$$\exp(-\tau)\,\frac{1}{4\pi R^2}\,[\gamma\exp(-\tau_1)\,\frac{L}{4\pi R^2}\,d\tau\,dA]$$

$$= \gamma\exp(-\tau_1)\,L\,dA\,\frac{1}{16\pi^2R^2r^2}\exp(-\tau)\,d\tau\ \mathrm{erg/cm^2\ sec} \tag{4.6}$$

Integrated over the length of the cylinder of total optical depth τ_0, this gives an intensity I_n:

$$I_n = \gamma[1-\exp(-\tau_0)]\,\exp(-\tau_1)\,\frac{L\,dA}{16\pi^2R^2r^2}\ \mathrm{erg/cm^2\ sec}$$

$$= \gamma[1-\exp(-\tau_0)]\,\exp(-\tau_1)\,\frac{L}{16\pi^2r^2}\,d\Omega \tag{4.7}$$

Ω denoting the solid angle subtended at the Earth by the area dA. The intensity I_* of starlight received at Earth being $(L/4\pi R^2)e^{-\tau_*}$ we have from (4.7)

$$\frac{I_n}{I_*} = \gamma[1-\exp(-\tau_0)]\,\exp(-\tau_1)\,\frac{R^2\,d\Omega}{4\pi r^2}\exp(\tau_*) \tag{4.8}$$

From Fig. 4.2, we have further

$$r\sin\alpha = Ra \tag{4.9}$$

where a is the angular separation in radians between the star and the patch of nebulosity which is the projection of the cylinder considered. Or in minutes of arc,

$$r\sin\alpha = Ra\left[\frac{2\pi}{360\text{x}60}\right] \tag{4.9a}$$

Further, the solid angle $d\Omega$ corresponding to 1 square second of arc is

$$d\Omega = \left[\frac{2\pi}{360\text{x}60\text{x}60}\right]^2 \tag{4.10}$$

and (4.8), (4.9a) and (4.10) then give

$$60^2\,.\,\frac{I_n}{I_*} = \gamma[1-\exp(-\tau_0)]\,\exp(-\tau_1)\,\frac{\sin^2\alpha}{4\pi a^2}\exp(\tau_*) \tag{4.11}$$

Expressed in magnitudes

$$m_n - m_* = 5\log a - 2.5\log(\gamma\sin^2\alpha)$$
$$+ 11.5 - 2.5\log\{[1-\exp(-\tau_0)]\exp(\tau_*-\tau_1)\} \tag{4.12}$$

At Hubble's limit of observability, (4.12) becomes (writing m_n = 23.25 mag/square second)

$$m_* + 5 \log a = 11.75 + 2.5 \log(\gamma \sin^2 \alpha) + 2.5 \log \{[1-\exp(-\tau_0)] \exp(\tau_* - \tau_1)\} \quad (4.13)$$

We can now compare this result with the observational relation (4.1). Ignoring the slight difference in the coefficient of log a, we have

$$\log(\gamma \sin^2 \alpha) = -0.30 - \log\{[1-\exp(-\tau_0)] \exp(\tau_* - \tau_1)\} \quad (4.14)$$

i.e.

$$\gamma \sin^2 \alpha = \frac{0.5}{[1-\exp(-\tau_0)] \exp(\tau_* - \tau_1)} \quad (4.15)$$

For all admissible values of τ_0, τ_* and τ_1, it is seen from (4.15) that

$$\gamma \geq 0.5 \quad (4.16)$$

It should be stressed, however, that the above condition holds good only provided: (a) the assumptions inherent in our cloud model are valid; and (b) Hubble's 1922 data are reliable. There is no reason to doubt the quality of Hubble's observations. One questionable assumption implied in the present model, however, is that the scattering phase function for a grain is spherically symmetric. The precise form of the scattering diagram of a spherical particle depends on its refractive index as well as its radius (see Chapter 2). For spherical dielectric grains of radii $\sim 0.3\mu m$ the scattering is strongly forward directed. In this case, one needs to know the relative position between star and nebula before any conclusions concerning the albedo can be reached. For values of the angle α greater than $\pi/2$ and with a strongly forward directed phase function, a somewhat lower value of γ could suffice to produce the observed limiting luminosity.

4.3. OBSERVATIONS OF NGC 7023

Although for the type of observation carried out by Hubble (1922) a large number of reflection nebulae were suitable, there are relatively few nebulae with sufficient surface brightness on which more detailed photometric measurement is possible. Among the more suitable cases is NGC 7023. The illuminating star in this case – HD 200775 – is of spectral type BV, and the nebula has an apparent angular size of about 19 minutes of arc in blue light.

Van Houten (1961) has discussed the brightness distribution and polarisation of this nebula, with a view to obtaining information on the scattering function and albedo of grains. Following van Houten (1961), we shall adopt a spherical cloud model around the star, with the density of grains assumed uniform throughout the cloud. As in section 4.2, consider a cylindrical column of cross–sectional area dA, with generators along the line of sight of a particular point of the nebulosity.

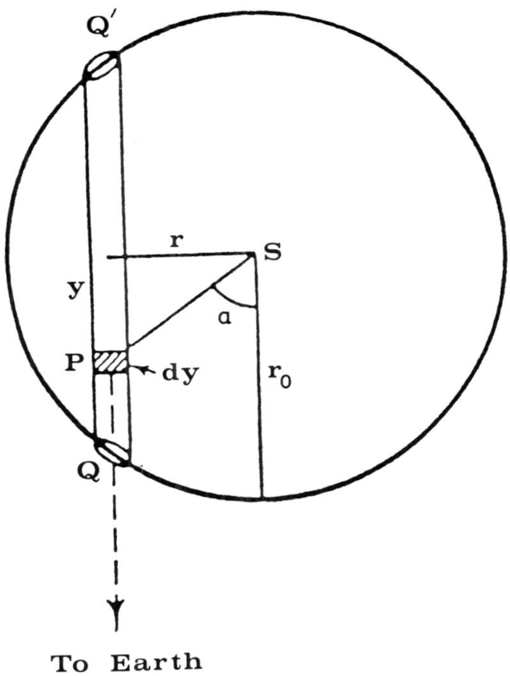

Fig. 4.3 *Section of a spherically symmetric reflection nebula*

The geometrical situation is shown in Fig. 4.3, where QQ′ represents the cylindrical column. Consider an element of cylinder of height dy containing the point P. Let k be the absorption coefficient, defined so that the optical depth corresponding to a geometrical distance dy through the cloud is k dy. The intensity of radiation received at the point P is

$$\exp(-kPS)\,\frac{L}{4\pi r^2\ \mathrm{cosec}^2\,\alpha}\ \mathrm{erg/cm^2\,sec} \tag{4.17}$$

and the flux removed by the grains in the element at P is (following (4.5))

$$\exp(-kPS)\,\frac{L}{4\pi r^2\ \mathrm{cosec}^2\,\alpha}\ k\ dy\ dA\ \mathrm{erg/sec} \tag{4.18}$$

If γ is the albedo, and $S(\alpha)$ is the scattering function, a fraction $\gamma S(\alpha)$ of this flux is scattered into unit solid angle about the direction PQ. The flux of scattered light leaving the nebula at Q per unit solid angle centred about PQ is therefore

$$\gamma \exp(-kPQ)\ S(\alpha)\ \exp(-kPS)\,\frac{L}{4\pi r^2\ \mathrm{cosec}^2\,\alpha}\ k\ dy\ dA\ \mathrm{erg/sec} \tag{4.19}$$

and the flux/cm^2 received at Earth is $1/R^2$ times this value. Noting that dA/R^2 is the solid angle (say $\delta\Omega$) subtended at Earth by the nebular area dA, the flux received at the Earth from element at P is

$$dI(r) = \gamma \exp[-k(PQ+PS)] \frac{L}{4\pi r^2 \operatorname{cosec}^2 \alpha} S(\alpha)\, k\, dy\, \delta\Omega \tag{4.20}$$

Making use of the geometrical relations

$$\begin{aligned} SP &= r \operatorname{cosec} \alpha \\ PQ &= \sqrt{(r_0^2-r^2)} - r \cot \alpha \\ y &= r \cot \alpha \\ -dy &= r \operatorname{cosec}^2 \alpha \, d\alpha \end{aligned} \tag{4.21}$$

and integrating over the length of the column from Q to Q′ we have an intensity I(r) in erg/cm^2 sec per unit solid angle of nebular area,

$$I(r) = \frac{\gamma k L}{4\pi r} \exp[-k\sqrt{(r_0^2-r^2)}] \int_{\arcsin(r/r_0)}^{\pi-\arcsin(r/r_0)} \exp[-kr(\operatorname{cosec} \alpha - \cot \alpha)] S(\alpha)\, d\alpha \tag{4.22}$$

For the phase function occurring in the integral, van Houten adopted an empirical function $S(\alpha)$ given by

$$S(\alpha) = \frac{(1-g^2)}{4\pi} (1+g^2-2g \cos \alpha)^{-3/2} \tag{4.23}$$

so that $g = 0$ corresponds to a spherically symmetric scattering diagram, and for $g = 1$ the scattering is entirely forward directed. The following observations are then interpreted in terms of (4.22) and (4.23):

(a) The surface brightness I(r) is closely proportional to r^{-1}, for r between 0.5 and 3.5 minutes of arc from star.

(b) The polarisation of nebular light is about 15% at the photographic wavelength, at r = 22 seconds of arc.

(c) I(r) has an observed value at a point 1 minute from star.

Van Houten obtained the best combination of values γ, g at optical wavelengths consistent with the above observations to be $\gamma \cong 0.8$, $0.3 \lesssim g \lesssim 0.5$. The albedo is high and the scattering function moderately forward throwing. These conclusions are again valid only on condition that our assumptions concerning the cloud geometry and the uniform grain density are correct.

A further quantity which we can derive for a given cloud model is a colour index difference between star and nebula, e.g.

$$E = (B-V)_* - (B-V)_n$$

as a function of angular distance r from the direction of the star. We shall indicate how this function may be obtained for the uniform spherical cloud model considered above. If the wavelength is left as a free parameter in the preceding calculation, we have to replace γ, k, $S(\alpha)$, I(r) and L in (4.22) by $\gamma(\lambda)$, $k(\lambda)$, $S(\alpha,\lambda)$, $I(r,\lambda)$ and $L(\lambda)$ respectively. We then have

$$E = E(r) = (B-V)-(B-V)_n = -2.5 \log \frac{I(\lambda_B,r)}{I(\lambda_V,r)} \frac{L(\lambda_V)}{L(\lambda_B)} \tag{4.24}$$

where the right-hand side is directly calculable from the modified form of (4.22) provided r_0, $k(\lambda)$, and the particle parameters $\gamma(\lambda)$ and $S(\alpha,\lambda)$ are known.

Measurements of the run of E(r) for different paths across NGC 7023 were made by Martel (1958), and later by Vanysek and Svatos (1962) and Zellner (1970, 1973). The results in all cases indicate that the nebula is bluer than the star nearby to the star and becomes increasingly red at greater distances from the star.

Vanysek and Svatos (1962) have also attempted to interpret their observations on the basis of spherical clouds models with a symmetrical location of the illuminating star. They considered the case of spherical dielectric grains, and come to the conclusion that a refractive index of about 1.2 and a particle diameter close to 0.3 micron would give a satisfactory agreement between theory and observations at optical wavelengths. Generally similar conclusions were obtained by Zellner (1973) who used more refined models including the adoption of single particle phase functions in place of equation (4.23).

In a more recent investigation Witt et al. (1982) combined earlier data on nebular brightness at optical wavelengths with measurements of nebular fluxes at far infrared and ultraviolet wavelengths. Fig. 4.4 shows the relative nebula/star flux curve for an offset of 22.5" from HD 200775 over the spectral range 1200 – 5500 Å, and Table 4.1 shows the observed total nebular to stellar flux ratio at several wavelengths.

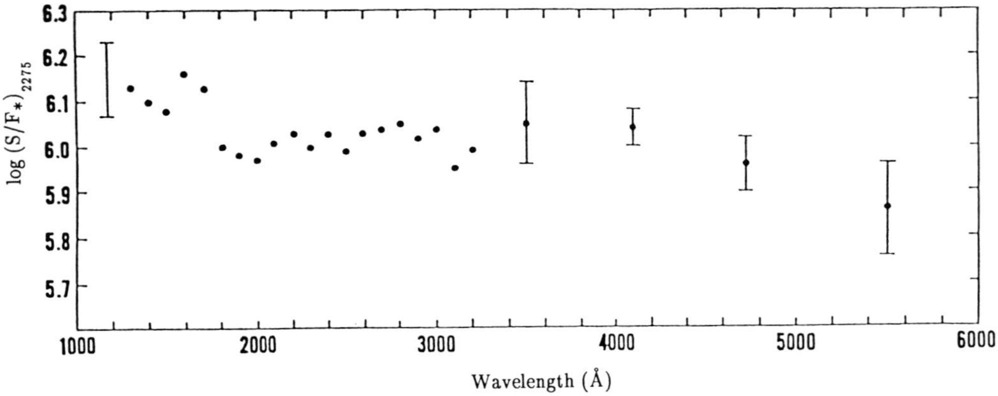

Fig. 4.4 *The average nebular surface brightness in NGC 7023 at an offset distance of 22.5" from HD 200775 as a function of wavelength. The error bar on the left represents an uncertainty of ± 20%, typical of the error in the ultraviolet measurements. (Witt et al., 1982)*

TABLE 4.1

Observed Ratio of Nebular to Stellar Flux in NGC 7023

λ(A)	F_N/F_*
5515	0·41
4733	0·43
4093	0·54
3470	0·56
2400–1300	0·85 ± 0·15

Using the spherically symmetric model of Fig. 4.3 and taking account of multiple scattering in a Monte Carlo calculation, Witt et al. (1982) derive the dust parameters listed in Table 4.2.

TABLE 4.2

Particle parameters for NGC 7023

λ(Å)	γ	g
5515	0·60	0·63
4733	0·56	0·63
4093	0·58	0·60
3470	0·55	0·56
2500	0·47	0·50
2000	0·42	0·40
1700	0·52	0·35
1400	0·60	0·25

4.4. OBSERVATIONS OF REFLECTION NEBULA AROUND MEROPE

The reflection nebula surrounding Merope is particularly suited for accurate surface photometry, on account of its high surface brightness and also because its distance is well defined, being in the Pleiades cluster. Spectrophotometric measurements of this nebula have been made by several investigators (O'Dell, 1965; Elvius and Hall, 1967; Roark, 1966; Hanner, 1971; Andriesse et al., 1977; Witt, 1977; Witt et al., 1986).

Merope, like NGC 7023, is a case where the difficulties of theoretical interpretation are reduced owing to known geometrical characteristics. Being in the Pleiades cluster, its distance may be set as ~ 126 pc. Further, it has been argued by White (1984a,b) that the illuminating star lies 0.1 – 0.01 pc behind a slab of dust and molecules that constitutes the reflection nebulosity.

Roark (1966) calculated star–nebula colour differences Δ(B–V) and Δ(U–B) for homogeneous, plane parallel slab models of reflection nebulae, with various values for the parameters defining the nebular particles and the star–nebula geometry. An advantage of Roark's models was that he used phase functions computed from Mie theory. His model with the star inside a plane slab is applicable to the case of Merope. A satisfactory fit with the Merope observations, on the basis of scattering by dielectric ice grains, is obtained for the following values of the relevant parameters: particle radius, 0.5μm; nebular thickness, 1 pc; distance of star behind front face of nebula, 0.5 pc; and angle of view of the nebula, 90°. The theoretical curve for this case is plotted as the solid line in Fig. 4.5, together with the observational points of Elvius and Hall (1965) and of Roark (1966).

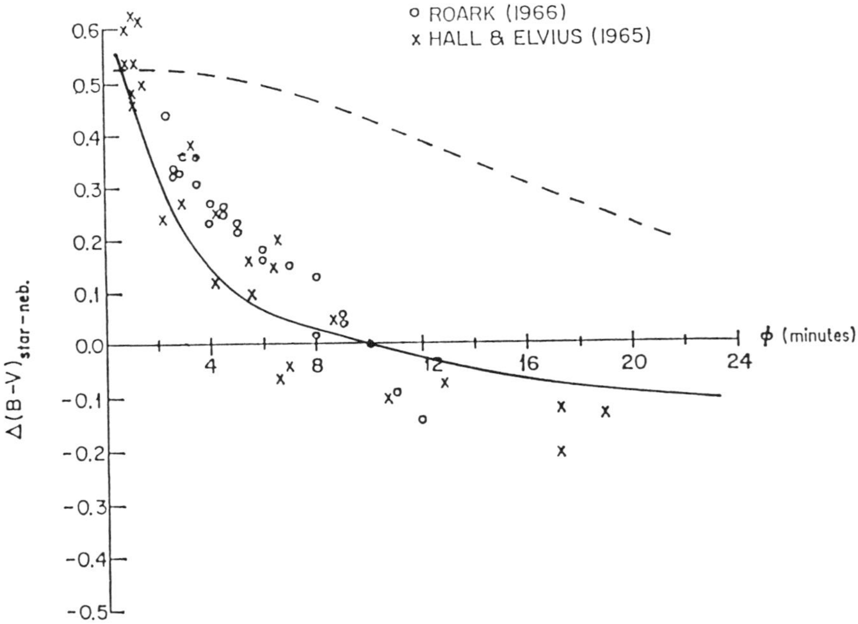

Fig. 4.5 *Star–nebula difference in (B–V) colour as a function of displacement angle ϕ. Open circles are observations of Roark (1966); crosses are observational points of Elvius and Hall (1966). Solid curve is for dielectric ice grains of radius 0.5μm; nebular thickness 1 pc; distance of star behind front face of nebula 0.05 pc angle of view 90°. Dashed curve replaces ice grain with a graphite sphere of radius 0.05μm.*

A typical theoretical curve for a graphite model of the nebula with particle radius 0.05 micron is plotted as the dashed line of Fig. 4.5. This is seen to be in marked disagreement with the observations, indicating that the nebular particles cannot be pure graphite, nor indeed any metal particle of radius $< 0.05\mu m$. Roark's computations further show that the situation for small graphite grains cannot be improved by varying the particle or nebular parameters. The colour differences always remain very positive (blue) at small angular displacements from the star, and become only slightly redder with increasing angle.

A striking feature of the dielectric particle model of the nebula is that a vast range of shapes of Δ(B–V) versus ϕ curves is possible by varying the available parameters. For metallic grains, typified by graphite, the variation is very restricted, the curve always remaining close to the dashed line of Fig. 4.5. This is essentially due to the highly anisotropic scattering diagrams appropriate for dielectric spheres, as compared with the more of less spherical scattering diagram for small metallic particles.

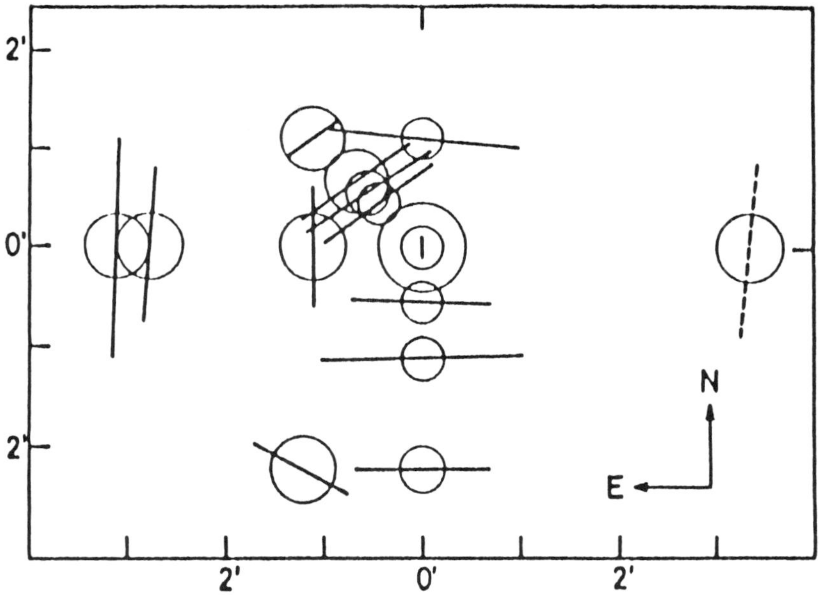

Fig. 4.6 *Polarisation of NGC 7023 showing large polarisation vectors oriented transverse to the direction to the illuminating star HD 200775. Scale: (for blue light) 10 per cent = ————————. (Elvius and Hall, 1966).*

Although the first studies of reflection nebulae concentrated on determining U, B and V magnitudes as a function of offset angle from the illuminating star, a wealth of observations have subsequently accumulated giving both polarisation and brightness as functions of wavelength. Fig. 4.6 shows a typical pattern of transverse polarisation vectors around the central star for the case of NGC 7023 (Elvius and Hall, 1966). Polarisations in the blue range from a few to several percent. The polarisation data impose severe constraints on grain properties, but their interpretation is straightforward only if the nebula is optically thin. It can be shown using the theory of Chapter 2 that if dielectric grains become larger than a certain critical value, the transverse polarisation vectors of Fig. 4.6 should not be seen. For larger dielectric grains the directions of maximum electric vector should become radial. Indeed it can be proved from direct calculations of the Mie formulae that for grains with $m = 1.6$, $a = 0.3\mu m$ polarisation vectors going radially from the star are expected. But if $m = 1.2$, $a = 0.5\mu m$, transverse vectors in agreement with observation are obtained. The significance of these conclusions is restricted because of our assumption of single scattering, however. A more realistic model of reflection nebulae must include the effects of multiple scattering.

Fig. 4.7 shows the polarisation of the Merope nebula as a function of offset angle for two effective wavelengths $\lambda^{-1} = 1.74\mu m^{-1}$ and $2.24\mu m^{-1}$. We note that the polarisation at a particular value of the wave number increases steeply from small offset angles to reach a flat plateau. The polarisation increases with wave number as seen in Fig. 4.7 for Merope and more strikingly for NGC 2068 as shown in Fig. 4.8.

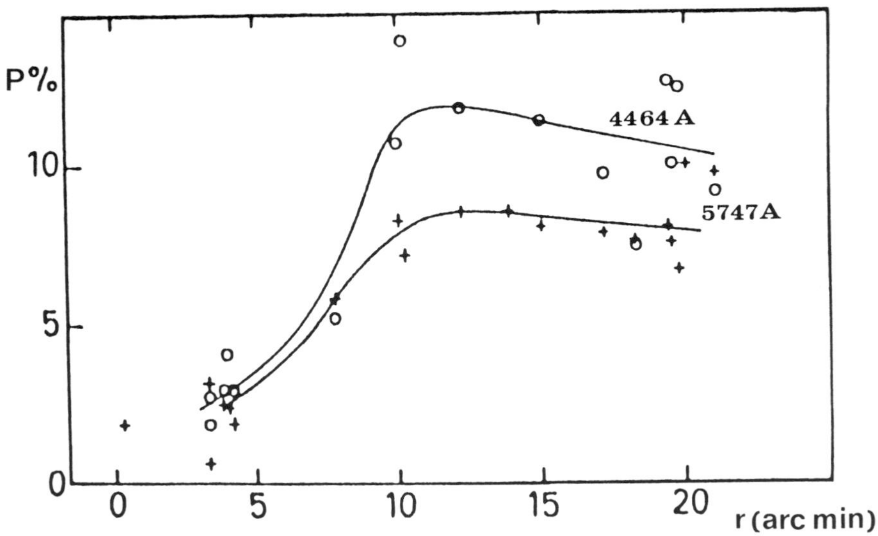

Fig. 4.7 *Polarisation of the Merope nebula as a function of offset in two bandpasses (λ_{eff} = 4464 Å (o) and λ_{eff} = 5747 Å (+). (Elvius and Hall, 1966).*

A dielectric model with m = 1.10, a = 0.38μm has been shown by Zellner (1973) to fit this data. However, since the scattering properties for particles with small values of $|m-1|$ scale with $2\pi a|m-1|/\lambda$, a particle with m = 1.2, a = 0.19μm would yield an identical fit to the data.

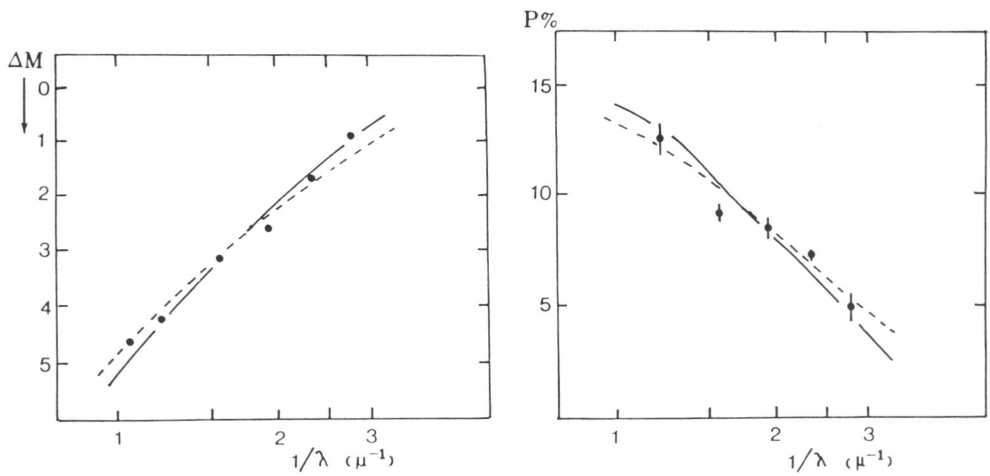

Fig. 4.8 *The brightness and polarisation data for a region in NGC 2068. ΔM is the magnitude of the nebula minus the magnitude of the illuminating star. Solid curves for m = 1.5, average radius ~ 0.06μm; broken curve for m = 1.1, average radius ~ 0.12μm. The range of scattering angles in the model was 10 – 50°. (Zellner, 1973).*

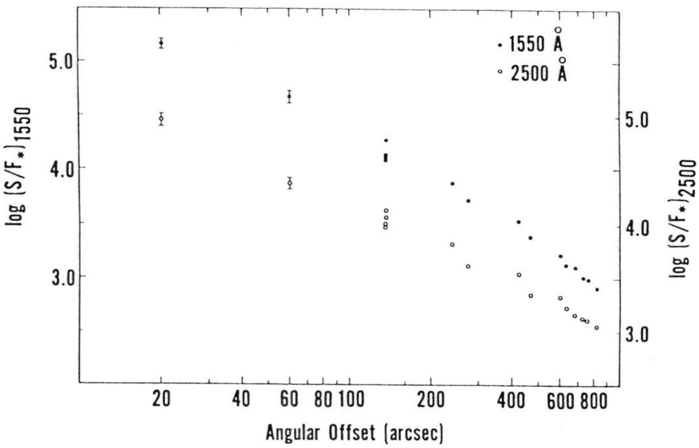

Fig. 4.9a *Ratio of nebular surface brightness to stellar flux (on a log scale) as a function of offset angle for the Merope (23 Tau) nebula for the effective wavelengths 1550 and 2500 Å. (Witt et al., 1986)*

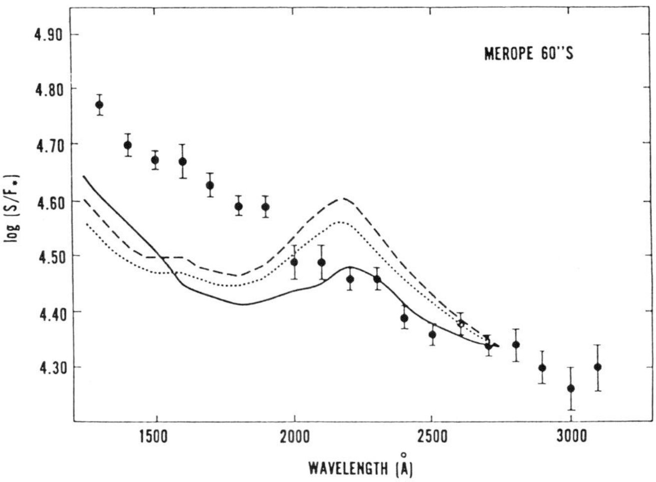

Fig. 4.9b *The points represents the same function as in Fig. 4.9a for the Merope nebula observed 60" south of 23 Tau. The curves are theoretically expected wavelength dependences of the colour difference nebula minus star, normalised at 2740 Å: The solid curve is based on the UV reddening curve observed for the Pleiades, while the dashed curve is based on the average galactic extinction law, with only dust in front of the stars taken into account. The dotted curve assumes the average galactic extinction law with equal amounts of dust in front of and behind 23 Tau involved in the scattering. (From Witt et al., 1986)*

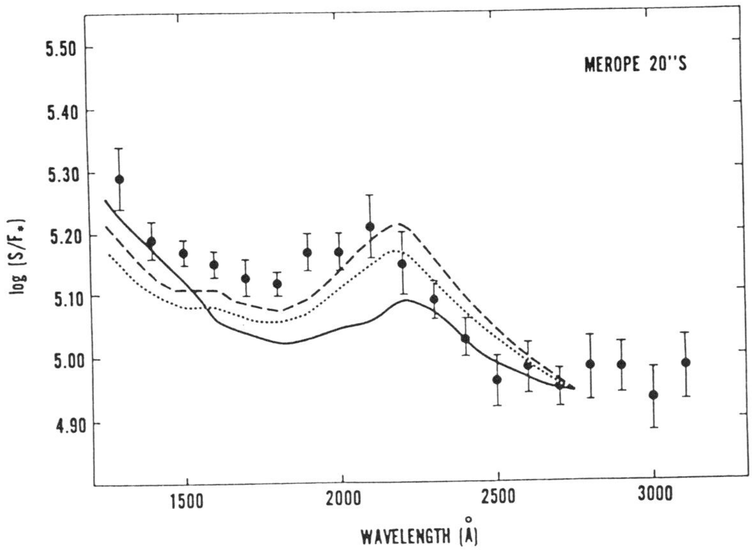

Fig. 4.9c *Same as Fig. 4.9b but for the location 20" south of 23 Tau. (From Witt et al., 1986).*

Lillie and Witt (1973), Andriesse et al. (1977) and Witt et al. (1986) extended the photometric measurements of the Merope nebula into the ultraviolet using OAO, ANS and IUE satellite data. The most recent results are shown in Fig. 4.9a,b,c.

From Fig. 4.9a it is seen that the ultraviolet 'reflectivity' of the dust increases with decreasing wavelength, and the relative nebula/star flux decreases with increasing offset angle. Of particular interest is the relative flux curve for 20"S offset which shows a clear peak near λ 2200 Å, similar to the interstellar extinction peak discussion in Chapter 2.

To analyse Figures 4.9b and c, we need to modify our analysis in equations (4.2 – 4.8) to take account of a possible anistropy in the scattering function. For the general star–nebula geometry illustrated in Fig. 4.2, we can show, following Witt (1985), that for any two wavelengths λ_1, λ_2 equation (4.8) when suitably modified, leads to

$$\log \frac{(I_n/I_*)_{\lambda_1}}{(I_n/I_*)_{\lambda_2}} \approx \log \frac{\gamma(\lambda_1)}{\gamma(\lambda_2)} \frac{[1-e^{-\tau_0(\lambda_1)}]\{\exp[\tau_*(\lambda_1)-\tau_1(\lambda_1)]\}}{[1-e^{-\tau_0(\lambda_2)}]\{\exp[\tau_*(\lambda_2)-\tau_1(\lambda_2)]\}p(\lambda_1,\lambda_2)} \tag{4.24a}$$

where $p(\lambda_1,\lambda_2) = \int_{\alpha_1}^{\alpha_2} \Phi(\alpha,\lambda_2)\sin^2\alpha d\alpha \Big/ \int_{\alpha_1}^{\alpha_2} \Phi(\alpha,\lambda_1)\sin^2\alpha d\alpha$

$\Phi(\alpha,\lambda)$ being the scattering phase function and α_1, α_2 the extreme values of α along the elementary cylinder (see Fig. 4.2). For a nearly isotropic phase function, appropriate to small particles, $p(\lambda_1,\lambda_2) \approx 1$.

Witt et al. (1986) have argued that the ultraviolet scattering data of Fig. 4.9a–c implies the dominance of scattering by particles of radius a ~ 10^{-6}cm, with $p(\lambda_1,\lambda_2) \simeq 1$ throughout the ultraviolet waveband 2740 Å $\geq \lambda \geq$ 1300 Å. The curves are calculated from equation (4.24a) with the star assumed to lie behind a dust layer of $\tau_V = 0.24$ and with $\gamma(\lambda)/\gamma(2740) = 1$ assuming a galactic extinction law $\tau(\lambda)$, and $e^{\tau_*-\tau_1} \approx 1$ for the small optical depths involved. Although many uncertain and possibly unrealistic assumptions are involved in these calculations, the general indication is that the ultraviolet scattering particles in the Merope nebula are similar to those that are responsible for interstellar extinction at ultraviolet wavelengths. These particles (graphite or otherwise) must constitute a population of grains distinct from the larger forward scattering dielectric particles that are required to account for the data in Fig. 4.4.

In another investigation involving IUE satellite observations of the reflection nebula CED 201 associated with the star BD+69°1231, Witt et al. (1986b) have found evidence of scattered light in the 2175 Å band implying grains with an albedo of ~ 0.5. Graphite spheres with such a high albedo are possible only for radii > 0.03μm, but such large particles have absorption maxima at wavelengths significantly longwards of 2200 Å. These results cast doubt on the validity of a graphite model for the λ 2175 Å absorption feature in reflection nebulae.

4.5. MULTIPLE SCATTERING MODELS OF REFLECTION NEBULAE

In modelling observational data on reflection nebulae, particularly polarisation measurements, single scattering calculations of the kind discussed earlier become unreliable for optical depths in excess of unity. There have been several attempts to deal with multiple scattering in the present context, notably by Mattilla (1970), Fitzgerald et al. (1976), Andriesse et al. (1977) and Warren–Smith (1983a). The Monte Carlo techniques developed by Warren–Smith (1983a) lead to the most accurate computational procedures with a maximum economy of computing time. The model makes allowance for polarisation effects by calculating the Stoke's vector along each trial photon path through the nebula. In practice only a few scatterings need to be considered, but even with two or three scatterings per light path, significant departures from a single–scattering model are found. Warren–Smith's computations take account of the Mie–scattering phase functions for any desired particle model and for a given star–nebula geometry.

A detailed set of observations of surface brightness and polarisation is available for the reflection nebula NGC 1999 in the Orion region associated with the star V380 Orionis (Warren–Smith, 1983b). The nebula lies 10 – 20 arcsec WSW of V380 Ori and is at a distance of 470 pc. The star appears to be embedded close to the edge of an opaque dust cloud, a section of which is shown in Fig. 4.10. Here d is the

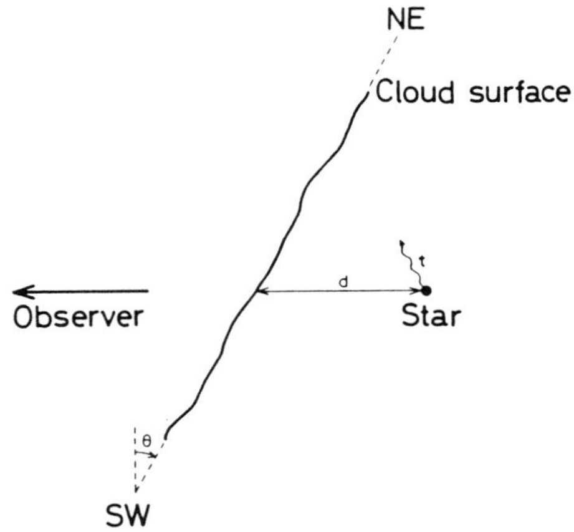

Fig. 4.10 *The model geometry: a star within a large dust cloud – Warren–Smith's model for NGC 1999.*

depth of the star in the line of sight behind the cloud, θ is the 'tilt angle' and t is the mean free path of photons in the cloud. Warren–Smith computes the behaviour of dielectric spherical grains according to three model distributions of radius:

(i) $n(a) \propto a^{\gamma}, \gamma < 0$

(ii) $n(a) \propto \exp(-a/a_0)$

(iii) $n(a) \propto \exp(-5(a/a_0)^3)$

where a is the grain radius and a_0 is a fixed value defining a size parameter.

The surface brightness and polarisation data at $\lambda = 4900$ Å for NGC 1999 shown in Figs. 4.11 and 4.12 were fitted by Warren–Smith (1983b) with models (i) – (iii) defined above using particle–nebula parameters as set out in Table 4.3 below.

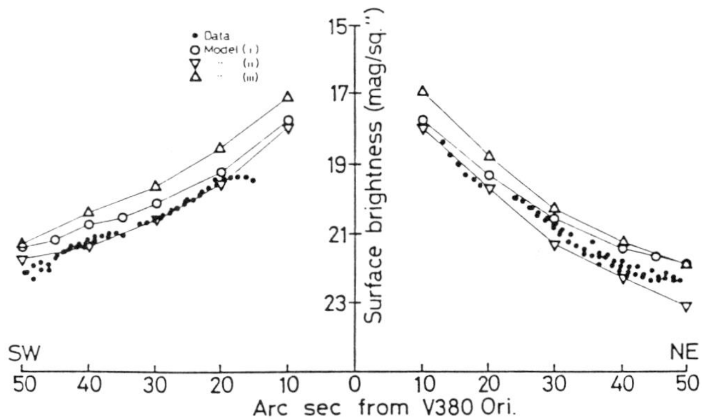

Fig. 4.11 *The observed surface brightness for NGC 1999 at 4900 Å compared with the fits obtained by Warren–Smith (1983) for each of the three models*

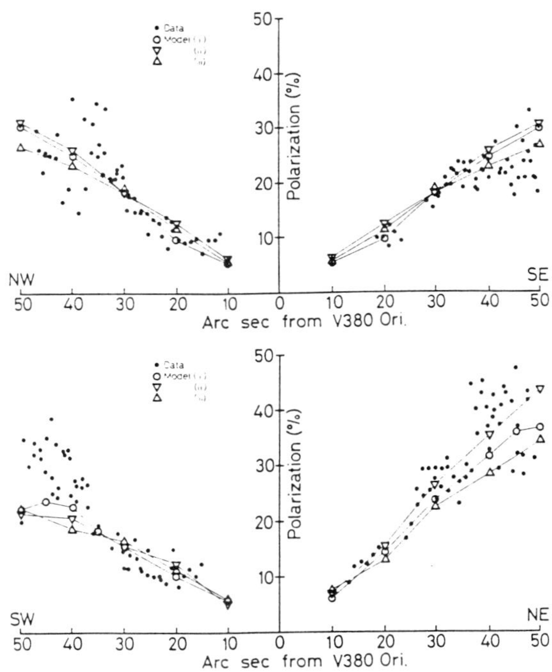

Fig. 4.12 *The linear polarisation data for NGC 1999 at 4900 Å compared with the fits obtained for each of the three models. (Warren–Smith, 1983)*

TABLE 4.3

Warren–Smith's summary of the parameters for each of his models which give the best fit to surface brightness and polarisation data at 4900Å

Model	(i)	(ii)	(iii)
Depth of star (arcsec)	45·9 ± 5·2	46·0 ± 5·1	36·3 ± 10·8
Nebular tilt (°)	15·1 ± 2·6	15·2 ± 4·0	9·5 ± 6·7
Mean free path (arcsec)	38·5 ± 3·6	43·7 ± 6·9	23·7 ± 15·0
Power–law index γ	–4·24 ± 0·18		
Size parameter a_0 (μm)		0·089 ± 0·019	0·55 ± 0·13
Real part of RI	1·19 ± 0·09	1·10	1·10
Imaginary part of RI	0·0	0·0	0·0
Internal A_v (mag)	1·12	0·90	1·30
$R = [A_v/E(B-V)]$	3·14	1·79	1·67

We note that in all three cases low values of $|m-1| \sim 0.1$–0.2 appear to be strongly favoured. Of these cases model (i) with $n = 1.19 \pm 0.09$ is to be preferred as a unified model for interstellar and nebular grains because of the value of $R \simeq 3.14$ that is implied. We saw in Chapter 3 that a ratio $R \cong 3$ is appropriate for normal interstellar grains.

4.6. DIFFUSE GALACTIC LIGHT

The diffuse galactic light (DGL) is the starlight that has been scattered diffusely by the interstellar dust grains. It is evident as a background sky brightness, and a study of this radiation may be expected to reveal important information concerning the scattering properties of grains. Measurements of this radiation, even with modern techniques, are difficult due to spurious effects – airglow, zodiacal light and light from unresolved faint stars which inevitably add to the sky brightness. Uncertain corrections have to be introduced to take account of these factors. The best hope for accurate results is to make measurements above the atmosphere and only a few such observations are available at the present time. The interpretation of diffuse galactic light data once obtained is also difficult. A radiative transfer model is required in order to deduce properties of grains, in particular to deduce values of the albedo γ and the phase parameter g(= <cos θ>).

Some idea of the asymmetry of the scattering diagram of individual grains may be obtained if observations are available of the diffuse light as a funtion of galactic latitude. The first detailed model that sought to interpret such data was due to Henyey and Greenstein (1941).

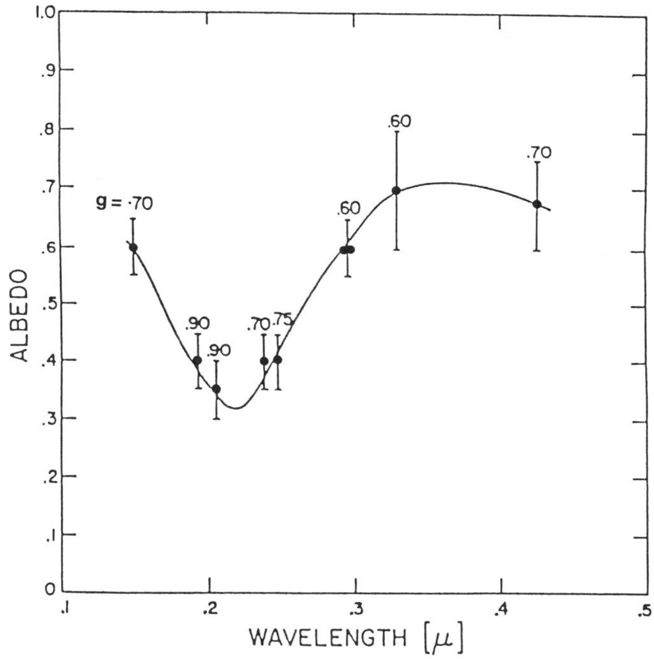

Fig. 4.13 *The albedo and phase parameter g (within brackets) of interstellar dust as a function of wavelength based on OAO−2 measurements of the diffuse galactic light (Witt and Lillie, 1973).*

Witt and Lillie (1971) and Lillie and Witt (1977) have also obtained diffuse light data in the waveband 1500 – 4200 Å using equipment carried on the Orbiting Astronomical Observatory OAO 2. Both the latitude dependence of the diffuse galactic light at several wavelengths, and the ratio of diffuse galactic light to direct starlight, have been measured in a large number of selected areas of the sky. Using this data in conjunction with the computations of van de Hulst and de Jong (1969), Witt and Lillie have obtained a wavelength dependence for the grain albedo and also for the phase parameter g = <cos θ>. Fig. 4.13 shows their results for the albedo which indicate a sharp minimum at 2200 Å, coinciding with the peak in the extinction curve. The scattering phase function is found to be strongly forward throwing at all wavelengths. The excessively high values of g ~ 0.9 found near the wavelength λ ~ 2200 Å would militate against a model of grains including an appreciable fraction of graphite spheres of radius 0.02μm for which g << 0.5.

References

Andriesse, C.D., Piersma, Th. R. and Witt, A.N., 1977, *Astron. Astrophys.*, **54**, 841.

Elvius, A. and Hall, J.S., 1967, in T.P. Roark and J.M. Greenberg (eds.) *Interstellar Grains*, (NASA SP–140).

Elvius, A. and Hall, J.S., 1966, *Lowell Obs. Bull.*, **135**, vi, No. 16.

Fitzgerald, M.P., Stephens, T.C. and Witt, A.N., 1976, *Astrophys. J.*, **208**, 709.

Hanner, M.S., 1971, *Astrophys. J.*, **164**, 425.

Henyey, L.G. and Greenstein, J.L., 1941, *Astrophys. J.*, **93**, 70.

Hubble, E.E., 1922, *Astrophys. J.*, **56**, 162 and 400.

Lillie, C.F. and Witt, A.N., 1969, *Astrophys. Lett.*, **3**, 201.

Lillie, C.F. and Witt, A.N., 1973, in J.M. Greenberg and H.C. van de Hulst (eds.) *Interstellar Dust and Related Topics*, D. Reidel.

Lillie, C.F. and Witt, A.N., 197, *Astrophys. J.*, **208**, 64.

Martel, T.M., 1958, *Ann. Astrophys.*, Suppl. 7.

Mattila, K., 1970, *Astr. Astrophys.*, **9**, 53.

O'Dell, C.R., 1965, *Astrophys. J.*, **142**, 604.

Roach, F.E. and Smith, L.L., 1964, *Nat. Bur. Stand. Tech. Not.*, p. 214.

Roark, T.P., 1966, *PhD Thesis*, Rensselaer Polytechnic Inst., Troy, New York.

van de Hulst, H.C., 1957, 'Light Scattering by Small Particles', Wiley.

van de Hulst, H.C. and de Jong, T., 1969, *Physica*, **41**, 151.

van Houten, C.J., 1961, *B.A.N.*, **16**, 509.

Vanysek, V. and Svatos, J., 1964, *Acta Univ. Carol.*, Ser. Math. No. 1.

Warren–Smith, R.F., 1983, *Mon. Not. Roy. Astr. Soc.*, **205**, 337.

Warren–Smith, R.F., 1983, *Mon. Not. Roy. Astr. Soc.*, **205**, 349.

White, R.E., 1984a, *Astrophys. J.*, **284**, 685.

White, R.E., 1984b, *Astrophys. J.*, **294**, 216.

Witt, A.N., 1968, *Astrophys. J.*, **152**, 59.

Witt, A.N., 1977, *Publ. Astron. Soc. Pacific*, **89**, 750.

Witt, A.N., 1985, *Astrophys. J.*, **294**, 216.

Witt, A.N., Walker, G.A.H., Bohlin, R.C. and Stecher, T.P., 1982, *Astrophys. J.*, **261**, 492.

Witt. A.N., Bohlin, R.C. and Stecher, T.P., 1986a, *Astrophys. J.*, **302**, 421.

Witt. A.N., Bohlin, R.C. and Stecher, T.P., 1986b, *Astrophys. J.*, **305**, L23.

Wolstencroft, R.D. and Rose, L.J., 1966, *Nature*, **209**, 389.

Zellner, B., 1970, *PhD Thesis*, University of Arizona.

5. Interactions Between Dust, Gas and Radiation

5.1. INTRODUCTORY REMARKS

As we have noted in Chapter 1, interstellar grains are intimately mixed with gas within interstellar clouds. They are also immersed in a dilute radiation field with which they interact. A 'standard' diffuse interstellar clouds has for many years been considered as one with the following average properties:

TABLE 5.1

Standard Cloud Properties

Hydrogen density (HI)	$n_H \cong 20\ cm^{-3}$
Temperature (gas)	$T_g = 80$ K
Radius	$d = 5$ pc
Stellar radiation energy density	$U = 7 \times 10^{-13}\ erg\ cm^{-3}$
Visual extinction	$\Delta m_v = 0{\cdot}2$ mag

Although the concept of a 'standard cloud' is useful for many purposes, it should be stressed that interstellar clouds with sharp boundaries as physical entities are unreal, and that the observations only indicate an interstellar medium with density fluctuations over distance scales ranging from hundredths of a parsec in dark globules and protostars, to ~ 50 parsecs in the case of the molecular clouds. The gas density and temperature, molecular composition and ionization state all vary from the hot 'intercloud' regions to the deep interiors of dark globules. The spectral distribution and the energy density of the ambient radiation field also varies sharply as one goes from diffuse clouds to the interiors of dense clouds.

One parameter that is relatively constant in interstellar space is the ratio of N_H/E_{B-V} where N_H, the total column density of hydrogen, is obtained from a combination of $L\alpha$ and H_2 observations (Bohlin, 1975) and E_{B-V} is the colour excess. Including the contribution of HII in ionized hydrogen regions around the O and B stars, from which colour excess measures are derived, Spitzer (1978) obtained the relation

$$N_H = 5.9 \times 10^{21}\ E_{B-V}\ mag^{-1}\ cm^{-2} \tag{5.1}$$

Remembering that $A_V = 3E_{B-V}$ for the general interstellar medium, and $A_V = 1.086\ \pi a^2 <Q_{ext}> n_g L$, where a is the average grain radius, $<Q_{ext}>$ is the average visual extinction efficiency, n_g is the number density of grains, n_H is the hydrogen density, and L is the column length, we have

$$\frac{\pi a^2 <Q_{ext}> n_g}{n_H} \cong 5 \times 10^{-22} \tag{5.2}$$

Since $<Q_{ext}> \simeq 2$ at $\lambda = \lambda_V$ for most grain models of interest, we have

$$\frac{4\pi a^2 n_g}{n_H} \cong 10^{-21} \tag{5.3}$$

as the almost invariant ratio of grain surface area per hydrogen atom in most directions that exclude substantial HII regions. For a given value of a, equation (5.2) also gives a ratio of n_g/n_H. For grains of radius 3×10^{-5} cm and $<Q_{ext}> \simeq 2$ we have

$$\frac{n_g}{n_H} \simeq 10^{-13} \tag{5.4}$$

For grains of radii 3×10^{-6} cm, with the same value of $<Q_{ext}>$ we have $n_g/n_H \simeq 10^{-11}$.

The first determinations of the dust to gas ratios in HII regions were made by O'Dell and Hubbard (1965) and O'Dell et al. (1966). They used an interference filter system to isolate the emission Hβ line from the scattered light continuum to obtain an effective dust to gas ratio in the form

$$x = \frac{\pi a^2 <Q_{sca}> n_g}{n_H} \tag{5.5}$$

Equating $<Q_{ext}>$ to $<Q_{sca}>$ for dielectric particles, it was found that x is close to the normal interstellar value given by equation (5.2) in most cases, although in the Orion nebula a distinct variation appears to take place from the inner to the outer regions. In the inner regions, x is an order of magnitude less than the normal value, whilst in the outer regions it is $\sim$ 4 times larger.

In a typical interstellar radiation field the flux of starlight energy reaching a grain will be $(c/4)\ 7 \times 10^{-13}$ erg cm^{-2} s^{-1}. The number of 4 eV photons impinging on a grain of radius a is thus $\sim 3.10^9\ a^2\ s^{-1}$, and the mean time interval between successive arrivals of optical photons would be

$$\tau \sim (3.10^9\ a^2)^{-1}\ s \tag{5.6}$$

For a grain of radius 3×10^{-5} cm we have $\tau \sim 1/3$ s, whereas for a grain of radius 3×10^{-7} cm we have $\tau \approx 3.10^3$ s.

A dust grain immersed in an interstellar gas environment experiences collisions at a rate

$$\sim 4\pi a^2\ n_H\ v_{th}\ s^{-1} \tag{5.7}$$

where v_{th} is the thermal speed. With $a = 3 \times 10^{-5}$ cm, $n_H \simeq 8$ cm^{-3}, $v_{th} \sim 10^5$ cm s^{-1} this gives a rate $\sim 10^{-2}$ s^{-1}. Thus the average time interval between successive collisions of a 'standard' grain with gas atoms in a 'standard' interstellar cloud is about ~ 100 sec. In this chapter we shall explore some aspects of the interaction between grains and the ambient gas, with regard to general considerations of dynamics, alignment and molecule formation at the grain surface. Since many of the gas–dust interaction properties are controlled by the value of the grain temperature, we begin now with a discussion of this topic.

5.2 GRAIN TEMPERATURES FOR STANDARD GRAINS

In a typical diffuse interstellar cloud, the radiation field may be approximated by a blackbody of temperature 10^4 K with a dilution factor $W \simeq 10^{-14}$. An average energy density of optical light may be taken as ~ 7×10^{-13} erg cm^{-3}. The rate of transport of energy from optical photons reaching a grain of radius a is thus

$$\sim 7 \times 10^{-13} \frac{c}{4} (4\pi a^2) \text{ erg s}^{-1}$$

If $<Q_{abs}>$ is an average absorption efficiency of visible light, F_s, the absorption rate of starlight energy, is

$$F_s \cong 2.1 \times 10^{-2} <Q_{abs}> \pi a^2 \text{ erg s}^{-1} \tag{5.8}$$

The rate of transport of gas thermal energy, F_g, using the incidence rate (5.7), is given by

$$F_g \cong n_H v_{th} 4\pi a^2 (\frac{1}{2} m_H v_{th}^2) \tag{5.9}$$

which yields

$$\frac{F_s}{F_g} \cong \frac{5.10^6 <Q_{abs}>}{n_H} \tag{5.10}$$

with $v_{th} \simeq 10^5$ cm s^{-1}. Thus gas collisions are unimportant for heating grains with the values of n_H appropriate to normal diffuse clouds. Only in situations of exceptionally high gas density can gas–heating compete with radiative heating. Even if one considers situations where H_2 molecules form at grain surfaces, giving up a fraction of the 4.48 eV energy release to the solid, this chemical energy input does not decrease F_s/F_g by more than ~ 10^2. The energy input from cosmic ray encounters is also totally negligible. For 1 MeV protons (low energy cosmic rays) which have the largest effect, the energy deposited per encounter is ~ 250 (a/cm) MeV assuming data on the stopping power of water ice. With a mean interstellar MeV proton flux estimated at ~ 1 cm^{-2} s^{-1} we thus have an energy deposition of

$$F_{cR} = 4 \times 10^{-4} (4\pi a^3) \text{ erg s}^{-1} \tag{5.11}$$

Using (5.8) we have

$$\frac{F_{cR}}{F_s} \cong 8.10^{-2} a <Q_{abs}> \tag{5.12}$$

and this is << 1 for $<Q_{abs}> \sim 1$, $a \sim 10^{-5}$ cm. A similar argument holds for absorption of energy from a background of soft X–rays in interstellar space.

In the foregoing discussion we have ignored a possible contribution from the cosmic microwave background to grain heating. Although the energy density in the microwave background $\sim 4 \times 10^{-13}$ erg cm^{-3} is over half that in starlight, spherical grains of ordinary sizes, $a \sim 10^{-6} - 10^{-4}$ cm, in the diffuse clouds absorb this radiation only very inefficiently due to the $2\pi a/\lambda$ factor. Only for long conducting whiskers, or particles with exceedingly low visual and ultraviolet extinction efficiencies, can the situation be appreciably different. We shall return to such matters in Chapter 10.

The only effective mechanism for the radiation of energy by a grain is thermal re–emission corresponding to the temperature T_g of the solid lattice. The total energy lost by radiation per second is thus

$$4\pi a^2 \int_0^\infty Q_{abs}(\lambda)\, B_\lambda(T_g)\, d\lambda \tag{5.13}$$

where a is the grain radius for a spherical particle. In accordance with our earlier remarks, we assume that energy gain occurs only through absorption of a dilute radiation field which we approximate by a radiation field corresponding to a blackbody of temperature $T_* = 10{,}000°$ K and a dilution factor $W = 10^{-14}$. The gain of energy per second is thus

$$4\pi a^2\, W \int_0^\infty Q_{abs}(\lambda)\, B_\lambda(T_*)\, d\lambda \tag{5.14}$$

The equilibrium grain temperature is calculated by assuming detailed balance of the absorption and re–emission of energy. If grains behave like blackbodies, i.e. $Q_{abs} = 1$ at all wavelengths, we obtain a grain temperature T_g given simply by

$$T_g^4 = WT_*^4 + T_b^4 \tag{5.15}$$

where T_b is the microwave background temperature. An actual spherical grain of radius $\sim 10^{-5} - 10^{-6}$ cm does not act like a black body, however. It absorbs optical and ultraviolet radiation relatively efficiently but emits and absorbs far infrared and millimeter waves highly inefficiently. The result is a 'greenhouse effect' which causes T_g to rise above the effective black body temperature, thus shifting the maximum photon emission towards shorter infrared wavelengths.

The procedure for calculating the temperature of a spherical grain of radius a is as follows, the second term on the left due to absorption of the cosmic microwave background being usually negligible.

$$W\int_0^\infty Q_{abs}(a,\lambda)\,B(\lambda,T_*)\,d\lambda + \int_0^\infty Q_{abs}(a,\lambda)\,B(\lambda,T_b)\,d\lambda = \int_0^\infty Q_{abs}(a,\lambda)\,B(\lambda,T_g)\,d\lambda\,. \qquad (5.16)$$

Strictly, the integrand in the first term of (5.14) should be replaced by a superposition of black body spectra with various dilution factors, in agreement with the actual star fields making up the radiation background, but the effect of this refinement on calculated grain temperatures is small. The values of Q_{abs} in (5.16) should be calculated from the theory of Chapter 2 using optical constants appropriate to the grain material in question. The main contribution to RHS of (5.16) arises in the infrared, and with the particle radius such that $2\pi a/\lambda << 1$, $Q_{abs}(a, \lambda)$ in this term may be approximated by

$$Q_{abs} \cong \frac{8\pi a}{\lambda}\,\mathrm{Im}\,\frac{1-m^2}{2+m^2}\,,\quad m^2 = K - 2i\sigma\lambda/c\,, \qquad (5.17)$$

K being dielectric constant and σ the conductivily. When the latter are known from laboratory measurements T_g can be computed without extrapolations being required.

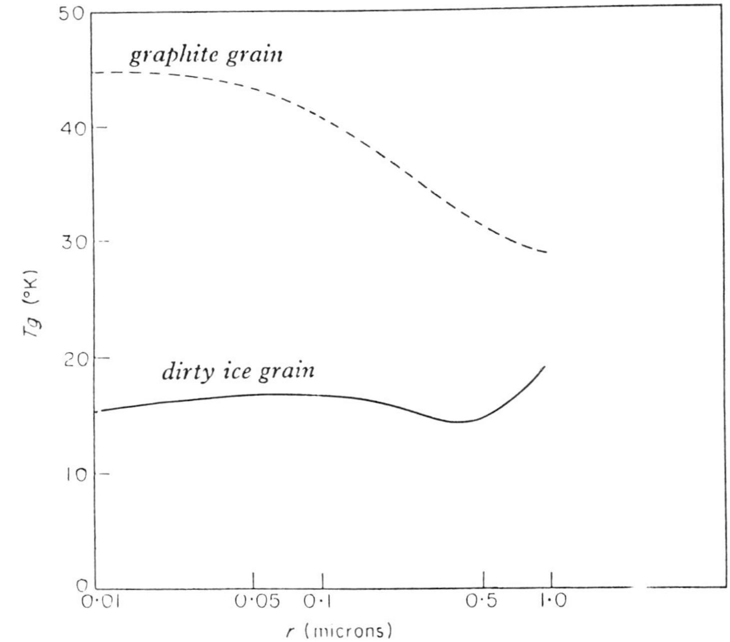

Fig. 5.1 *Computed temperatures of spherical grains in an interstellar radiation field characterised by $T_* = 10{,}000°$ K and $W = 10^{-14}$. Solid curve is for 'dirty ice' spheres (with $k = 0.05$ for $\lambda < 6000$ Å) and the dashed curve is for graphite.*

The solid curve in Fig. 5.1 shows the computed temperature curves for dirty ice spheres with m = n(λ)–ik(λ) taken directly from experimental data, except that k = 0.05 is assumed for λ < 6000 Å. The dashed curve is for graphite spheres. In this case we note that temperatures ~ 45° K are appropriate for grains of radius 0.02μm. The situation for grains with a siliceous composition is less certain because the absorptive index in the far ultraviolet is not well defined and depends sensitively on the particular type of silica, as well as on the impurity content and the presence of lattice defects. The value of k in the range 3000–1800 Å could vary from ~ 10^{-5} for pure quartz, to ~ .05 for disordered silica. Impurities and radiation damage could also enhance the far infrared emissivity of grains up to maximum values consistent with the particle size. Ignoring this latter complication, we calculate T_g for quartz grains of radius 0.1μm for three different values of k in the ultraviolet and visual waveband. The results of this calculation are set out in Table 5.2 below.

TABLE 5.2

The temperature in degrees Kelvin of a silica grain of radius 0.1μm in a diffuse cloud

UV absorptive index	T_g
5×10^{-2}	10·5
5×10^{-3}	6·7
5×10^{-4}	4·4

The temperatures of ice or organic grains are also dependent on the UV absorptive index which could be modified due to impurities or defects. A temperature in the range 10–15° K may be reasonable for organic or icy grains of radius 0.3μm with an ultraviolet absorption index k ≃ 0.01.

These results (shown in Fig. 5.1 and Table 5.2) are consistent with similar calculations reported by Greenberg (1971). Greenberg and Shah (1971) have further discussed the effect of shape on the temperature of interstellar grains. They show that infinite dielectric cylinders can take up temperatures ~ 10% lower then spheres of the same radii.

Calculations are also available for the temperatures of grains in the interiors of dense interstellar clouds, where the effect of optical depth reduces the fluxes of incident ultraviolet light (Solomon and Wickramasinghe, 1969; Werner and Salpeter, 1969; Greenberg, 1971; Leung, 1975). For a dielectric grain model temperatures could fall from 15° K at a visual optical depth $\tau = 0$ to 7° K at $\tau = 5$ in a typical case. In clouds with very large optical depths, T_g could reach a value close to 3° K.

5.3. TEMPERATURE SPIKES IN VERY SMALL GRAINS

The discussion of grain temperature in the preceding section presupposed that the grain was of a size that its total heat energy (enthalpy) was much larger than the energy of a single optical or ultraviolet photon. This assumption breaks down for particles significantly smaller than 0.01μm (Purcell, 1976). When a grain of this type absorbs a typical interstellar photon of energy E ~ 4 eV, its temperature will experience a spike rise. Duley (1973) discussed such an effect in general terms, and a quantitative discussion was given later by Purcell. A neglect of this effect can be shown to introduce large errors in estimations of temperatures based on the use of equation (5.16). Usually, though not always, the error is in the direction of giving overestimates of the average grain temperature.

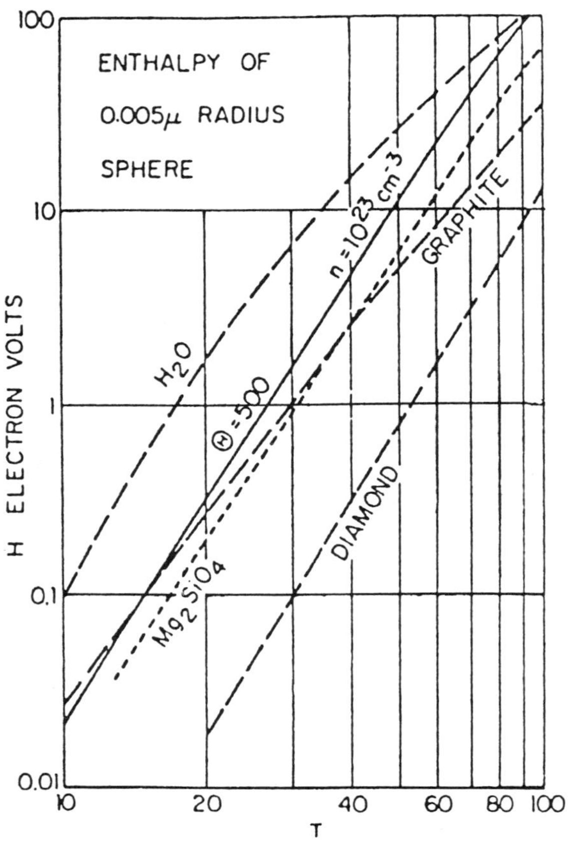

Fig. 5.2 *Purcell's calculations of the enthaply of small spheres of radius $a = 0.005\mu m$ comprised of various pure substances. The solid curve is for an ideal crystal with a Debye temperature $\Theta = 500° K$ and an atom density 10^{23} cm^{-3}.*

Fig. 5.2 shows Purcell's calculations of the enthalpy of small spheres of radius a = 0.005μm comprised of various pure substances. The solid curve is for an ideal crystal with a Debye temperature $\Theta = 500°$ K and an atom density 10^{23} cm^{-3}. It is clear from this figure that the total heat energy exceeds 2 eV for most materials with T above about 45° K, which is the equilibrium temperature calculated by equation (5.16) for a graphite particle of radius 0.01μm (see Fig. 5.1). The energy of an absorbed optical photon would then raise the temperature to a value consistent with the new total energy in a timescale of $\sim 10^{-9}$ s, the time required to redistribute the energy between the chemical bonds. The timescale involved in thermal re–emission of this energy at temperature T is τ given by

$$Q_{abs}(4\pi a^2 \sigma T^4)\tau \approx 4 \text{ eV} \qquad (5.18)$$

where Q_{abs} is the absorption efficiency at the re–emission wavelengths. With a = 5 x 10^{-7} cm, equation (5.17) gives

$$\tau = \frac{3.6 \text{ x } 10^{-4}}{Q_{abs}(T/100)^4} \text{ sec} \qquad (5.19)$$

For T = 100° K, $Q_{abs} \approx 10^{-4}$ which appear to be representative values, we get cooling in seconds following each temperature spike. This value of τ is to be compared with the mean time interval between spikes, reckoned in tens of minutes, as shown by equation (5.6). A schematic temperature plot as a function of time is shown in Fig. 5.3.

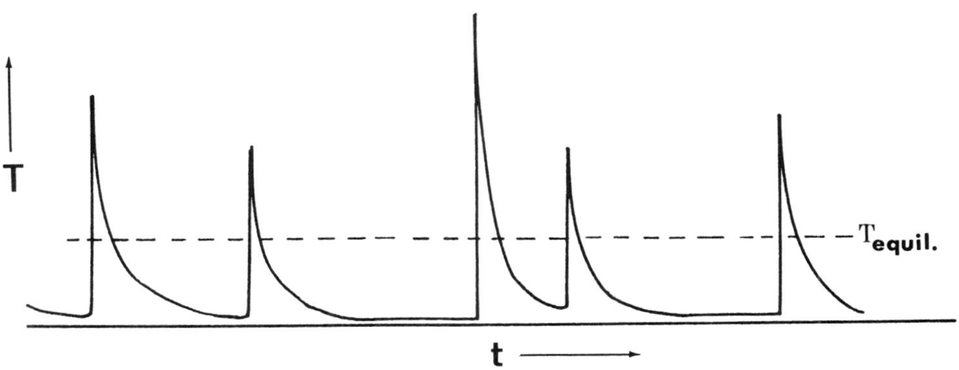

Fig. 5.3 *Schematic temperature profile of a very small grain.*

These considerations apply with little change to the case of clusters of covalently bound atoms such as occur in large aromatic molecules. If such a molecule contains N atoms, the temperature in a spike is given approximately by the condition

$$3NkT \simeq E \qquad (5.20)$$

where E is the energy of the incident photon. This leads to the relation

$$T \simeq 77 \frac{E/4 \text{ eV}}{N/200} \text{ K} \qquad (5.21)$$

Temperature spikes can reach higher values on occasions when photons of greater energies – say $\sim$ 10 eV – are absorbed. With $Q_{abs} \propto \lambda^{-2}$, equation (5.19) gives $\tau \propto T^{-6}$, showing that the higher temperature spikes cool off more quickly. We shall return to matters concerning temperature spikes in Chapter 9 when we deal with the behaviour of polyaromatic hydrocarbon molecules in space.

5.4. ELECTROSTATIC CHARGE ON GRAINS

Over any sizeable region, the interstellar medium is electrical neutral. Thus in the vicinity of a grain we expect the number density of positive ions n_i to equal the electron number density n_e. Equipartition of energy between ions and electrons yields the condition

$$\frac{1}{2} m_i <v_i>^2 \cong \frac{1}{2} m_e <v_e>^2 \tag{5.22}$$

where $<v_i>$, $<v_e>$ are their respective average velocities, and m_i, m_e are their respective masses. In diffuse interstellar clouds, the main source of ions is carbon with $(m_i/m_e)^{1/2} \cong 149$, and in HII regions with protons the dominant ion this ratio is $\sim$ 43. The electrons therefore approach a grain faster than the ions in all cases. It is thus clear that grains tend to acquire a net negative charge, provided photoemission can be neglected as a competing process. To follow Spitzer's (1941) original treatment of this problem, we begin by neglecting photoemission.

The collision cross–sections for an electron and ion (single charged) with a grain of radius a carrying x units of positive charge are

$$\sigma_e = \pi a^2 \left[1 + \frac{2xe^2}{m_e v_e^2 a}\right], \tag{5.23}$$

$$\sigma_i = \pi a^2 \left[1 - \frac{2xe^2}{m_i v_i^2 a}\right], \tag{5.24}$$

where v_e, v_i are their respective velocities at large distances from the grain. The rate of capture of electrons by a grain is then

$$n_e \sigma_e v_e \ \mathrm{sec}^{-1} \tag{5.25}$$

and the rate of accumulation of negative charge due to these captures is

$$e\, n_e \sigma_e v_e \ \mathrm{sec}^{-1} . \tag{5.26}$$

Similarly, the rate at which positive charge is accumulated is

$$e\, n_i \sigma_i v_i \ \mathrm{sec}^{-1} . \tag{5.27}$$

The steady–state grain charge may be estimated by using (5.23) – (5.27) to equate the rates of positive and negative charge accumulation. Including an integration of (5.26) and (5.27) over a Maxwellian distribution of electron and ion velocities, we obtain:

$$1 - y = \sqrt{\frac{m_i}{m_e}}\, e^y \tag{5.28}$$

with

$$y = \frac{xe^2}{akT}\,. \tag{5.29}$$

The solution of equation (5.28) gives

$$\begin{aligned} y &= -3.5, \text{ for an HI region} \\ y &= -2.5, \text{ for an HII region} \end{aligned} \tag{5.30}$$

the difference arising from taking the ions to be CII and HII respectively. We then have

$$|x| = 3.5\ akT/e^2 \ \text{ or } \ |x| = 2.5\ akT/e^2 \tag{5.31}$$

depending on whether the analysis refers to an HI or HII region. Here $|x|$ is the excess number of electrons on the grain. For grains of radii of 3×10^{-5} cm immersed in a diffuse cloud of kinetic temperature of $\sim 100°$ K we find x = 6 electron; in HII regions with temperatures $\sim 10^{4°}$ K we have $|x| \approx 400$. Both these estimates require modification to take account of photoelectric emission due to ultraviolet light; for HI regions they would hold good in clouds with $\tau_V \geq 0.5$ and for HII regions they would be relevant in the outer parts where the ultraviolet radiation is greatly attenuated.

When the rate of emission of photoelectrons from the grain surface exceeds the rate at which electrons are captured, positive charge builds up to an equilibrium value x, determined by

$$\alpha\, n_e\, \sigma_e\, v_e = F_e\, 4\pi a^2 \tag{5.32}$$

with σ_e, v_e defined as before, α is sticking coefficient for electrons, and F_e is the photoelectron emission flux per unit area. The quantity F_e depends on the ambient ultraviolet flux and the photoelectric yield which varies with the grain material and with photon energy. If W is the work function of the solid, typically ~ 4 eV for ices and organics, photons of minimum energy in eV

$$W = \frac{xe^2}{a} + 4\ \text{eV}$$

would be required to cause photoemission. The efficiency of photoemission is negligible at visible wavelengths, but generally increases in the far ultraviolet, with a value of ~ 0.1 common to most substances near $\lambda \approx 1000$ Å. With an estimate of $\sim 3 \times 10^7\ \text{cm}^{-2}\ \text{s}^{-1}$ for the flux of $\lambda = 1000$ Å photons in a diffuse cloud (with no extinction) the equilibrium grain potential is ~ 0.07 eV (Watson, 1972) assuming $n_e = 5 \times 10^{-2}\ \text{cm}^{-3}$ and $a = 10^{-5}$ cm. This corresponds to a grain charge $x \cong 5$. If $n_e = 5 \times 10^{-3}\ \text{cm}^3$ we find $x \cong 42$.

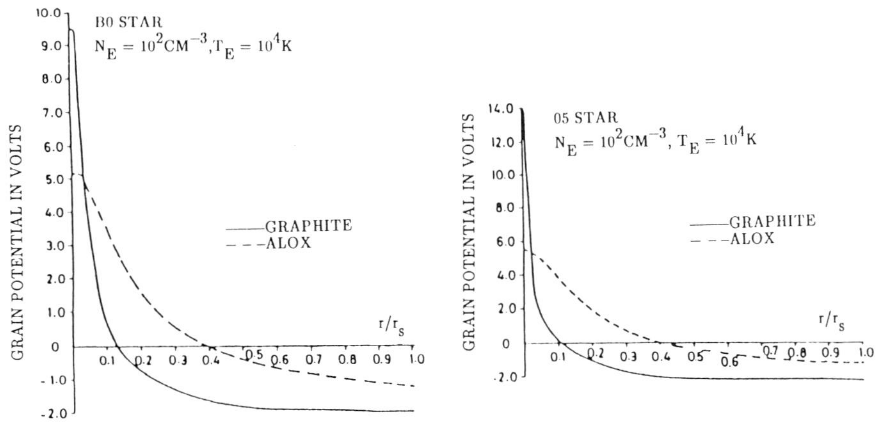

Fig. 5.4 *Equilibrium grain potentials calculated for graphite and Al_2O_3 grains as functions of distance from an early-type star (Moorwood and Feuerbacher, 1976).*

For HII regions detailed calculations of Moorwood and Feuerbacher (1976) have shown that grain potentials are sensitively dependent on the grain material. Fig. 5.4 shows tha potentials calculated for graphite and Al_2O_3 grains as function of distance (in units of the Strömgren radius). We note that the potential tends asymptotically to the 'Spitzer values' at large values of r/r_S.

5.5. ROTATIONS OF GRAINS

A gas atom striking a grain surface may either be specularly reflected or be adsorbed and re-evaporated from the grain surface. (We ignore molecule formation in this context.) In either case an angular momentum increment $\sim$ mav is transferred to the grain, where m is the mass of the impinging atom, v is its mean velocity, and a is the grain radius. The square of any component of the grain's angular momentum begins to grow linearly with time by a random walk process until it is finally maintained at an equilibrium value due to the resistive torque of the gas. The equilibrium angular velocity of a grain is determined by a simple equipartition relation:

$$\frac{1}{2} I\omega^2 = \frac{3}{2} kT , \tag{5.34}$$

where I is the grain's moment of inertia, and T is the ambient gas kinetic temperature. Taking an average moment of inertia

$$I \cong \frac{2}{5}\left[\frac{4}{3}\pi a^3 s\right] a^2 \tag{5.35}$$

where s is the grain density, appropriate to a spherical grain, and writing $\omega = 2\pi\nu$, equation (5.34) gives for s = 1

$$\nu \cong 3 \times 10^7 \left[\frac{10^7}{a}\right]^{2.5} \left[\frac{T}{100}\right]^{\frac{1}{2}} \text{Hz} . \tag{5.36}$$

We thus see that small grains with $a \cong 10^{-6}$ cm rotate at frequenties of ~ 30 MHz in interstellar clouds with gas temperature ~ 100° K; larger particles with $a \cong 10^{-5}$ cm rotate at frequenties of 0.1 MHz.

The timescale required for a grain to acquire its equipartition angular velocity ω is

$$\tau_{eq} \approx \frac{(I\omega)^2}{\langle mva\rangle^2 \; n \langle v\rangle \; (4\pi a^2)} \tag{5.37}$$

where n is the ambient gas density. Using equation (5.34) together with suitable values of the averages indicated in the denominator of (5.37) we find that equipartition is reached in about 10^5 years or less in most cases.

We note that the estimate of ν according to equation (5.36) would be too low if we admit the possibility of H atoms recombining to from H_2 on the surfaces of grains and being ejected with translational energies ε ~ 1 eV (an appreciable fraction of the chemical energy release). In such cases the equipartition condition will be $\frac{1}{2} I\omega^2 \approx \varepsilon$ where ε is the mean energy released on recombination that is converted to translational motion of the molecule. Frequencies of rotation may then be ~ 1000 MHz for smaller grains with $a \simeq 10^{-6}$ cm. The timescale for equipartition on the other hand will not be significantly altered because of the invariance of the ratio $\langle\omega^2\rangle / \langle\frac{1}{2} m\, v^2\rangle$ in equation (5.37).

5.6. RADIO WAVES FROM GRAINS

Erickson (1957) and the present authors (Hoyle and Wickramasinghe, 1970) have discussed the possibility that charged spinning grains emit radio waves. For grains of radii in the range $10^{-7} - 10^{-5}$ cm equation (5.36) shows that the rotation frequencies cover the radio waveband from 10^{10} Hz – 10^6 Hz. In Fig. 5.4 we saw that over a major part of an ionized hydrogen cloud surrounding a hot star, grains are charged to a potential

$$\Phi \cong -2 \text{ Volts} . \tag{5.38}$$

In general, the charge distribution of a non–spherical grain would have a centroid which is appreciably distinct from the centre of mass of the grain. The dipole moment arising from such a charge distribution is

$$\langle P\rangle \cong eax \tag{5.39}$$

where x is the excess number of electrons on the grain. For a rotation frequency ν the rate of emission of electromagnetic waves of frequency ν is given by

$$\frac{dE}{dt} = \frac{16\pi^4\nu^4}{3c^3} <P>^2 . \tag{5.40}$$

Using equations (5.38) – (5.40) with $\Phi = ex/a$, we obtain

$$\frac{dE}{dt} \cong 8.5 \times 10^{-34} a^4 \nu^4 \text{ erg s}^{-1} . \tag{5.41}$$

The efficiency of this radiation process is easily determined from equation (5.41). The mass of a grain of radius a is $\frac{4}{3}\pi a^3 s \approx 10a^3$, where s is the density grain material. Hence the emission per gram of material is

$$\sim 10^{-34} a\nu^4 \text{ erg s}^{-1} \text{ gm}^{-1} . \tag{5.42}$$

This result may be compared with the rates of emission of radio waves from dust–filled nebulae associated with hot stars. The inner part of the radio source Sagittarius A at the centre of our galaxy has an output of $\sim 10^{35}$ erg s^{-1} at $\nu = 2.10^9$ Hz. To obtain this output we require

$$\sim 10^5 M_\odot \tag{5.43}$$

of grains of radii 2.10^{-7} cm which probably represents a fraction of the total mass associated with this source. As we shall see in a later chapter, particles of these sizes may be identified with interstellar polyaromatic molecular clusters. The possibility therefore arises that non–thermal radio sources associated with hot dust filled nebulae may well arise from this process, rather than by an electron synchrotron process as is usually believed.

5.7. MOLECULE FORMATION

A variety of interstellar molecules and radicals is known to exist in interstellar space. CH, CH^+, CN were detected several decades ago by their characteristic bands in optical stellar spectra, while H_2 was detected optically more recently (Carruthers, 1970). A host of other molecules of a predominantly organic composition have been detected over the past few years by their spectral features in the radio and millimetre waveband. Table 5.3 gives a list of the interstellar molecules that have been discovered to date. The list continues to grow as new transitions are predicted and instruments deployed to search for these predictions.

TABLE 5.3

Interstellar Molecules Listed According to Number of Atoms

2	3	4	5	6	7	8	9	10	11	13
H_2	H_2O	H_2CO	H_2C_2O	$HCONH_2$	HC_5N	$HCOOCH_3$	HC_7N	CH_3C_5N	HC_9N	$HC_{11}N$
CH	H_2	H_2CS	H_2CNH	CH_3CN	$HCOCH_3$	CH_3C_3N	$(CH_3)_2O$	CH_3COCH_3		
CH^+	HCN	$HCNH^+$	H_2NCN	CH_3NC	CH_3C_2H		CH_3CH_2CN			
C_2	HNC	HNCO	HC_3N	CH_3OH	CH_2CHCN		CH_3CH_2OH			
CN	HCO	HNCS	HCOOH	CH_3SH	NH_2CH_3		CH_3C_4H			
CO	HCO^+	$HOCO^+$	CH_2CN	C_5H	C_6H					
CS	HOC^+	C_3H	C_3H_2							
NO	HCS^+	C_3N	C_4H							
NS	HNO	C_3O	SiH_4							
OH	N_2H^+	C_3S	CH_4							
SiO	C_2H	NH_3								
SiS	OCS	H_3O^+								
SO	SO_2	C_2H_2								
SO^+	SiC_2	C_2H_4								
HCl	C_2S									

The process of molecule formation in interstellar space in the gas phase tends to be generally inefficient in the absence of three–body collisions and of grain surfaces. The role of grain surfaces in assisting molecule formation has been discussed by several authors, particularly in relation to interstellar H (van de Hulst, 1949; McCrea and McNally, 1960; Gould and Salpeter, 1963; Knaap et al., 1966; Solomon and Wickramasinghe, 1969). Radiative association of two neutral H atoms to form H_2 is effectively ruled out because vibration–rotation transitions are forbidden for electric–dipole radiation. If an interstellar H atom hitting a grain resides on the surface for long enough to encounter and combine with another H atom, an H_2 molecule could form. The excess energy of the newly formed excited molecule is distributed into the grain lattice and the molecule subsequently re–evaporates.

Van de Hulst (1949), McCrea and McNally (1960) and Gould and Salpeter (1963) argued that almost every impinging atom could reside on the grain long enough to combine with another atom. For a total mass density of grains of $\sim 2 \times 10^{-26}$ g cm^{-3} and a mean grain radius of $\sim 10^{-5}$ cm, the total grain surface per cm^3 is

$$\frac{2 \times 10^{-26}}{\frac{4}{3}\pi a^3 s} \cdot 4\pi a^2 \cong 3 \times 10^{-21} \text{ cm}^2 \text{ grain surface per cm}^3 . \qquad (5.44)$$

taking the density of grain material to be $s \approx 2$ g cm^{-3}. The rate of molecule formation per unit surface (assuming every impact leads to 0.5 molecule) is $0.5 n_H v_{th}$, where n_H is the number density of ambient hydrogen, and v_{th} is their thermal speed; and the rate per unit cloud volume is

$$\sim 10^{-8}\, n_H \text{ cm yr}^{-1} . \tag{5.45}$$

However, Knaap et al. (1966) correctly pointed out that equation (5.45) is valid only with certain reservations. Including a zero–point energy which had formerly been neglected, one could show the residence time for an adsorbed H atom on an ice grain to be

$$\tau \cong 10^{-12} \exp[200/T_g] \tag{5.46}$$

where T_g is the grain temperature. (Here k x 200° K is the depth of the potential well in which the atom lies minus the quantum mechanical zero point energy.) The mean time interval between successive H atom arrivals is

$$\tau_{coll} \approx [\pi a^2 (kT/m_H)^{\frac{1}{2}}\, n_H]^{-1} \tag{5.47}$$

where T is the gas kinetic temperature. With $a \cong 10^{-5}$ cm and $T \simeq 100°$ K equation (5.47) gives

$$\tau_{coll} \approx 3.10^4\, n_H^{-1} . \tag{5.48}$$

A necessary condition for H_2 formation is that the mean residence time τ exceeds the time between successive H–atom arrivals, τ_{coll}. From (5.46) and (5.48) we then obtain

$$T_g < T_{crit};\; T_{crit} = \frac{87}{16.5 - \log_{10} n_H} . \tag{5.49}$$

With $n_H = 10$ cm^{-3} the critical grain temperature for recombination is $T \simeq 6°$ K on an ice grain of radius 10^{-5} cm. We have already seen in Section 5.2 that in diffuse clouds such low temperatures may occur only for a limited number of grain species with exceptionally low ultraviolet absorptive properties, e.g. pure quartz. They could, however, occur in the interiors of dark clouds with modest amounts of visual extinction (Solomon and Wickramasinghe, 1969).

A stringent limitation on the amount of H_2 in interstellar space arises on account of the rapid photodissocation of this molecule (Stecher and Williams, 1967). Considering the balance between the H_2 formation on grain surfaces and photodestruction, it is readily shown that dark clouds with optical extinction exceeding a few magnitudes would consist almost wholly of molecular hydrogen (Solomon and Wickramasinghe, 1969). Similar conclusions have been reached in investigations by Hollenbach and Salpeter (1971) and Hollenbach, Werner and Salpeter (1971).

The preceding discussion was confined to the case of H_2. Grain surfaces could also be effective in the production of other diatomic and more complex interstellar molecules. In addition to the formation of molecules by the physical adsorption process discussed here, interstellar molecules may also form by chemical exchange reactions and by chemisorption on particular types of grain.

5.8. GROWTH AND DESTRUCTION OF GRAINS

An interstellar grain immersed in a dense molecular cloud would tend to accrete a mantle from molecules in the ambient gas. The precise composition of such a mantle remains uncertain, but since the molecular gas is organic in character, it would be reasonable to expect a mantle of organic 'ice'.

Suppose n is the number density of condensible molecules in the cloud and r the radius of a grain at time t. The number of condensible molecules striking a grain per unit time is

$$4\pi r^2 n\left[\frac{m}{2\pi kT}\right]^{3/2}\int_0^{\infty}\int_{-\infty}^{\infty}\int_{-\infty}^{\infty} u \exp\left[\frac{-m(u^2+v^2+w^2)}{2kT}\right] du\, dv\, dw$$

$$= 4\pi r^2 n(kT/2\pi m)^{\frac{1}{2}}, \tag{5.50}$$

where m is the mean molecular weight, and u, v, w are orthogonal velocity components. The rate of increase of mass of a grain is therefore

$$\frac{dM}{dt} = \alpha . 4\pi r^2 n\left[\frac{kT}{2\pi m}\right]^{\frac{1}{2}} m, \tag{5.51}$$

where α is a sticking coefficient.

Writing $M = \frac{4}{3}\pi r^3 s$ where s is the density of the condensed material equation (5.51) gives

$$\frac{dr}{dt} = \frac{\alpha n}{s}\left[\frac{kTm}{2\pi}\right]^{\frac{1}{2}}. \tag{5.52}$$

Equation (5.52) integrates to give

$$r = r_0 + \frac{\alpha n}{s}\left[\frac{kTm}{2\pi}\right]^{\frac{1}{2}} t. \tag{5.53}$$

With $T \simeq 50^\circ$ K, $n_H \simeq 10^4$ cm^{-3}, $n/n_H \simeq 10^{-3}$, $m = 30\ m_H$ (appropriate for H_2CO) for a grain in a dense cloud, we have

$$r = r_0 + 7.32 \times 10^{-11}(t/yr) \quad \text{cm} \tag{5.54}$$

where we have set $\alpha = 1$ and $s = 1$ gm/cm^3, and t is expressed in years. If the initial radius r_0 is assumed to be small, the time of growth to radius $r >> r_0$ is

$$t \simeq 1.4 \times 10^{10}\ r \ \text{yr} \tag{5.55}$$

In a timescale of $\sim 10^6$ years (the lifetime of the cloud against gravitational collapse) grains would grow to radii of $\sim 1.3 \times 10^{-5}$ cm. If the initial grain radius was 3×10^{-5} cm accretion under these conditions would lead only to relatively thin molecular mantles. Higher density conditions (which persist for shorter timescales) would be needed for growth of much larger grain mantles.

Grain mantles of molecular or polymeric composition are easily disrupted when their temperature rises above a critical value. The rate of decrease in radius is given by an equation similar to (5.52):

$$\frac{dr}{dt} = \frac{p}{skT_g}\left[\frac{kT_g m}{2\pi}\right]^{\frac{1}{2}} \tag{5.56}$$

where p is the vapour pressure of the solid at the temperature T_g. Since to a good approximation

$$p \propto T^{5.2} \exp{-\chi/kT_g} \tag{5.57}$$

where χ is the binding energy in the solid, (5.56) is seen to give a negligible rate of decrease of r for $T_g << \chi/k$. Swift evaporation begins when $T_g \approx \chi/k$. For an icy composition a temperature $T_g \sim 100°$ K may be needed for destruction, while for organic polymers the temperature for destruction is $T_g \sim 800°$ K.

Conditions for the disruption of icy grains by thermal evaporation would be expected to prevail throughout most HII regions, whereas destruction of organic polymers or other refractory grain materials may be confined to more restricted regions around early–type stars.

Sputtering and Related Matters. Sputtering – the ejection of target atoms due to irradiation by energetic atoms or ions – has been the subject of active investigation in the laboratory. Experimental data have been mainly confined to measurements of sputtering yield as a function of incident ion energy for a variety of ion–target combinations. Some data is also available on the energy distribution and angular spread of the sputtered particles. An excellent review of work in this field is provided by Kaminsky (1965). We present here only a summary of those results which are of direct relevance to interstellar dust grains. The relative energies of ions encountered under interstellar conditions range from ~ 10 keV down to 0.01 eV, the highest energies being relevant to collisions of high speed grains with gas.

For most refractory substances (e.g. iron, graphite, silicates) the displacement threshold of a lattice atom or molecule may be taken as ~ 25 eV. An impinging ion which produces a recoil in a grain atom close to the surface with energy greater than this value has a significant probability of causing an ejection of such an atom. The impinging ion should in general have an energy which is several orders of magnitude above the threshold value E_t in order to produce a significant yield of sputtered atoms. The qualitative situation appears to be as follows: an ion in the range 1–100 keV colliding with or traversing a target crystal causes a cascade of moving target atoms. The cascade may extend over a considerable region of the target but sputtering results only from atoms within a few atomic layers of the crystal surface.

As one might expect, the sputtering yield – number of sputtered atoms per impinging ion – depends upon the ion/target parameters and upon the incident energy in a rather complicated way. Several energy regimes could be distinguished:

(1) $E < E_t$ – no–sputtering regime,
(2) $E_t \lesssim E \lesssim E_A$ – hardsphere collision regime,
(3) $E_A \lesssim E \lesssim E_B$ – weak–screening regime,
(4) $E > E_B$ – Rutherford collision regime.

Determinations of E_t have been somewhat unreliable on account of the difficulty of measuring very low sputtering yields. Values of E_t in the range 20–30 eV would appear appropriate for sputtering silicate, iron and graphite particles by protons, with lower values in the range 3–20 eV for ices and organic materials. This follows if we fit the best available data (Kaminsky, 1965) to an equation of the form

$$E_t \cong 10[(M+m)^2/4mM]L \tag{5.58}$$

where m is the mass of the impinging ion, M is the atomic mass, and L is the vaporization energy of the target crystal (Wehner, 1957; Langberg, 1958). The energies E_A, E_B are defined as follows:

$$E_A = 2E_R Z_1 Z_2 (Z_1^{2/3}+Z_2^{2/3})^{1/2} \frac{m+M}{M} \tag{5.59}$$

$$E_B = 4E_R{}^2 Z_1^2 Z_2^2 (Z_1^{2/3}+Z_2^{2/3})^{1/2} \frac{m}{M} \cdot \frac{1}{E_t} \tag{5.60}$$

where E_R = 13.68 eV is the Rydberg energy for hydrogen, Z_1, m, Z_2, M are respectively the charge mass combinations of the impinging ion and target atom. Table 5.4 shows the values E_A, E_B for iron, silicate and graphite particles irradiated with protons.

TABLE 5.4

	E_A/keV	E_B/keV
Graphite	0·367	0·382
Iron	2·251	3·499
Silicate ($MgSiO_3$)	0·849	1·109

Experimental data which is most nearly applicable to situations of astrophysical interest pertain to the sputtering of metal targets, including iron, by H_2^+, H_3^+ over the energy range 1–5 keV (KenKnight & Wehner, 1964). The quantity measured is the sputtering yield: S(E) = number of sputtered atoms/number of impinging ions of energy E. We can deduce from the data of KenKnight & Wehner (1964) that S(E) has a value ~ 0.03 over the fairly extended range E ~ 1–5 keV. Lower values of S(E) obtain for both higher and lower energies. Experiments on the sputtering of rocks by beams of H and He indicate sputtering yields similar to those appropriate for iron (Wehner, KenKnight & Rosenberg, 1963). From an inspection of available data we can derive the following semi-empirical expressions for S(E):

$$\begin{aligned} S(E) &= S_0,\ E_A \leq E \leq 10E_B \\ &= S_0(E/E_A),\ E_t \leq E \leq E_A \\ &= S_0(10E_B/E),\ E > 10E_B \\ &= 0,\ E < E_t\,. \end{aligned} \tag{5.61}$$

The rate of decrease of grain radius a due to sputtering is given by equation (5.52) where α is replaced by $-S(E)$. Since $E \sim 0.5$ eV in a typical HII region, we find that sputtering is an inefficient grain destruction mechanism even for the most volatile grains with $E_t \sim 3$ eV. ($n \cong 10^4$ cm^{-3}, $T = 10^{4\circ}$ K, $t = 10^5$ s). Only in cases where grains are propelled through gas clouds with speeds in excess of 1000 km/sec could gas atom impacts be sufficiently energetic for sputtering to be important.

In addition to sputtering, energetic collisions with atoms of energies exceeding about 100 eV, including low-energy cosmic rays, lead to the production of displacement defects – the displacement of atoms into interstitial sites (Dienes & Vineyard, 1957). Although these effects might lead to atrophysically interesting consequences the fluxes of such particles remain uncertain at the present time. However, it is known that the concentration of displacement defects in a solid tends to saturate at high values of the irradiation dose. This occurs due to interactions between groups of dispaced atoms and vacancies, and the maximum possible defect concentration that could persist in low-temperature interstellar grains would appear to constitute ~ 3 percent of the total density of atoms in the solid (Kinchin & Pease, 1955).

5.9. EFFECTS OF RADIATION PRESSURE

Light carries momentum as well as energy. The direction of the momentum is that of photon propagation and the amount is c^{-1} times the energy. Of the incident radiation I_0, the absorbed part $I_0\pi a^2 Q_{abs}$, where a is the radius of a spherical grain, is totally lost to the incident beam, whereas a fraction of the scattered energy is resupplied to the forward beam. This fraction is simply $\overline{\cos\theta}$, so that the total flux of energy removed from the forward beam is

$$I_0\pi a^2(Q_{abs} + Q_{sca} - (\overline{\cos\theta})\, Q_{sca})$$

$$= I_0\pi a^2(Q_{ext} - (\overline{\cos\theta})\, Q_{sca})$$

The quantity $Q_{ext} - (\overline{\cos\theta})\, Q_{sca}$ is the efficiency factor for radiation pressure and is usually denoted by Q_{pr}. The forward momentum removed from the beam per unit time is

$$\frac{I_0}{c}\,\pi a^2\, Q_{pr} \tag{5.63}$$

and this is the radiation pressure force on the grain.

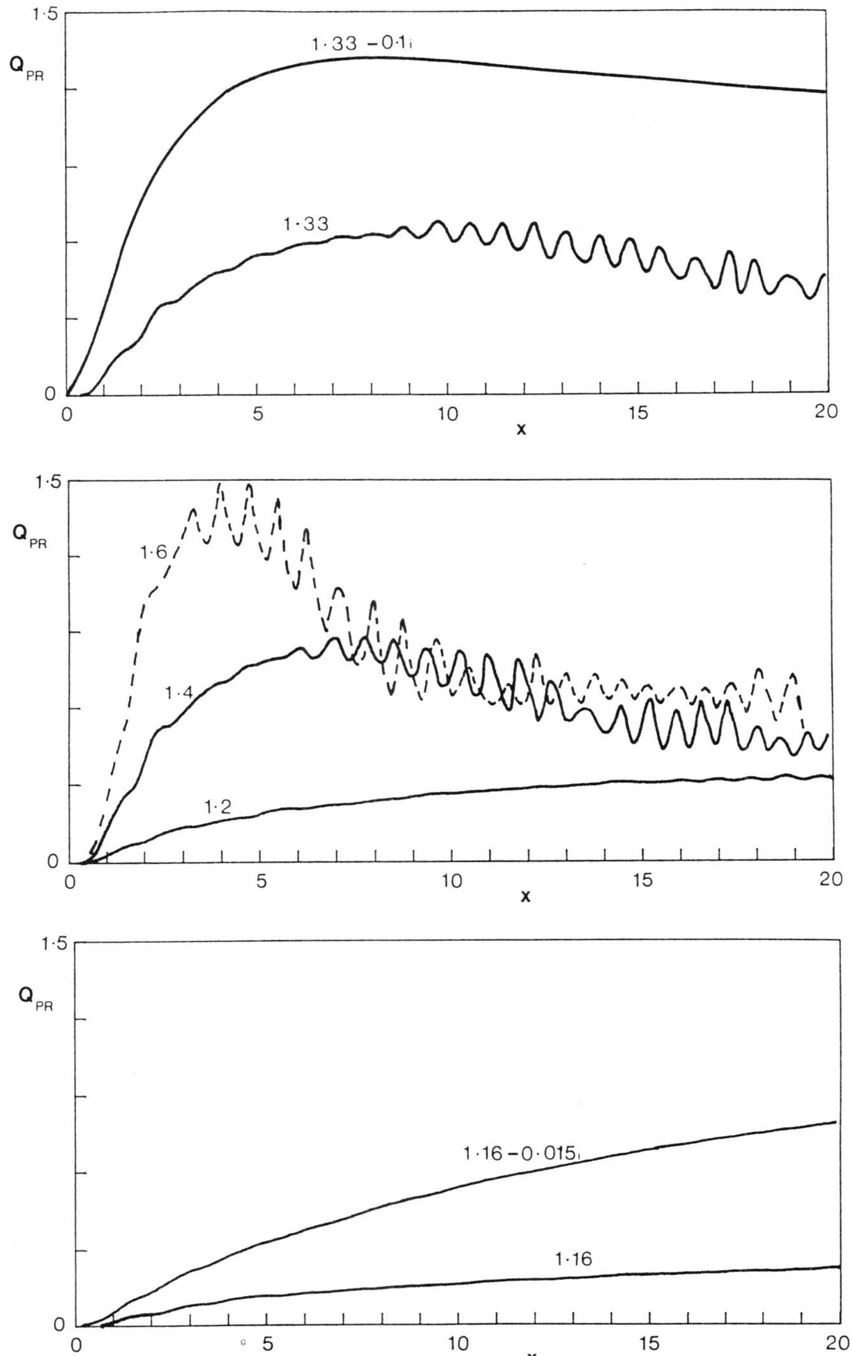

Fig. 5.5 *Plots of $Q_{pr}(x)$ for various values of the complex refractive index m.*

Fig. 5.5 shows Q_{pr} as a function of $2\pi a/\lambda = x$ calculated from the Mie formulae for spheres of various refractive index values m. In general, dielectric particles with radius $a \simeq 3 \times 10^{-5}$ cm have Q_{pr} values in the range ~ 0.2 to 1 at $\lambda \approx 5000$ Å. The radiation force F on a grain given by (5.63) often exceeds the local gravitational attraction G by an appreciable amount. For instance in the vicinity of a star P/G values exceeding unity arise in most cases of interest. Whenever P >> G large velocities can be generated in the direction of asymmetry of the local radiation field.

For grains in a stellar environment (for example in a situation where grain growth occurs) the radiation incident on them exerts a pressure tending to push them away. For a spherical grain of radius a and complex refractive index m the radiation force due to monochromatic light of intensity I_λ is

$$F_\lambda = Q_{pr}(a, m, \lambda)\, \pi a^2 \frac{I_\lambda}{c} \tag{5.64}$$

where Q_{pr} is the efficiency factor for radiation pressure given by the Mie formulae, and c is the velocity of light. For a star of effective temperature T* and radius R* the radiation force on a grain at distance $R \geq R*$ is

$$P = \left[\frac{R*}{R}\right]^2 \int_0^\infty Q_{pr}(a, m, \lambda)\, \pi a^2 \frac{1}{c} B(\lambda, T*)\, d\lambda \tag{5.65}$$

where $B(\lambda, T)$ is the Planck function. The oppositely directed gravitational attraction on the grain at a distance R from the star is

$$G = \frac{4}{3} \pi a^3 s \frac{\gamma M}{R^2} = g \cdot \left[\frac{R*}{R}\right]^2 \cdot \frac{4}{3} \pi a^3 s \tag{5.66}$$

where s is the density of grain material, γ is the gravitational constant, M is the mass, and g the surface gravity of the star. From (5.65) and (5.66) the ratio P/G may be calculated for given values of grain and stellar parameters. Representative values of P/G for the cases $g \cong 1$ cm s^{-2}; T* = 3000, 2500° K are set out in Table 5a for graphite, silicate and iron grains of various radii. Laboratory determinations of $m(\lambda)$ are used for graphite and iron particles (Wickramasinghe, 1973), but for silicates we use m = 1.66–0.005i at all optical and near infra–red wavelengths.

TABLE 5a

Ratio of Radiation Pressure to Gravity, P/G

($T_* = 3000$ K; $g \cong 1$ cm s^{-2}; $R = 3 \times 10^{13}$ cm)

a/μ	Iron	Silicate	Graphite
0·01	$5{\cdot}59\text{x}10^2$	$1{\cdot}57\text{x}10$	$1{\cdot}62\text{x}10^3$
0·03	$9{\cdot}18\text{x}10^2$	$2{\cdot}32\text{x}10$	$1{\cdot}83\text{x}10^3$
0·05	$1{\cdot}70\text{x}10^3$	$5{\cdot}04\text{x}10$	$2{\cdot}46\text{x}10^3$
0·08	$2{\cdot}84\text{x}10^3$	$1{\cdot}44\text{x}10^2$	$3{\cdot}90\text{x}10^3$
0·10	$3{\cdot}25\text{x}10^3$	$2{\cdot}32\text{x}10^2$	$4{\cdot}72\text{x}10^3$
0·30	$1{\cdot}96\text{x}10^3$	$9{\cdot}55\text{x}10^2$	$3{\cdot}49\text{x}10^3$
0·50	$1{\cdot}02\text{x}10^3$	$9{\cdot}44\text{x}10^2$	$1{\cdot}84\text{x}10^3$
1·0	$3{\cdot}34\text{x}10^3$	$4{\cdot}20\text{x}10^2$	$6{\cdot}24\text{x}10^2$

($T_* = 2500$ K; $g = 1$ cm s^{-2}; $R_* = 3 \times 10^{13}$ cm)

a/μ	Iron	Silicate	Graphite
0·1	$7{\cdot}81\text{x}10$	$6{\cdot}21$	$5{\cdot}88\text{x}10^2$
0·03	$1{\cdot}08\text{x}10^2$	$8{\cdot}07$	$6{\cdot}41\text{x}10^2$
0·05	$1{\cdot}93\text{x}10^2$	$1{\cdot}49\text{x}10$	$8{\cdot}36\text{x}10^2$
0·08	$3{\cdot}52\text{x}10^2$	$3{\cdot}96\text{x}10$	$1{\cdot}32\text{x}10^3$
0·10	$4{\cdot}54\text{x}10^2$	$6{\cdot}54\text{x}10$	$1{\cdot}66\text{x}10^3$
0·30	$4{\cdot}88\text{x}10^2$	$3{\cdot}45\text{x}10^2$	$1{\cdot}64\text{x}10^3$
0·50	$2{\cdot}72\text{x}10^2$	$4{\cdot}01\text{x}10^2$	$8{\cdot}90\text{x}10^2$
1·0	$9{\cdot}66\text{x}10$	$2{\cdot}19\text{x}10^2$	$3{\cdot}17\text{x}10^2$

Once clear of the high density photospheric and atmospheric gases – within a few stellar radii – grains may be shown to attain very high speeds. Ignoring the effect of gas friction, which is unimportant over most of the distance scale involved in the acceleration, the equation of motion of a grain is

$$\frac{4}{3}\pi a^3 s\ddot{R} = \frac{P}{G}\cdot g\cdot\frac{4}{3}\pi a^3 s\left[\frac{R_*}{R}\right]^2$$

that is,

$$\ddot{R} = \frac{P}{G} g \left[\frac{R*}{R}\right]^2 . \tag{5.67}$$

Multiplying by R and integrating, we have

$$\dot{R}^2 = 2\left[\frac{P}{G}\right] g \left[\frac{1}{R*} - \frac{1}{R}\right] R_*^2 \tag{5.68}$$

so that the ejection velocity which is effectively reached within a couple of stellar radii is

$$V = \left[2\left(\frac{P}{G}\right) gR*\right]^{1/2} = \left[2\left(\frac{P}{G}\right) \gamma \frac{M}{R*}\right]^{1/2} . \tag{5.69}$$

The maximum ejection velocities of iron, graphite and silicate grains of various radii from a giant star with $T* = 3000°K$, $g = 1$ cm s^{-2} and $R* = 3 \times 10^{13}$ cm (appropriate to a giant star) have been calculated and are set out in Table 5b.

TABLE 5b

Ejection velocities in km s^{-1}

a/μ	Graphite	Silicate	Iron
0·01	$3{\cdot}11 \times 10^3$	$3{\cdot}07 \times 10^2$	$1{\cdot}17 \times 10^3$
0·03	$3{\cdot}31 \times 10^3$	$3{\cdot}73 \times 10^2$	$1{\cdot}45 \times 10^3$
0·05	$3{\cdot}84 \times 10^3$	$5{\cdot}50 \times 10^2$	$1{\cdot}96 \times 10^3$
0·08	$4{\cdot}83 \times 10^3$	$9{\cdot}30 \times 10^2$	$2{\cdot}60 \times 10^3$
0·10	$5{\cdot}32 \times 10^3$	$1{\cdot}18 \times 10^3$	$2{\cdot}86 \times 10^3$
0·30	$4{\cdot}58 \times 10^3$	$2{\cdot}39 \times 10^3$	$2{\cdot}51 \times 10^3$
0·50	$3{\cdot}32 \times 10^3$	$2{\cdot}38 \times 10^3$	$1{\cdot}84 \times 10^3$
1·00	$1{\cdot}94 \times 10^3$	$1{\cdot}59 \times 10^3$	$1{\cdot}08 \times 10^3$

It is seen from Table 5b that initial speeds of several thousand km s^{-1} are appropriate to all three grain species. The characteristic stopping distance of grains is easily estimated if grains have a constant radius throughout their motion. For drift velocities $v >> v_{th}$, the thermal velocity of gas atoms, the drag force on a grain of radius a projected into a gas cloud of density n_H is

$$F \approx \pi a^2 (m_H v)\, n_H v \,. \tag{5.70}$$

Equating F to $-(4/3)\,\pi a^3 s(dv/dt)$ the equation of motion is

$$\frac{dv}{dt} \approx -\frac{3 m_H n_H}{4as} v^2 \tag{5.71}$$

and this may be written as

$$\frac{dv}{dx} \cong -\frac{3 m_H n_H}{4as} v \tag{5.72}$$

where x is the path length traversed through cloud. The distance traversed before slowing down the gas thermal speed of $v_{th} \sim 1$ km s^{-1} is thus

$$\frac{L}{(1 \text{ pc})} \approx 3s \left[\frac{a}{10^{-5} \text{ cm}}\right] \left[\frac{1 \text{ cm}^{-3}}{n_H}\right] \ln \left[\frac{v_i}{\text{km s}^{-1}}\right] \tag{5.73}$$

where v_i is the initial speed in km s^{-1}. The time for slowing to this speed is (by integrating equation (5.71))

$$\frac{4as}{3 m_H n_H} \left[\frac{1}{v_{th}} - \frac{1}{v_i}\right] \cong 3s \left[\frac{a}{10^{-5} \text{ cm}}\right] \left[\frac{1 \text{ cm}^{-3}}{n_H}\right] 10^6 \text{ yr} \,. \tag{5.74}$$

A grain of radius $a \cong 3 \times 10^{-6}$ cm, density $s \cong 3$ g cm^{-3} and initial speed $v_i \approx 1000$ km s^{-1} is stopped within ~ 1 pc inside a typical interstellar cloud of number density $n_H \approx 10$ cm^{-3}. The time scale for stopping is $\sim 3 \times 10^5$ yr.

The velocity of grains in an optically thin cloud will not be controlled by gas collisions, but rather by the asymmetric component of the interstellar radiation field. This asymmetry could arise stochastically due to proximity to a 'nearest' star or due to excess radiation from the galactic bulge. If ζ is the fractional anisotropy in a typical cloud, the radiation pressure force acting on a grain of radius a, with efficiency factor for radiation pressure Q_{pr} is

$$F_{rad} \simeq Q_{pr}\, \pi a^2 U \zeta \,, \tag{5.75}$$

where U is the energy density of the stellar radiation field. The drag force on a spherical grain of radius a moving with velocity u through a cloud of density n, temperature T, is given by

$$F = 4\pi a^2 n_H T \frac{u}{v}(1 + \frac{u^2}{5v^2}), \quad u \lesssim v$$

$$= 4\pi a^2 n_H T \frac{u^2}{v^2}(\frac{3}{4} + \frac{3}{4}\frac{v^2}{u^2} - \frac{v^3}{2u^3} + \frac{v^4}{5u^4}), \quad u \gtrsim v \tag{5.76}$$

where $v = (3kT/m_H)^{\frac{1}{2}}$. For most practical purposes the following orders of magnitude estimates suffice

$$F \approx \pi a^2 m_H n_H uv \qquad u << v$$

$$\approx \pi a^2 m_H n_H u^2 \qquad u >> v \tag{5.77}$$

which we already used in equation (5.70). Equating (5.77) for $u >> v$ to (5.76) we obtain a terminal velocity

$$u_{min} \cong \left\{\frac{Q_{pr} U \zeta}{m_H n_H}\right\}^{\frac{1}{2}} \tag{5.78}$$

Setting $Q_{pr} = 0.5$, $U = 1$ eV cm^{-3}, $\zeta \cong 0.2$, as representative values we obtain

$$u_{min} \cong 10^5 \left\{\frac{n_H}{10\ cm^{-3}}\right\}^{-\frac{1}{2}} cm\ s^{-1} \tag{5.79}$$

We shall adopt (5.80) as a reasonable estimate for the velocity of 'thermal' cloud grains in an asymmetric radiation field. Assuming a diameter of $D \simeq 10$ pc for such a cloud the lifetime of thermal grains against dispersal from the cloud is

$$t_d \simeq \frac{D}{v} \simeq 10^7 \left\{\frac{n_H}{10\ cm^{-3}}\right\}^{\frac{1}{2}} yr\ . \tag{5.80}$$

Intercloud grains. Grains escaping from clouds are likely to be accelerated to much higher velocities in the intercloud medium. The intercloud gas density being $n_H \sim 0.1$ cm^{-3} equation (5.78) already implies a terminal grain velocity $\sim 10^6$ cm s^{-1}. Strictly for intercloud grains the resistive force is increased slightly due to grain charging and the dominance of collisions with protons. The force is now (Spitzer, 1968):

$$F = \frac{4}{3}\pi a^3 su/\tau_s$$

where

$$\tau_s = \frac{11.7(\frac{4}{3}\pi a^3 s/m_H)\ T^{3/2}}{A^{1/2} n_i Z_g^2 Z^2\ (\ln\Lambda)}$$

Here T is the gas kinetic temperature, A, Z, n_i are respectively the atomic weight, charge and number density of the ambient ions, Z_g is the grain charge, and ln Λ is a slowly varying logarithmic function with a value close to 25 for conditions appropriate to the intercloud medium. The use of this expression for the resistive force does not introduce a major modification to our velocity estimate from equation 5.79. Velocities in excess of 10^7 cm s^{-1} could arise due to stochastic variations of U during encounters with stars as are bound to occur. We thus assume that grains in the intercloud medium, irrespective of their mode of origin, have suprathermal speeds imparted to them by interaction with starlight.

5.10. GYRATION ABOUT THE MAGNETIC FIELD

Since grains are charged (Z_g = 1 to 10 in typical cases) their coupling to the magnetic field of the galaxy cannot be ignored. The Larmor radius of a grain of radius a in a magnetic field of intensity B is

$$R_L = \frac{m_g u\ c}{Z_g e B} \tag{5.81}$$

where m_g is the mass of the grain and u is its transverse velocity. Writing $m_g = \frac{4}{3}\pi a^3 s$ and $Z_g = |x| = 3.5\ akT/e^2$ from equation (5.31), equation (5.81) yields

$$R_L = 4 \times 10^{-3} s \left[\frac{a}{10^{-5}\text{cm}}\right]^2 \left[\frac{u}{10^5\text{cm s}^{-1}}\right] \left[\frac{10^{-6}\text{G}}{B}\right] \left[\frac{100\text{K}}{T}\right] \text{pc}\ . \tag{5.82}$$

A particle of radius 10^{-5} cm and density s ≈ 1 g cm^{-3} projected at a speed of 10^5 cm s^{-1} has a Larmor radius of 4.10^{-3} pc in an HI cloud with T ≃ 100° K, B ≃ 10^{-6} G. It is clear from (5.82) that under a wide range of conditions appropriate to interstellar space R_L << the dimensions of a diffuse cloud. Grains are thus effectively tied to magnetic field lines and motions relative to the gas are mainly confined to drifts along field lines. The quantity ζ in the earlier discussion must be interpreted as representing an asymmetry component of the local interstellar radiation field *in* the direction of the local galactic magnetic field.

5.11. ALIGNMENT OF GRAINS

We have already seen that interstellar polarisation is caused by selective extinction due to anisotropic (elongated or flattened) grains that are systematically aligned in the galaxy. If the long axes of such grains lie preferentially in planes normal to some fixed direction, the polarisation of starlight is maximum when the line of sight is perpendicular to this direction and the electric vector of polarisation is parallel to the same direction. Observations of the position angle of interstellar polarisation indicate that this preferred direction lies close to the galactic plane and that it also coincides approximately with the direction of the mean magnetic field (Hiltner, 1965; Hall, 1958; Davis and Berge, 1968).

The main processes discussed for the alignment of grains fall into two categories: (1) processes involving magnetic relaxation in spinning grains (Spitzer and Tukey, 1951; Davis and Greenstein, 1953; (b) dynamical alignment involving the streaming of gas relative to dust (Gold, 1952).

Consider the grains to be elongated prolate spheroids with moments of inertia I_1, I, I, I >> I_1, about principal axes at the grain centroid, with $\omega = (\omega_1, \omega_2, \omega_3)$ the angular velocity about these axes. Taking the grain to be spur–up by gas collisions the equipartition situation is given by

$$I_1\omega_1^2 = I\omega_2^2 = I\omega_3^2 = kT \;, \tag{5.84}$$

which implies

$$I_1\omega_1 = (kTI_1)^{1/2}, I\omega_2 = I\omega_3 = (kTI)^{1/2} \,. \tag{5.85}$$

For I >> I_1, (5.84) requires the spin to be dominantly about the axis of grain symmetry, whereas (5.85) requires the angular momentum to be dominated by components at right angles to the axis of symmetry. It is implicit in these statements that the axes are fixed with respect to the grain material, not in space. Thinking of the grain as a cylinder of radius a and length l composed of material of density s

$$I_1 \simeq \frac{1}{2}\,\pi\, a^4\, l\, s \tag{5.86}$$

Putting a = 3.10^{-5} cm, l = 10^{-4} cm, s = 1g cm^{-3}, (5.86) gives I $\simeq 1.27 \times 10^{-22}$ g cm^2, while (5.84) for a gas temperature T = 50 K gives $\omega_1 \simeq 7.4 \times 10^3$ s^{-1}. The grain rotates of the order of 10^3 times per second about its axis of symmetry, provided the main contribution to its rotational energy comes from gas collisions.

It was seen earlier that a grain situated in a typical diffuse cloud will acquire a nett charge. The charge is negative, corresponding to deficit of $|x|$ electron, with $|x| \approx 10$. To examine the effect of this charge when the grain is considered to be in a uniform magnetic field **B**, let the charge be distributed uniformly around a circle of radius a, taken in the plane through the centroid perpendicular to the axis of symmetry. Because of the main rotation ω_1, the charge can be considered to generate an electric current $|x|e\omega_1/2\pi$ flowing in the circle, on which the magnetic field exerts a torque. It is not hard to obtain the components of this torque with respect to axes fixed in space and thence to obtain the equations of motion of a symmetrical spinning top subject to such a torque. From the equations of motion it follows that, neglecting the small component of angular momentum $I_1\omega_1$, the grain precesses about the direction of the magnetic field with angular velocity Ω given by

$$\Omega = \frac{ea^2|x\mathbf{B}|}{2cI} \tag{5.87}$$

Putting I = 1.27×10^{-22} g cm^2, $|x| = 6$, $|\mathbf{B}| = 2.5 \times 10^{-6}$ G, a = 3.10^{-5} cm, gives $\Omega = 8.50 \times 10^{-13}$ s^{-1}. Thus the grain precesses extremely slowly compared the spin $\omega_1 \simeq 7.4 \times 10^3$ s^{-1}. Nevertheless, the precessional period $2\pi/\Omega \simeq 2.10^5$ years is not long compared to the characteristic time, t_{rel}, for the equipartitional condition (5.84) to become established for a grain situated in a diffuse cloud with $n_H \simeq 10$ cm^{-3}. The latter is given by

$$t_{rel} \simeq \frac{4\,\pi a^3 s/3}{\pi a^2.n_H m_H(v_{th}/3)}\,, \tag{5.88}$$

which is to say the time required for a grain to encounter a mass of gas equal to itself. For $a = 3.10^{-5}$ cm, $s = 1$ g cm^{-3}, $n_H = 10$ cm^{-3}, $v_{th} = 10^5$ cm s^{-1}, $t_{rel} = 2.3 \times 10^6$ years, an order of magnitude greater than the precessional period.

The angle, Θ say, between the direction of the axis of symmetry of the grain and the magnetic field is a constant of motion when dissipation is neglected. With Θ distributed randomly for an ensemble of grains there is no systematic alignment and no polarisation arises for light passing through such an ensemble. Dissipation is present, however, leading to a slow decrease in the initial value of Θ, an effect that in principle is capable of producing polarisation. Thus for $\Theta \neq 0$ the magnetic field considered instantaneously has a component parallel to the symmetry axis of a grain and a component transverse. The former stays constant with respect to the grain material as it spins but the latter oscillates with frequency $\omega_1/2\pi$. The induced magnetisation arising in the dielectric material of a grain is thus subject to oscillation with this frequency, a process that is dissipative. The problem in the theory of Davis and Greenstein is to calculate this dissipation rate. After extensive investigations it was found that the time scale, t_{mag}, for an appreciable reduction of Θ is given by

$$t_{mag} \simeq 3.10^{11}\, T_g\, a^2\, s \,/\, B^2 \sin\Theta \tag{5.89}$$

For the theory to be viable it is necessary that (5.89) should yield a reduction time for Θ not much longer that the relaxation time given by (5.88), otherwise the situation is repeatedly randomised before dissipative effects can become important. When the theory was first proposed it was anticipated the measurements of the galactic magnetic field would give $B \simeq 10^{-5}$ G, but measurements eventually gave $B \simeq 10^{-6}$ G, for which (5.89) is much too long. Thus for $T_g = 50$ K, $a = 3.10^{-5}$ cm, $s = 1$ g cm^{-3}, $B = 10^{-6}$ G, $\sin\Theta \simeq 1$, (5.89) leads to $t_{mag} \simeq 4.10^8$ years, more than a thousand times longer than t_{rel}.

The coefficient in (5.89) depends inversely on ω_1, so that by increasing the spins of grains sufficiently t_{mag} could be reduced suitably towards t_{rel}. An ingenious suggestion for greatly increasing ω_1 has been made by Purcell (1969, 1979). When H atoms adsorbed at a grain surface combine to H_2 it is likely that the hydrogen molecule leaves the grain surface with an energy ~ 1 ev. The recoil energy imparted to the grain is then some two orders of magnitude greater than it is for a gas atom collision. If a grain is considered to spin up through the random occurrence of such events then the equipartition angular velocities would be changed to the extent of replacing kT in (5.84) by ~ 1 ev, i.e. the angular velocity ω_1 is increased by an order of magnitude, thereby increasing the dissipation rate considered above correspondingly. But this increase is not sufficient to overcome the difficulty that t_{mag} remains much greater that t_{rel}. To increase ω_1 still further by a large factor Purcell suggested that H_2 molecule formation does not occur randomly on grain surfaces but at particular active sites, when a grain is spun–up proportionately to the number of molecules formed at its surface rather than proportionately to the square root of that number, so yielding appreciably higher values of ω_1. However, the conditions needed for this process to work succesfully appear to be rather stringent. As pointed out by Spitzer and McGlynn (1979) changes at a grain surface, such as the transient saturation of active sites, lead to reversals of the spin–up torques. At each reversal the spin of the grain dips to a minimum, with these "cross over" events occurring on a significantly shorter time scale than the alignment time. Monte Carlo simulations of the H_2 formation process appear to confirm this objection.

There remains the possibility discussed by Jones and Spitzer (1967), namely that grains contain some material with a far higher dissipation rate then ordinary dielectric materials. A ferromagnetic substance such as magnetite, if there were enough of it, would yield a suitably high dissipation rate to reduce t_{mag} appropriately in relation to t_{rel}. The difficulty was that not enough iron is present in the interstellar medium to provide sufficient magnetite for all grains to become aligned. But if there were the concomitant difficulty would be that the resulting polarisation of starlight by the grains would be far too large. In order to restrict the polarisation of starlight to only a few percent only a small fraction of grains need to be efficiently aligned, and for this there is enough interstellar iron. But the question that still remains is why the available iron should be especially concentrated in only a small fraction of grains. In relation to a biological model for grains, which is shown in Chapters 8 and 9 to give excelent correspondences to observations in other respects, we have noted that a small fraction of bacterial species do in fact contain domains of magnitite, and we have suggested that it could be this fraction that take up the available iron and which then become aligned to yield the few percent polarisation of starlight shown by observation (Jabir, Hoyle and Wickramasinghe, 1983).

Gold (1952) proposed a dynamical alignment mechanism which would operate whenever there is a net streaming motion of gas relative to grains. Gas atoms colliding with a grain contribute angular momentum chiefly transverse to the streaming velocity $\mathbf{v}_s$ and the effect is to make the long axes of grains preferentially parallel to $\mathbf{v}_s$. This alignment process is particularly strong when $|v_s|$ exceeds the thermal velocity of the interstellar gas.

Although the Gold alignment process does not depend directly on the magnetic field, each grain must carry a net electric charge and gyrates many times about the galactic magnetic field during a single 'relaxation period' of the stream alignment. Hence only the component of the streaming velocity $\mathbf{v}_s$ parallel to the magnetic field **B** can contribute systematically to any alignment. The sense of this alignment is therefore to make the long axes of grains parallel to **B**, resulting in a direction of polarisation which is in conflict with the observations.

A necessary condition for the operation of magnetic alignment in an appropriate sense, however, is that the grain temperature T_g is less than the gas temperature T (Jones and Spitzer, 1967). The rotational modes of the grain energised by the paramagnetic relaxation effect will tend to have energies that are in equipartition with the thermal energy of grain atoms $\sim kT_g$, whereas the rotation of the grain transverse to **MxB** comes into equipartition with the gas with energy $\sim$ kT. Alignment in the desired sense demands $T_g < T$. If $T_g > T$ alignment will occur in a sense that is inconsistent with observations.

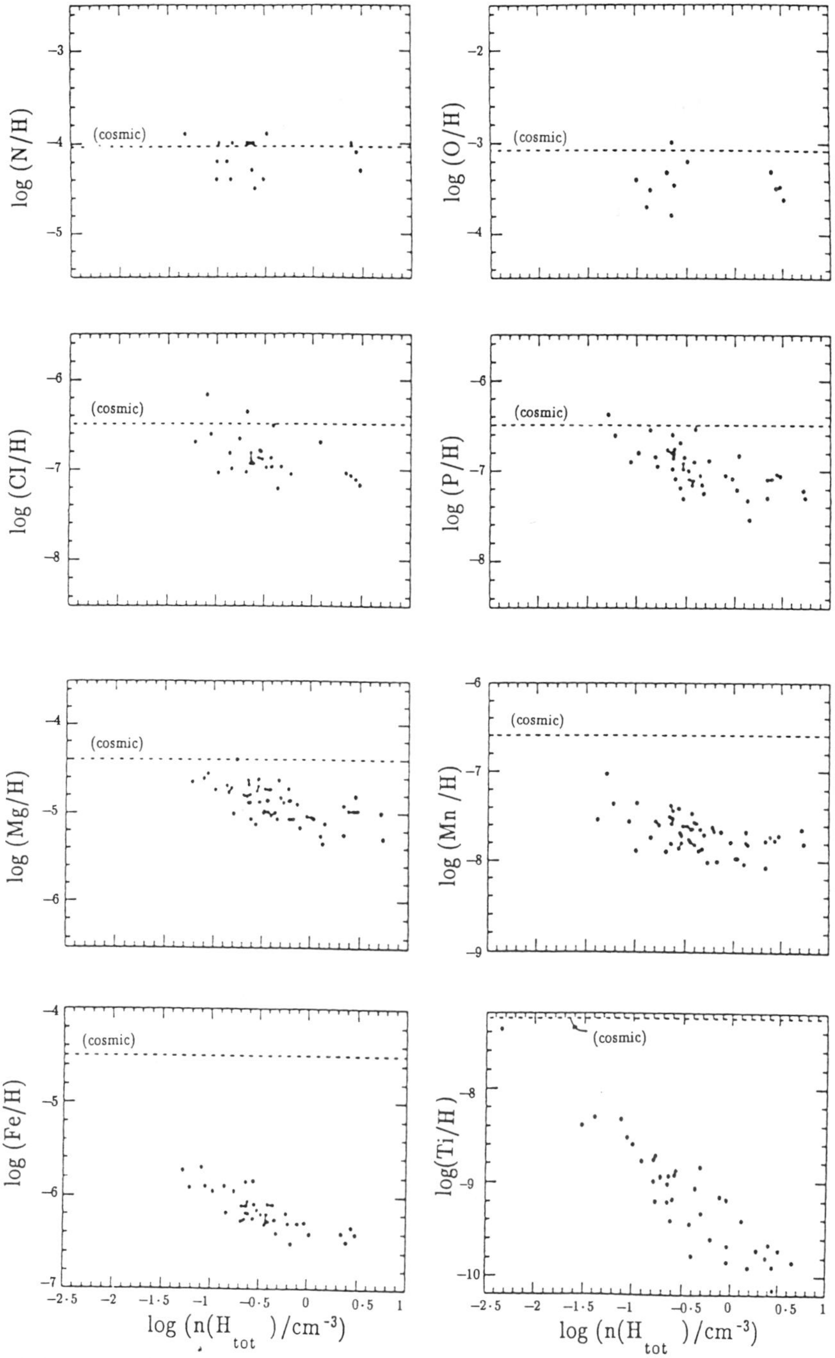

Fig. 5.6 ***Logarithmic abundances A relative to H plotted as functions of average line of sight densities $\overline{n}(H_{tot})$ (after Jenkins, 1986).***

5.12. DEPLETION OF ELEMENTS FROM THE GAS PHASE

Satellite observations of ultraviolet lines have led to important clues relating to the chemical composition of interstellar gas. It has been known for some years that the distribution of elements inferred from ultraviolet studies differs from cosmic abundances (Morton, 1974; Spitzer and Jenkins, 1975; Cowie and Songaila, 1986; Jenkins et al., 1986; Snow et al., 1987). The depletion of an element X is measured using the parameter

$$\delta_X = \log \left[\frac{N_X}{N_H}\right]_{\text{star}} - \log \left[\frac{N_X}{N_H}\right]_{\text{cosmic}} \tag{5.90}$$

where N refers to column densities and the cosmic abundances ratios are determined mainly from the solar spectrum and meteorite studies. The star ζ Oph, which is associated with a relatively dense cloud, was among the first to be studied in relation to depletion measurements. The quantity δ_X for a given element is found to vary somewhat from one star to another with a general trend to increase with hydrogen column density or E_{B-V}. Fig. 5.6 shows the trend for several elements.

TABLE 5.6

Depletion Factors for Elements

Element X	Cosmic Abundance	δ_X (ζ Oph)	δ_X (α Sco)
H	12·0	0	0
Li	3·2	-1·5	–
C	8·6	–0·7	–0·41
N	8·0	–0·7	–0·30
O	8·8	–0·6	–0·22
Na	6·3	–0·9	
Mg	7·5	–1·5	–1·04
Al	6·4	–3·3	–3·09
Si	7·5	–1·6	–1·53
P	5·4	–1·1	
S	7·2	–0·3	–0·05
Ca	6·4	–3·7	–3·70
Fe	7·4	–2·0	–2·10

Table 5.6 shows the depletions estimated for ζ Oph and σ Sco. We note that elements such as Fe, Ca, Al, Si, Mg are drastically depleted, whereas C, N, O, appear to be depleted only to a moderate extent. The depletions of C as given in Table 5.6 are likely to be an underestimate. We note that the depletion of this element cannot yet be directly observed because it is mainly singly ionized. The available CII lines tend either to be saturated, or they are too weak to be detected by currently available observational techniques (Jenkins, 1986). It is safe to suppose that δ_C is so poorly determined as to be virtually unknown.

References

Bethe, H. van Ashkin, J., 1953, in E. Segre (ed.), *Experimental Nuclear Physics*, Wiley, N.Y.

Bohlin, R.C., 1975, *Astrophys. J.*, **200**, 402.

Burbidge, E.M. and Burbidge, G.R., 1972, in A. and M. Sandage (eds.) *Galaxies and the Universe, Stars and Stellar Systems*, Vol. 9.

Carruthers, G., 1970, *Astrophys. J. Lett.*, **161**, L81.

Chiao, R.Y. and Wickramasinghe, N.C., 1972, *Mon. Not. Roy. Astr. Soc.*, **159**, 361.

Cowie, L.L. and Songaila, A., 1986, *Ann. Rev. Astr. Ap.*, **24**, 499.

Davis, L. and Berge, G.L., 1968, in B.M. Middlehurst and L.A. Aller (eds.) *Stars and Stellar Systems*, Vol. 7, Chicago.

Davis, L. and Greenstein, J.L., 1951, *Astrophys. J.*, **114**, 206.

Dienes, G.J. and Vineyard, G.H., 1957, *Radiation Effects in Solids*, Interscience, NY.

Duley, W.W., 1973, *Astrophys. Sp. Sci.*, **23**, 43.

Erickson, W.C., 1957, *Astrophys. J.*, **126**, 480.

Gold, T., 1952, *Mon. Not. Roy. Astr. Soc.*, **112**, 215.

Gould, R.J. and Salpeter, E.E., 1963, *Astrophys. J.*, **138**, 393.

Greenberg, J.M., 1971, *Astron. and Astrophys.*, **12**, 240.

Greenberg, J.M. and Shah, G.A., 1969, *Physica*, **41**, 92.

Hall, J.S., 1958, *Publ. US Naval Observatory*, **17**, 275.

Hiltner, W.A., 1965, in A. Beer (ed.) *Vistas in Astronomy*, 2, Pergamon, NY.

Hollenbach, D. and Salpeter, E.E., 1971, *Astrophys. J.*, **163**, 155.

Hollenbach, D.J., Werner, M.W. and Salpeter, E.E., 1971, *Astrophys. J.*, **163**, 165.

Hoyle, F. and Wickramasinghe, N.C., 1970, *Nature*, **227**, 473.

Jabir, N.L., Hoyle, F. and Wickramasinghe, N.C., 1983, *Astr. and Sp. Sci.*, **91**, 327.

Jenkins, E.B., Savage, B.D. and Spitzer, L., 1986, *Astrophys. J.*, **301**, 355.

Jenkins, E.B., 1986, in D.A. Hollenbach and H.A. Thronson (eds.) *Interstellar Processes*, D. Reidel.

Jones, R.V. and Spitzer, L., 1967, *Astrophys. J.*, **147**, 943.

Kaminsky, M., 1965, *Atomic and Ionic Impact Phenomena on Metal Surfaces*, Springer Verlag.

Ken Knight, C.E. and Wehner, G.K., 1964, *J. Appl. Phys.*, **35**, 322.

Kinchin, G.H. and Pease, R.S., 1955, *Rep. Prog. Phys.*, **18**, 1.

Knaap, H.F.P., van den Meidenberg, C.J.N., Beenakker, J.J.M. and van de Hulst, H.C., 1966, *Bull. Astr. Soc.*, Netherlands, **10**, 137.

Langberg, E., 1958, *Phys. Rev.*, **111**, 91.

Leung, C.M., 1975, *Astrophys. J.*, **199**, 340.

McCrea, W.H. and McNally, D., 1960, *Mon. Not. Roy. Astr. Soc.*, **121**, 238.

Moorwood, A.F.M. and Feuerbacher, B., 1976, in N.C. Wickramasinghe and D.J. Morgan *Solid State Astrophysics*, D. Reidel.

Morton, D.C., 1974, *Astrophys. J. Lett.*, **193**, L35.

O'Dell, C.R. and Hubbard, W.B., 1965, *Astrophys. J.*, **142**, 591.

O'Dell, C.R., Hubbard, W.B. and Peimbert, M., 1966, *Astrophys. J.*, **143**, 743.

Oort, J.H., 1965, *Stars and Stellar Systems*, **5**, 455.

Page, T., 1972, in A. and M. Sandage (eds.) *Galaxies and the Universe, stars and Stellar Systems*, Vol 9.

Purcell, E.M., 1976, *Astrophys. J.*, **206**, 685.

Purcell, E.M., 1969, *Physica*, **100**.

Purcell, E.M., 1979, *Astrophys. J.*, **231**, 404.

Roberts, M.S., 1972, in A. & M. Sandage (eds.) *Galaxies and the Universe, Stars and Stellar Systems*, Vol. 9.

Rubin, V.C. and Ford, W.K. Jr., 1970, *Astrophys. J.*, **159**, 379.

Sandage, A. 1972, *Ann. Report of the Directors, Hale Observatories*, 1970 – 1971, p. 417.

Snow, T.P., Buss, R.H., Gilra, D.P. and Swings, J.P., 1987, *Astrophys. J.*, **321**, 921.

Solomon, P.M. and Wickramasinghe, N.C., 1969, *Astrophys. J.*, **159**, 449.

Spitzer, L., 1941, *Astrophys. J.*, **93**, 369.

Spitzer, L., 1978, *Physical Processes in the Interstellar Medium*, J. Wiley & Sons.

Spitzer, L. and Jenkins, E.B., 1975, *Ann. Rev. Astron. Astrophys.*, **13**, 133.

Spitzer, L. and McGlynn, T.A., 1979, *Astrophys. J.*, **231**, 417.

Spitzer, L., 1968, *Physics of Fully Ionized Gases*, Interscience, NY.

Stecher, T.P. and Williams, D.A., 1967, *Astrophys. J. Lett.*, **149**, L29.

van de Hulst, H.C., 1949, *Rech. Astron. Obs. Utrecht*, XI, Part 2.

Watson, W.D., 1972, *Astrophys. J.*, **176**, 103.

Wehner, G.K., 1957, *Phys. Rev.*, **108**, 35.

Wehner, G.K., Ken Knight, C.E. and Rosenberg, D., 1963, *Planet. Sp. Sci.*, **11**, 1257.

Werner, M.W. and Salpeter, E.E., 1969, *Mon. Not. Roy. Astr. Soc.*, **145**, 249.

Wickramasinghe, N.C., 1971, *Nature Phys. Sci.*, **234**, 7.

Wickramasinghe, N.C., 1973, *Light Scattering Functions for Small Particles with Applications in Astronomy*, Adam Hilger.

6. Inorganic Theories of Grain Formation

6.1. INTERSTELLAR CONDENSATION

The idea of nucleation of grains from low density interstellar gas clouds owes its origin to the classic work of B. Lindblad (1935). Lindblad's appreciation of the need for interstellar condensation stemmed mainly from the observation that parameters characterising the gas and the dust correlated strongly one with another. This has become evident to an even greater degree in recent times, as can be seen from the correlation of the hydrogen column density to E_{B-V} (Fig. 2.1). Furthermore, we have seen in Chapter 5 that the depletions of heavy elements from the gas phase increase with E_{B-V} indicating an incorporation of gas atoms into grains. The present evidence does not, however, support the old ideas entirely in demanding that solid particles nucleated and condensed from the gas phase. The minimum requirement is that interstellar atoms become somehow incorporated into grains, possibly through a cycle involving star formation, or simply by atoms and molecules sticking onto pre-existing grains. The latter possibility has a measure of support in view of the high values of A_V/E_{B-V} that are found to be associated with dense clouds and star-forming regions.

Ice Grain Theory. Lindblad's (1935) suggestion led to a provisional grain composition based on stable inorganic combinations of the elements O,C,N with H, and with trace quantities of other materials included within a basically icy matrix. In view of the large abundance of H in interstellar space, hydrides of the C,N,O elements appeared to have some *prima facie* plausibility. H.C. van de Hulst (1949) who developed these ideas in a classic thesis, argued for the composition set out in Table 6.1.

TABLE 6.1

van de Hulst's Suggested Composition of Interstellar Grains

	Refractive Index
100 molecules H_2O	m = 1·31
30 molecules H_2	1·10
20 molecules CH_4	1·26
10 molecules NH_3	1·32
5 molecules MgH, etc	complex

It should be noted, however, that the above composition was obtained on the basis of a simplified scheme. An assumption implicit here is that every heavy atom hitting the growing grain resides on the surface long enough to encounter and combine with adsorbed H atoms, and then freezes down as a stable hydride.

Nucleation in Interstellar Space? The detailed physical processes that might lead to the condensation of interstellar gas atoms into grains (in the absence of pre-existing grains) are yet somewhat obscure. We state briefly the ideas that have been proposed and indicate points in the chain of argument where difficulties have subsequently arisen. Broadly speaking, the following sequence of steps was proposed by ter Haar (1943), van de Hulst (1949) and others:

(a) The formation of diatomic molecules, such as CH, CH^+, CN, CN^+, from the interstellar gas atoms.

(b) The growth from diatomic molecules to large molecules containing 10 to 50 atoms which serve as condensation nuclei.

(c) The growth from condensation nuclei to grains.

The first two steps in this sequence have insecure features, however.

Starting from an interstellar composition of monatomic gases with a density of a few atoms per cubic centimetre, the rate of molecule formation will be very slow. The crucial question is whether condensation nuclei, consisting of aggregates of 10 to 50 atoms, could form at a sufficiently fast rate to produce the observed interstellar grain density in a time-scale of $\sim 3 \times 10^9$ years. In 1941, the University of Leiden offered a prize for the best solution to this important astronomical problem. The prize-winning contribution of ter Haar, which was later published (B.A.N., 361) was perhaps the most significant step towards a solution.

The formation of diatomic molecules is probably the first step in any grain formation process that might conceivably occur in interstellar space. The usual ideas of chemical kinetics unfortunately do not apply in this case. The laws of chemical kinetics are applicable only when molecules form predominantly by three-body collisions. An encounter between two atoms could result directly only in the formation of a diatomic molecule in a high vibrational state. Unless the excited molecule encounters a third molecule or a solid surface within a timescale of the order of the vibrational period, $\sim 10^{-13}$ sec, it would instantly dissociate. Under interstellar conditions, however, the gas densities are far too low for three-body collisions to play an important role.

Excluding three-body collisions, and also in the absence of available solid surfaces, the only other way in which a diatomic molecule might form is when a two–body collision is accompanied by the emission of radiation. The energy loss by radiation has the effect of bringing the excited molecule to the stable lower state. Such a radiative association between two atomic species A and B may be written schematically as

$$A + B \rightarrow AB + h\nu \tag{6.1}$$

The number of two-body collisions between species A and B per unit volume per unit time is given by

$$n(A)n(B)v\sigma \tag{6.2}$$

where n(A), n(B) are the number densities of the species, v is their mean speed and σ is their collision cross-section. The rate of formation of molecules of type AB is then given by

$$\frac{dn(AB)}{dt} = pn(A)n(B)v\sigma \tag{6.3}$$

where p is the probability that the emission $h\nu$ takes place, and n(AB) is the number density of the molecule AB at time t. Writing $\gamma = pv\sigma$, we have from (6.3)

$$\frac{dn(AB)}{dt} = \gamma n(A)n(B) \tag{6.4}$$

γ is the rate constant for the reaction (6.1) and may either be determined experimentally or obtained by theoretical calculation. Unfortunately, it is the case that the former procedure is not available for the type of reaction we shall be concerned with, so that we have to rely solely on theoretical estimates.

Two atoms approaching each other in a two-body collision possess a varying dipole moment and may therefore be expected to emit radiation. The probability for the emission of such radiation leading to the formation of a stable molecule may be computed by assuming a suitable potential for the interaction forces between the two atoms.

ter Haar (1943) and Kramers and ter Haar (1946) considered the formation of the molecules CH, CH^+ to be the first step in the process of grain building. This would appear to be a fair assumption, since the rates of formation of molecules such as NH, OH by radiative two-body collisions are believed to be considerably slower.

Observations of interstellar bands also indicate a considerable abundance of CH, CH^+, adding to about 10^{-8} cm^{-3}. Much of this observed abundance is, however, likely to have been formed by reactions taking place either on grain surfaces or by grain destruction rather than by direct two-body radiative associations.

The early estimates of rate constants for reactions leading to CH, CH^+ formation were very tentative, based on simple models for the interaction potential. Kramers and ter Haar also underestimated the importance of the reverse photodissociation reactions which are simultaneously taking place, and further assumed an interstellar gas temperature of $\sim 10^{4\circ}$ K for an HI cloud.

The question of CH and CH^+ formation was later re-examined by Bates and Spitzer (1951) using more reliable computations of rate constants, including the effect of photodissociations, and assuming an interstellar gas temperature of 100° K appropriate for HI clouds. The reactions they considered were essentially as follows:

$$\begin{aligned}
&C + H \rightarrow CH + h\nu \\
&C^+ + H \rightarrow CH^+ + h\nu \\
&CH + h\nu \rightarrow C + H \\
&CH^+ + h\nu \rightarrow C^+ + H \\
&CH + h\nu \rightarrow CH^+ + e \\
&CH^+ + e \rightarrow CH + h\nu \\
&CH^+ + e \rightarrow C + H
\end{aligned} \tag{6.5}$$

Equilibrium concentrations n(CH), n(CH^+) were determined for typical interstellar radiation fields and for given values of n(H), n(C), n(C^+), assuming a gas temperature of ~ 100° K. For n(C) = 3.10^{-3} cm^{-3}, n = 10 cm^{-3}, n(C^+) = 3.10^{-4} cm^{-3}, the computed value of n(CH), according to Bates and Sptzer, is in the general region of 10^{-10} to 10^{-11} cm^{-3}; the value of n(CH^+) may be higher by about an order of magnitude but a value close to 10^{-10} seems to be likely. The computed values of both n(CH) and n(CH^+) were considerably lower than the observed abundance of these molecules, ~ 10^{-8} cm^{-3}.

In the years between 1951 and the present time, much progress has been made in understanding the problems of molecule formation in interstellar clouds. For instance, the development of ion–molecule reaction networks where one of the reactant atoms is charged has shown that reaction rates could be increased by several orders of magnitude above the rates associated with radiative associations between uncharged species. Reactions of this type are

$$\begin{aligned}
&H_2^+ + H_2 \rightarrow H_3^+ + H \quad (\gamma \cong 2 \times 10^{-9}\ cm^3\ s^{-1}) \\
&O^+ + H_2 \rightarrow OH^+ + H \quad (\gamma \cong 10^{-9}\ cm^3\ s^{-1}) \\
&CO^+ + H_2 \rightarrow CO^+ + H \quad (\gamma \cong 1.4 \times 10^{-9}\ cm^3\ s^{-1})
\end{aligned} \tag{6.6}$$

On account of the greatly increased rate constants that are relevant when charged species are involved, ion-molecule chemistry in interstellar space looks at first sight promising. Indeed successes of this theory have been considerable with regard to correct predictions of some of the abundances of the smaller organic molecules and radicals set out in Table 5.1. However, with regard to using a pure gas phase chemistry to account for the nucleation of grains one is faced with a chicken and egg problem. Ion–molecule reaction networks invariably depend on the pre–existence of H_2, a molecule that can form effectively only on grain surfaces. (See section 5.7).

It is perhaps ironical that although the optical detection of the radicals CH, CH^+, CN, CN^+ is nearly half a century old, difficulties still remain in relation to their formation in interstellar clouds by gas phase reactions. For instance, there is a major problem that has persisted for over a generation to explain the observed ratio

$CH^+/CH \sim 1$. Although the formation rate of CH^+ is rapid through the reaction

$$C^+ + H \rightarrow CH^+ + h\nu \tag{6.7}$$

with a rate constant of $\sim 10^{-7}$ cm^3 s^{-1}, the destruction of CH^+ proceeds too fast through the reactions

$$CH^+ + H_2 \rightarrow CH_2^+ + H \tag{6.8}$$

and

$$CH^+ + e \rightarrow C + H\,, \tag{6.9}$$

the former having a rate constant of $\sim 10^{-9}$ cm^3 s^{-1} and the latter a rate constant of $\sim 10^{-7}$ cm^3 s^{-1}. The fact that the theoretical expectation $CH^+/CH << 1$ is not borne out by the observational data is already quite disturbing for the theory of molecule formation in the gas phase in interstellar clouds. Dalgarno (1976) has reviewed the subject of interstellar CH, CH^+, and we refer the reader to this review for more details.

Even if one can show that molecules such as CH, CH^+ are able to form at an adequate rate, it does not follow that grain formation proceeds sufficiently fast. The question is: what fraction of these diatomic molecules build up into condensation nuclei, and ultimately into grains?.

To explain the observed interstellar absorption on the basis of the most efficient grain sizes, we have already seen that the smoothed-out mass density of grains has to be $\sim 10^{-26}$ gm/cm^3. With a typical grain radius of $\sim 10^{-5}$ cm and a grain material of density 1 gm/cm^3, this gives a number density of grains

$$n_g = \frac{10^{-26}}{\frac{4}{3}\pi(10^{-5})^3} = 2 \times 10^{-12}\ \mathrm{cm}^{-3} \tag{6.10}$$

We therefore require the formation of this number density of condensation nuclei in the timescale of the age of the galaxy.

If we accept the computations of Bates and Spitzer (1951) the equilibrium densities of CH and CH^+ add up to $\sim 10^{-10}$ cm^{-3}, so that about 1/50 of these molecules are required to grow into condensation nuclei and thence to grains.

The growth and survival of diatomic species into polyatomic molecules comprised of 10 – 50 atoms is fraught with considerable difficulty. In the early discussions of this subject, the precise processes that could lead to such growth were, however, left somewhat obscure.

It has been argued that every C, O and N atom that impinged on an incipient condensation nucleus resulted in a stable addition, whereas an He or Ne atom was quickly re-evaporated. An impinging heavy atom such as C, O, or N may be expected, in the first instance, to be physically adsorbed; it then diffuses over the surface until it forms a chemical bond with an atom of the condensation nucleus. The newly formed molecule would be in an excited vibrational state, but a molecular aggregate consisting of 10 to 50 or more atoms was expected to have a sufficient number of internal degrees of freedom among which the excess energy could be shared, and a stable addition was therefore expected to ensue. The reason why an He or Ne atom was not expected to add on in this way is because they are chemically inert, and can form only very weak chemical bonds with other atoms. An impinging H atom may also stick, provided it forms a stable chemical bond with an atom already attached to the nucleus.

Destructive Effects on Nuclei. For a cluster of 20 atoms forming an incipient ice nucleus, the absorption of an individual optical photon would raise the temperature to value $T \cong 1060^\circ K$ according to equation (5.21). Such a temperature spike would certainly lead to the destruction of a volatile grain nucleus. From equation (5.6) we see that such events take place with a characteristic time interval of 10^4 s. This time is considerably shorter than the time required for regeneration of an ice grain to a radius $a = 10^{-7}$ cm (e.g. equation (5.54). Thus the occurrence of thermal spikes essentially kills off any process whereby grain nuclei might be generated form single atoms in interstellar space.

A similar argument casts doubt on theories of forming polyaromatic hydrocarbon molecules from a process of building up from smaller units in interstellar clouds. Such macro-molecules as were originally proposed by Platt (1956) and Platt and Donn (1956) can, however, arise from the break–up or degradation of larger organic grains.

6.2. CONDENSATION OF GRAPHITE GRAINS

A different approach to the problem of the formation of interstellar grains was first proposed by us in 1962 (Hoyle and Wickramasinghe, 1962). Various aspects of this problem have subsequently been re-investigated by other workers (Donn et al.,1968; Kamijo, 1969; Fix, 1969a,b; Friedemann and Schmidt, 1967; Blander and Katz, 1967; Tabak et al., 1975; Zettlemoyer, 1977; Salpeter, 1974a,b; Draine and Salpeter, 1977; Degushi, 1980; Draine, 1979). We now review the original graphite formation theory, taking account of later developments where necessary.

The basic proposal was that graphite particles could form in the atmospheres of cool carbon stars and subsequently be expelled into interstellar space. The difficulties of molecule formation and nucleation encountered in interstellar condensation theories are then overcome. For a giant star, the photospheric density is in the region of 10^{15} to 10^{16} atoms/cm^3 ; and at such densities three-body collisions between atoms are sufficiently frequent for the usual ideas of chemical kinetics to apply.

The obvious place to look for graphite grain condensation in the galaxy would seem to be the surfaces of cool giant stars. Of these, the so-called carbon stars, whose spectra are dominated by bands of C_2, CH, CH, etc., are likely to prove the strongest candidates. Giants showing bands of TiO or ZrO presumably have an excess concentration of oxygen over carbon. Since the dissociation energy of CO (11.2 eV) is considerably higher than that of O_2 (5.25 eV), it is likely that in such stars the carbon is largely combined with oxygen as CO, and is therefore not available to form graphite grains.

Carbon Stars. The carbon stars are red giants characterised by spectra showing strong bands of carbon compounds, and extremely weak metallic-oxide bands. In these stars, carbon is likely to be appreciable more abundant than oxygen with typical C/O ratios in the range 5 – 10.

Carbon stars are classified as CO–C9 according to the criteria of Keenan and Morgan (1941); or as RO–R9 and NO–N9 in the Henry Draper system. The R stars (RO–R9) correspond approximately to classes CO–C4, and the N stars (NO–N9) to classes C5–C9. The mean effective temperatures of these classes deduced from colour measurements are given in Table 6.2 below;

TABLE 6.2

Temperatures and Types of Carbon Stars

	R Stars			N Stars			
Spectral Class	C0	C2	C4	C5	C6	C7	C9
T_{eff}	4500	4000	350(	3000	2500	2000	1500

The carbon stars are variable stars which pulsate with periods of ~ 100 days. During a pulsation cycle, the effective temperature varies through several hundred degrees, covering a range of 2 or 3 spectral classes.

Thermodynamic Considerations. In order to discuss the formation of graphite particles during a pulsation cycle, we require to carry out molecular equilibrium calculations for the conditions prevailing in the stellar atmosphere. The gas phase atomic and molecular abundances in a mixture of the elements H, C, N, O, Si... in given proportions are functions of density and temperature. Thermodynamic equilibrium is required in each of a set of reactions such as:

$$\begin{aligned} &H + O \leftrightharpoons OH\,, \\ &OH + H \leftrightharpoons H_2O\,, \\ &C + O \leftrightharpoons CO\,, \\ &CO + O \leftrightharpoons CO_2\,, \\ &H + CO_2 \leftrightharpoons HCO_2\,, \\ &C + N \leftrightharpoons CN\,, \\ &H + CN \leftrightharpoons HCN\,, \end{aligned} \tag{6.11}$$

$$C + C \leftrightharpoons C_2$$

$$C_2 + C \leftrightharpoons C_3$$

If n(H), n(O), n(OH),... are the number densities of H,O,OH..., when thermodynamic equilibrium is reached, we have equations of the form:

$$\frac{n(H)\ n(O)}{n(OH)} = \frac{Z_H Z_O}{Z_{OH}} e^{-\chi/kT}, \tag{6.12}$$

where χ refers to the heat of formation of a molecule from its constituents on the left–hand side of the equations in (6.11) and Z refers to the relevant partition functions.

For each element we also have a constraint that the total number of atoms per unit volume in all the molecular species is constant. For example, in the case of H

$$n(H)+n(OH)+2n(H_2O)+n(HCO_2)+n(HCN) = [n_H] \text{ (given)} \tag{6.13}$$

Proceeding in this way gives N non–linear equations for N unknowns which are solved numerically

For a specified temperature T and specified total pressure $p = n_H kT$, the calculation gives the number density of free carbon particles $n \simeq n(C) + n(C_2) + n(C_3)$.

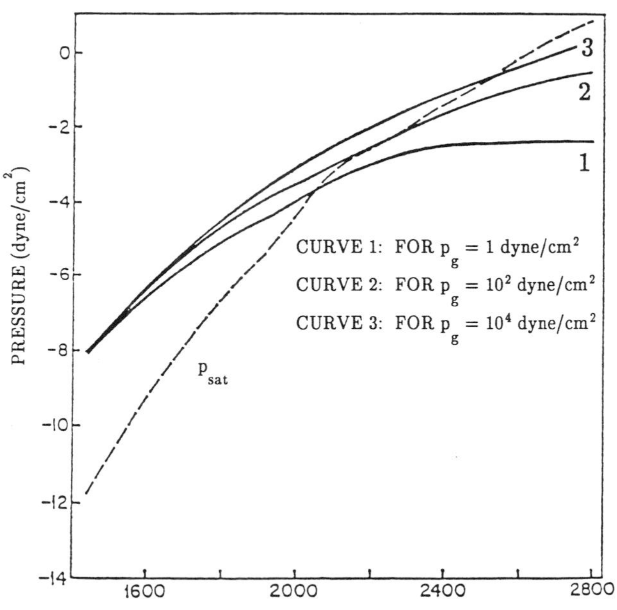

Fig. 6.1 *Partial pressure of free (uncombined) carbon in an N star atmosphere as a function of temperature; p_g is the total gas pressure.*

For the case H : C : N : O = 1 : 5 x 10^{-3} : 10^{-3} : 10^{-3} the carbon pressure $p_C = nkT$ is plotted in Fig. 6.1 as a function of T for the cases p_g = 1, 10^2, 10^4 dyne cm^{-2} (see also Table 6.3). The dashed curve shows the vapour pressure of bulk graphite. The photospheric value of p_g for a giant star, assuming an H^- source of opacity, is ~ 10 dyne/cm^2. For such a case, it is seen that since the free carbon pressure equals the vapour pressure of graphite at T ≅ 2300°K two possibilities are open. Either the carbon vapour would become increasingly supersaturated with respect to the bulk phase as the temperature falls below 2300°K or graphite particles would condense, lowering the carbon pressure to the vapour pressure of graphite. For reasons to be discussed below we shall find that the latter alternative is more likely.

Let T_{sat} denote the temperature at which the partial pressure of free carbon exceeds the saturation vapour pressure of bulk graphite, which we express in the form $p_{sat} = n_{sat}kT$. A tabulation of p_{sat} is also given in Table 6.1. For a given value of $T < T_{sat}$ we can re–calculate equilibrium abundances, including condensed graphite, by replacing the earlier condition of constraint on the number of C atoms in the gas phase by the new condition $n(C) = n_{sat}$, and re–calculating equilibrium in all reactions. In this way we can find the fraction of the carbon atoms which is

TABLE 6.3

T (°K)	Free Carbon Pressure ($C+C_2+C_3$) (dyne/cm^2)			Bulk Graphite Vapour Pressure
	$p_g \simeq 1$	$p_g \simeq 10^2$	$p_g \simeq 10^4$	p_{sat} (dyne/cm^2)
1008	4·6 x 10^{-16}	1·0 x 10^{-16}	1·7 x 10^{-21}	1·5 x 10^{-23}
1120	1·3 x 10^{-13}	1·1 x 10^{-13}	4·2 x 10^{-16}	7·4 x 10^{-20}
1260	3·6 x 10^{-11}	3·5 x 10^{-11}	2·7 x 10^{-11}	3·6 x 10^{-16}
1440	9·4 x 10^{-9}	1·0 x 10^{-8}	2·6 x 10^{-8}	1·7 x 10^{-12}
1680	1·5 x 10^{-6}	2·7 x 10^{-6}	8·8 x 10^{-6}	8·7 x 10^{-9}
1800	8·6 x 10^{-6}	2·3 x 10^{-5}	8·5 x 10^{-5}	2·7 x 10^{-7}
1938	5·0 x 10^{-5}	1·7 x 10^{-4}	8·5 x 10^{-4}	8·5 x 10^{-6}
2100	3·8 x 10^{-4}	1·1 x 10^{-3}	7·0 x 10^{-3}	8·1 x 10^{-4}
2291	2·2 x10^{-3}	8·4 x 10^{-3}	5·1 x 10^{-2}	7·8 x 10^{-3}
2520	3·6 x 10^{-3}	7·7 x 10^{-2}	3·3 x 10^{-1}	2·3 x 10^{-1}
2800	3·9 x 10^{-3}	3·3 x 10^{-1}	2·5	7·08
3150	4·0 x 10^{-3}	3·8 x 10^{-1}	18	2·1 x 10^2
3600	4·5 x 10^{-3}	4·0 x 10^{-1}	37	6·2 x 10^3

condensible as graphite at any temperature below T_{sat}. Detailed calculations have shown that for $T < T_{sat}$, when graphite may be formed under thermodynamic conditions, essentially all C except that tied up as CO goes into solid form.

Nucleation and Growth. Provided there is enough time within a single pulsation cycle of a carbon star for nucleation and growth of solid particles, the relevant temperature to be applied in the above considerations is that appropriate to the minimum of the cycle, the temperature when the star is reddest. It is usually the case that a modest degree of supersaturation is required in order for nucleation to proceed. We now calculate the degree of supersaturation necessary to complete nucleation in the available timescale of $\sim 10^7$ sec (Donn et al., 1968) and for a total carbon concentration in the atmosphere of

$$[n_c] \sim 10^{12}\ \mathrm{cm}^{-3}$$

Assuming that essentially all C condenses into graphite spheres of radii $r = 2 \times 10^{-6}$ cm, the number of atoms per grain is 10^6. We then require to form $\sim 10^6$ stable nuclei per cm^{-3} in $\sim 10^7$ sec, implying a nucleation rate

$$J = 10^{-1}\ \mathrm{cm}^{-3}\ \mathrm{s}^{-1} \qquad (6.14)$$

This requirement may be turned into a condition on p/p_{sat} using classical nucleation theory.

In thermodynamic equilibrium at temperature T the number density of clusters comprised of $\underline{i}$ carbon atoms is

$$n_i = n \exp(-\Delta G_i/kT)\ , \qquad (6.15)$$

Where $-\Delta G_i$ is the free energy of formation of the cluster. For the case of homogeneous nucleation (i.e nucleation in absence of foreign species)

$$\Delta G_i = 4\pi r^2 \sigma - \frac{4\pi r^3}{3} \frac{kT}{v} \ln\left[\frac{p_c}{p_{sat}}\right]\ , \qquad (6.16)$$

where r is the radius of cluster, σ is the surface energy per unit area, and v is the volume per C atom. As r increases, ΔG_i first increases with r, reaches a maximum, and then decreases. The cluster radius which makes ΔG_i a maximum is given by

$$r^* = \frac{2\sigma}{\frac{kT}{v} \ln\left[\frac{p_c}{p_{sat}}\right]} \qquad (6.17)$$

and the maximum value of ΔG_i is

$$\Delta G_i^* = \frac{16\pi\sigma}{3\left[\frac{kT}{v} \ln\left[\frac{p_c}{p_{sat}}\right]\right]^2} \qquad (6.18)$$

According to standard nucleation theory the rate of nucleation J (the rate at which single C atoms add on to clusters of critical radius r*) is:

$$J \cong \alpha(4\pi r^{*2}) \frac{p_c}{(2\pi mkT)^{1/2}} Zn_C \exp(-\Delta G_i^*/kT) \tag{6.19}$$

where α is a sticking factor and n_C is the number density of single carbon atoms. Here Z is a non–equilibrium factor which takes into account that nuclei of critical radii r* are removed from gas.

Equations (6.14) – (6.18) together with $Z = 10^{-2}$, $\alpha = 1$, $\sigma = 1000$ erg cm^{-2} can be solved for our assumed stellar conditions to give

$$p_c/p_{sat} \approx 2\text{–}10 \,. \tag{6.20}$$

From Fig. 6.1 we see that a temperature drop of only a few hundred degrees below T_{sat} produces supersaturation rates in this range. Such a temperature will certainly be achieved during the pulsation cycle of a carbon star. The above discussion using homogeneous nucleation theory is based on the work of Donn et al. (1968). Later workers have proposed modifications that ultimately affect the required supersaturation ratio by a small margin. In view of the steep dependence on temperature of the vapour pressure of bulk graphite, only small alterations of condensation temperature are involved. In the following discussion we shall assume that particle formation is ensured when the ambient temperature falls to T = 1900° K.

The rate of growth of a spherical particle of radius r is given by

$$\frac{dr}{dt} = \alpha \,.\, \frac{p_c}{s} \left[\frac{m}{2\pi kT}\right]^{1/2} \tag{6.21}$$

where α is the sticking coefficient, p_c is the partial pressure of free carbon, m is the mass of a C atom, T is the temperature, and s is the density of graphite. With $\alpha \cong 0.5$ (Thorn and Winslow, 1957) $s \cong 2.2$ gm/cm^3, $T \cong 2 \times 10^{3}$° K, we have

$$\frac{dr}{dt} \cong 10^{-6}\, p_c \text{ cm/sec} \,. \tag{6.22}$$

Thus for a particle to be able to grow to a radius $r \cong 2 \times 10^{-6}$ cm, or larger, in a timescale of $\sim 10^7$ sec, we have the following condition on the free carbon pressure

$$p_c \gtrsim \sim 2 \,.\, 10^{-7} \text{ dyne/cm}^2 \,. \tag{6.23}$$

It is seen from Fig. 6.1 that this condition is satisfied for a total gas pressure $\sim$ 1 dyne/cm at temperatures exceeding 1600° K If growth takes place at higher temperatures, the condition (6.23) will be satisfied as a strict inequality, and the resulting particle radius could exceed 2×10^{-6} cm, provided there is unlimited supply of carbon. The size to which particles may grow is, however, likely to be modulated by the number of available condensation nuclei (possibly ions) and the limited supply of C atoms. Before condensation begins, the total photospheric carbon density is $\sim 10^{12}$ cm^{-3} and the ion number density from ionisation of potassium is probably close to 10^6. If this number of ions become condensation nuclei, and if 30% of the total carbon is available for condensation, the number of C atoms per grain is 3 x 10^5, setting the particle radius at ~ 10^{-6} cm. However, a number density of condensation nuclei less than the number density of ions is likely to be established during the available timescale. With one percent of the ion density becoming effective as condensation nuclei, the resulting particle radius would become a few times 10^{-6} cm.

The Escape of Grains from the Stellar Atmosphere. As grains begin to grow the radiation incident on them exerts a pressure tending to push them away from the source of radiation. For a spherical grain of radius a and complex refractive index m, the radiation force due to monochromatic light of intensity I_λ is

$$F_\lambda = Q_{pr}(a,m,\lambda)\ \pi a^2 \frac{I_\lambda}{c} \tag{6.24}$$

where Q_{pr} is the efficiency factor for radiation pressure given by the Mie formulae, and c is the velocity of light. For a star of effective temperature T_* and radius R_* the radiation force on a grain at distance $R \geq R_*$ is

$$P = \left[\frac{R_*}{R}\right]^2 \int_0^\infty Q_{pr}(a,m,\lambda)\pi a^2 \frac{1}{c} B(\lambda,T_*)\ d\lambda\ , \tag{6.25}$$

where $B(\lambda, T*)$ is the Planck function. The oppositely directed gravitational attraction on the grain at a distance R from the star is

$$G = \frac{4}{3}\pi a^3 s \frac{\gamma M}{R^2} = g\left[\frac{R*}{R}\right]^2 \frac{4}{3}\pi a^3 s \tag{6.26}$$

where s is the density of the grain material, γ is the gravitational constant, M is the stellar mass, and g is the surface gravity of the star. From equations (6.25) and (6.26) the ratio P/G may be calculated for various values of grain and stellar parameters. Table 6.4 shows calculations of P/G for graphite spheres for the case g = 1 cm s^{-1} (appropriate for a giant star) and for the two cases T* = 2500, 2000° K. The data on the $m(\lambda)$ values for graphite are taken from the laboratory measurements of Taft and Phillip (1964).

TABLE 6.4

P/G ratio for g = 1 cm s^{-2}

a/μ	$T_* = 2500$ K	$T_* = 2000$ K
0·01	$5{\cdot}88 \times 10^2$	$1{\cdot}00 \times 10^2$
0·02	$6{\cdot}23 \times 10^2$	$2{\cdot}12 \times 10^2$
0·03	$6{\cdot}41 \times 10^2$	$2{\cdot}18 \times 10^2$
0·05	$8{\cdot}36 \times 10^2$	$2{\cdot}84 \times 10^2$
0·08	$1{\cdot}32 \times 10^3$	$4{\cdot}36 \times 10^2$
0·10	$1{\cdot}66 \times 10^3$	$5{\cdot}48 \times 10^2$
0·30	$1{\cdot}64 \times 10^3$	$5{\cdot}41 \times 10^2$

From Table 6.4 we note that P/G $\simeq$ 200 for grains of radius 0.02μm at the stage when the temperature of the star has fallen to ~ 2000° K. The grains are thus expelled outward. However, the grains are not at first repelled freely. They are immersed in photospheric gases which produce a drag as the grains move outwards. At photospheric densities $n_H \sim 10^{12} - 10^{15}$ cm^{-3}, the viscous drag force F on a spherical grain of radius a is

$$F = 6\pi a \eta u , \tag{6.27}$$

where u is the relative speed between the grain and the gas and η is the viscosity, given by

$$\eta = \frac{1}{2} \, a \, m_H \, n_H \left[\frac{kT}{\pi m_H}\right]^{1/2} \tag{6.28}$$

Equations (6.27) and (6.28) lead for T = 2000° K to

$$F = 3.6 \times 10^{-18} \, a^2 \, n_H \, u \tag{6.29}$$

The grain accelarates so long as P exceeds F. Eventually, however, u becomes so large that $F \cong P$. At this stage the frictional resistance of the gas prevents a further increase of u, so long as the grain remains within the photospheric layers. The maximum speed with which the grain can move upwards through the first scale–height is therefore given by equating F and P,

$$\left[\frac{P}{G}\right]\left[\frac{4}{3}\,\pi a^3 s\right] g = 3.6 \times 10^{-18} \, a^2 \, n_H \, u \,,$$

i.e.

$$u = 10^{15} \, n_H^{-1} \,, \tag{6.30}$$

where we have set $g = 1$ cm s^{-2}, $a = 2 \times 10^{-6}$ cm, $P/G = 2 \times 10^2$, and the time τ required for this is $\sim kT/2m_H g u$

$$\tau = 5 \times 10^{-5} \, n_H \tag{6.31}$$

The time required for a grain to move upward through the second scale–height is less than (6.31) by a factor e, since the hydrogen density n_H falls off exponentially with height above the photosphere. It is the first scale–height therefore that determines the timescale required for the grains to diffuse upward out of the star's atmosphere.

If the value of τ, computed for the smallest value of n_H arising in the pulsation cycle, were greater than the pulsation period of the star, then any appreciable escape of grains would be prevented by the frictional drag. This follows because towards maximum temperature phase of the cycle the grains must be wholly, or largely, evaporated. Grains condensing near the minimum phase must escape at that phase if they are to leave the star. Since the stars in question have pulsation periods of several hundred days, we can reasonably take 10^7 sec as the timescale allowed for the escape – i.e. we require $\tau \lesssim 10^7$ sec. From (6.31) therefore, we obtain the following condition on the photospheric hydrogen density

$$n_H < 2.10^{11} \text{ atom/cm}^3 \tag{6.32}$$

if escape is to take place. For an ordinary giant star with the opacity in the photosphere arising mainly from H^- absorption, the photospheric hydrogen density $n_H \approx 10^{15}$ cm^{-3}. The lower value of n_H required for escape would not be attained but for the opacity produced by the grains themselves.

It is easy to show that when condenstation proceeds the grain opacity causes n_H to fall significantly below 10^{11} cm^{-3}. The reason is that the atmosphere is forced to distend in order to permit the escape of radiation. This feature also ensures that grains move through the first scale–height in less than a pulsation period (Hoyle and Wickramasinghe, 1962). In extreme cases the high opacity of condensed grains may cause a wind of material, gas as well as grains, to flow entirely out of the star.

Once clear of the photospheric gases, a grain will begin to evaporate with a rate of decrease of radius given by equation (5.56) with p set equal to the saturation vapour pressure of graphite at $T \simeq 1900°$ K. Using the data in Fig. 6.1 and Table 6.1, it is easily shown that the evaporation time into a vacuum for a grain of radius $a = 2 \times 10^{-6}$ cm is $\sim 10^6$ sec. If the grain can be propelled by radiation pressure to, say, double the stellar radius in less than this time, the equilibrium grain temperature will fall to low enough values that would ensure their escape without further evaporation.

We now consider the outward motion of a grain once it is clear of photosperic gases. The outward force is

$$P - G \approx 200 \, \frac{4}{3} \, \pi a^3 s g \tag{6.33}$$

Here g is the surface gravity which we assumed above to be 1 cm s^{-2}. At a distance R from the centre of the star the gravitational force per unit mass is $g(R*/R)^2$. Ignoring the effect of gas friction, the equation of motion is

$$\frac{4}{3} \, \pi a^3 s \, \ddot{R} = \frac{P}{G} \, g \, \frac{4}{3} \, \pi a^3 s \left[\frac{R*}{R}\right]^2 , \tag{6.34}$$

that is

$$\ddot{R} = \frac{P}{G} \, g \left[\frac{R*}{R}\right]^2 \simeq 200 \left[\frac{R*}{R}\right]^2 \tag{6.35}$$

Multiplying by R and integrating, we have

$$\dot{R}^2 = 400 \left(\frac{1}{R*} - \frac{1}{R}\right) R*^2 \tag{6.36}$$

and the time taken for R to increase to 2R* is

$$\frac{1}{20 \, \sqrt{R*}} \int_{R*}^{2R_*} \left[\frac{R}{R - R*}\right]^{1/2} dR = \frac{R*^{1/2}}{20} \, [\sqrt{2} + \ln(1+\sqrt{2})] \tag{6.37}$$

Taking $R* = 3 \times 10^{13}$ cm for the photospheric radius, we find this time is an order of magnitude less than the typical pulsation period of a carbon star. The limiting value of R as $R \rightarrow \infty$ is seen from equation (6.36) to be $\sim 10^3$ km/s for $R* = 3.10^{13}$ cm. Grains are therefore expelled at high speed. Such grains are slowed down when they encounter a mass of gas comparable with but exceeding their own mass in interstellar clouds. For a grain of radius $a = 2 \times 10^{-6}$ cm and density 2.2 g cm^{-3} this gives a stopping distance of $\sim$ 1 pc in a typical cloud of density $n_H \sim 10$ cm^{-3}.

6.3. CONDENSATION OF GRAINS IN COOL OXYGEN–RICH GIANT STARS

The main difference between these stars and the carbon stars discussed above relates to the C/O ratio. In the oxygen-rich giant stars (M–type stars) we have a C/O ratio less than unity, whereas in the case of carbon stars the C/O ratio exceeds unity. For the case of an M-giant star with solar system relative abundances we may compute equilibrium molecular abundances as a function of T. Figure 6.2 shows the results of such a calculation assuming a total photospheric hydrogen density $[n_H] \cong 10^{15}$ cm^{-3}, effective temperature ~ 3000° K, solar abundances and a density law $\rho \propto R^{-3}$. This calculation is relevant both for the solar nebula and also for a mass flow from a typical Mira star. The computational procedure is based on equations (6.11) – (6.13), as in the above discussion of carbon stars. As above, condensation of a solid species is taken to occur at modest supersaturation values of the vapour pressure.

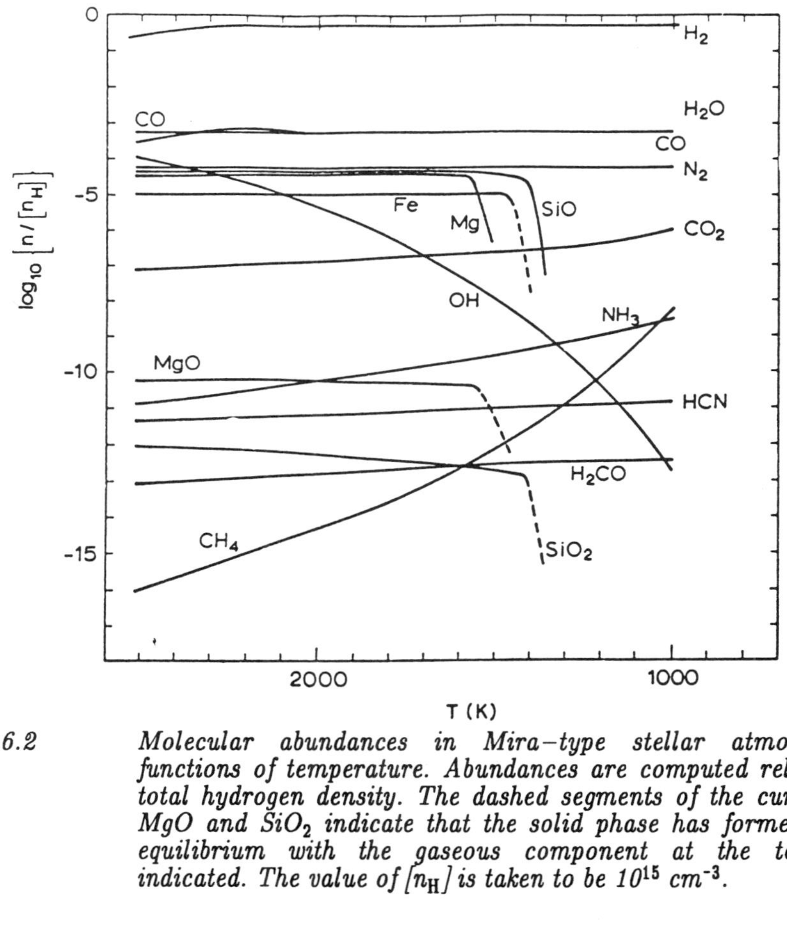

Fig. 6.2 *Molecular abundances in Mira–type stellar atmospheres as functions of temperature. Abundances are computed relative to the total hydrogen density. The dashed segments of the curves for Fe, MgO and SiO_2 indicate that the solid phase has formed and is in equilibrium with the gaseous component at the temperatures indicated. The value of $[n_H]$ is taken to be 10^{15} cm^{-3}.*

The dashed segments of the curves for Fe, MgO and SiO_2 indicate that the solid phase of these materials has formed and is in equilibrium with the gaseous component at the temperatures indicated (Hoyle and Wickramasinghe, 1968). We find from these calculations that the three species Fe, MgO and SiO are able to condense in a region of the stellar atmosphere where the temperature is in the range of $\sim 1400 - 1500°$ K. Since such temperatures occur fairly close to stars of effective temperatures in the range $2000 - 3500°$ K, extensive condensation of these particles takes place in the outer regions of appreciably distended stellar atmospheres. The particles could grow to radii $\sim 10^{-6}$ cm in a period of $\lesssim 1$ year and be ejected into the interstellar medium by radiation pressure.

Radiation pressure can produce striking differences in the way different grain species are expelled. The ratio P/G for iron and silicate grains of spherical shape around a Mira star is given in Table 6.5. The Q_{pr} values were computed from the Mie formulae using laboratory data for n and k for iron, and laboratory data for silica for $n(\lambda)$, assuming $k = 0.005$ in the visual and near infrared spectral region.

TABLE 6.5

Ratio of Radiation Pressure to Gravity P/G

($T = 3000$ K, $g = 1$ cm s^{-2}, $R_* = 3 \times 10^{13}$cm)

a/μ	Iron	Silica
0·01	$5{\cdot}59 \times 10^2$	$1{\cdot}46 \times 10$
0·02	$7{\cdot}98 \times 10^2$	$1{\cdot}93 \times 10$
0·03	$9{\cdot}18 \times 10^3$	$2{\cdot}16 \times 10$
0·05	$1{\cdot}70 \times 10^3$	$4{\cdot}69 \times 10$
0·08	$2{\cdot}84 \times 10^3$	$1{\cdot}34 \times 10^2$
0·10	$3{\cdot}25 \times 10^3$	$2{\cdot}16 \times 10^2$
0·30	$1{\cdot}96 \times 10^3$	$8{\cdot}88 \times 10^2$

Since asymptotic ejection speeds are given by

$$v = [2\left[\frac{P}{G}\right]g\ R*]^{1/2} = 7.75 \times 10^6\ (P/G)^{1/2}\ ,\ \text{for}\ g = 1\ \text{cm s}^{-2},\ R* = 3.10^{13}\ \text{cm}, \tag{6.38}$$

we have $v = 3 \times 10^7$ cm s^1 for silica grains and $v = 1.8 \times 10^8$ cm s^{-1} for iron grains of radius $a = 0.01\mu$m.

Although it is generally accepted that solid grains can condense in the atmospheres of oxygen-rich cool stars, there is some disagreement as to the precise composition of these particles. The possibility that quartz (SiO_2) particles may condense in these stars was first discussed by Kamijo (1963). More extensive thermodynamic calculations of the type described above indicated the condensation of SiO_2, MgO and Fe as separate particles (Hoyle and Wickramasinghe, 1968). In subsequent calculations, Gilman (1969) has reported that silicate particles (e.g. $MgSiO_3$, $MgFeSiO_4$) may form under similar conditions. The difference between the latter two sets of results might arise due to small differences in the adopted binding energies of solid silicate matter which are yet somewhat uncertain (Hoyle and Wickramasinghe, 1969). The kinetics of the nucleation process of solid particles, differential expulsion velocities already mentioned, the possibility of the fusion of core-mantle particles (particles of iron cores and silicate mantles if they are able to form) in stars of variable luminosity may, among other factors, determine the final composition of particles injected into interstellar space. It is likely that particles composed of quartz, iron and magnesium oxide are formed both singly and in mixtures in the range of physical conditions that may occur in distended stellar atmospheres or in mass flows from stars.

Comments on the Grain Expulsion Process. It is now generally accepted that the grain formation processes discussed in this section must operate in the case of most giant stars. Evidence for circumstellar dust exists in abundance in both oxygen-rich and carbon-rich red giant stars. Evidence also exists for gaseous mass flows from these stars and in such mass flows dust must also be formed to escape and provide a source of grains in interstellar space. Our original estimates of grain supply from carbon stars suggested that $\sim 10^{-27}$ g cm^{-3} of interstellar graphite grains might be supplied in this way in a time of $\sim 10^9$ years if some 10^4 N stars were present in the galaxy at any time. This estimate involved the somewhat extreme assumption that essentially all the carbon in the atmosphere is converted into graphite grains and replaced in each pulsation cycle. It also assumed that graphite grains reach interstellar clouds without any destruction taking place. More pessimistically, one might suppose that 10% of the atmospheric carbon is replaced in each cycle, and 10% of the grains survive injection into interstellar clouds. One would then have a graphite grain density that is a factor $\sim 10^2$ short of what would be necessary to explain the interstellar extinction.

Grain destruction becomes relevant particularly for grains ejected from stars at high speeds. The ejection speeds $\sim$ 1000 km/s calculated for graphite particles implies that H atoms colliding with them as they are slowed would have relative kinetic energies of $\sim$ 5 keV. It seems likely that grain radii are reduced due to sputtering in the course of slowing to thermal speeds, thus returning some fraction of the grain mass into the gas phase (Wickramasinghe, 1972).

For a gas density of $n_H \sim 10$ cm^{-3}, particles of radii 0.01μm will be stopped within a distance of $\sim$ 1 pc. If, on the other hand, particles are exposed to an intercloud medium with a much lower density, $n_H \sim 0.1$ cm^{-3}, the stopping distance is $\sim$ 100 pc, which is of the same order as the mean separation of diffuse clouds. It follows, therefore, that grains expelled from stars are stopped selectively in clouds, thereby explaining the well–known n_H/E_{B-V} correlation. On the assumption that the diffuse clouds have lifetimes that are more or less independent of dimension, a cloud of radius r will capture grains with a cross-section πr^2, and the mass of the captured grains will vary from cloud to cloud as $\sim r^2$. It follows that the total extinction through clouds of varying dimensions will be constant. This is indeed borne out by the observational data as compiled by Spitzer (1968) and set out in Table 6.6 below:

TABLE 6.6

Visible Dark Nebulae (after Spitzer, 1968)

Type	Mass	Radius	A_V
Small globule	$>0{\cdot}1\ M_\odot$	0·03 pc	>4mag
Large globule	3	0·25	1·4
Intermediate cloud	8×10^2	4·0	1·4
Large cloud	$1{\cdot}8 \times 10^4$	20·0	1·4

Correlations of this type will not follow if grains grow in clouds, but would occur in any formation model that involves injection of grains into clouds.

A controversy has arisen as to the speed of ejection of grains from cool stars. Our original estimate of $\sim 10^3$ km/s or more depended on the assumption that the gas and dust are *not* momentum coupled. Radiation pressure acts primarily on the grains, but the drag force due to relative grain-gas motion acts on the gas, producing under conditions when there is extensive grain formation, a large escape rate of gas. In such situations the final grain escape velocities may be reduced to 10^2 km/s (Gilman, 1973). The high escape velocities discussed earlier would apply under conditions where a small enough fraction of the atmospheric mass is in grains.

The question of gas-dust momentum coupling in stellar atmospheres and grain expulsion for different conditions has been investigated by Salpeter (1974a,b). There are three critical luminosities involved in this analysis designated $L_{cr,p}$, $L_{cr,i}$, $L_{cr,z}$ which are defined by

$$L_{cr,p} = z^{2/3}L_{cr,i} = zL_{cr,z} \sim L_\odot \tag{6.39}$$

where z is the mass fraction of material condensible into grains, and $L_\odot$ refers to the solar luminosity. For $z \simeq 10^{-3}$ this gives

$$L_{cr,p} = L_\odot,\ L_{cr,i} = 10^2L_\odot,\ L_{cr,z} = 10^3L_\odot \tag{6.40}$$

For stars with $L > L_{cr,z}$ both gas and grains are expelled and the final velocity of grains relative to gas is likely to be below $\sim 5 \times 10^6$ cm s^{-1}. If $L_{cr,p} \lesssim L << L_{cr,z}$, grains can be expelled leaving behind most of the gas. Grain velocities are not suppressed in this case. Asymptotic velocities of decoupled grains are given by

$$v_{dg}(\infty) \simeq (L/L_{cr,p})^{1/2}\, v_0\ , \tag{6.41}$$

where $v_0 = (2GM/R)^{\frac{1}{2}}$ is the velocity of escape from the star. The upper limit for the luminosity in this case may be taken as $L_{cr,i} \approx 10^2$ L⊙. A star with this luminosity is a giant with $v_0 \sim 100 - 300$ km s^{-1}, so that $v_{dg}(\infty)$ could be in the range $10^8 - 3 \times 10^8$ cm s^{-1}. Giant stars with luminosities in the range $\sim 30 - 100$ L⊙ provide, however, significantly lower rates of grain production than the more luminous stars. The grain flow rate is given (cf. Salpeter, 1974b) by

$$\phi_d \simeq 4 \times 10^{-10} \frac{Q(L/100\ L_\odot)}{(u/5\ \mathrm{km\ s^{-1}})}\ M_\odot\ \mathrm{yr}^{-1}\ , \tag{6.42}$$

where Q is the mean efficiency factor for radiation pressure on a grain, u is the thermal velocity, and z is set equal to 10^{-3}. For $L = L_{cr,i} \simeq 10^2$ L⊙, $Q \sim 0.25$, $u \sim 5$ km s^{-1} we obtain $\phi_d \sim 10^{-10}$ M⊙ yr^{-1}. For $z = 10^{-3}$ this is 10 times less than the grain flow associated with supergiants such as α–Ori with an observed total mass loss $\phi_{gas} \sim 10^{-6}$ M⊙ yr^{-1} (Deutsch, 1960; Weymann, 1963). A determination of the mean injection speed of grains into the interstellar medium requires a knowledge of the luminosity distribution function amongst giants and supergiants, which is at present uncertain. However, the analysis presented by Salpeter indicates grain injection velocities from giants and supergiants in the range 3×10^8 cm s^{-1} to $\sim 3 \times 10^6$ cm s^{-1} for $L \sim 10^2 - 10^3$ L⊙, the more luminous stars producing grains at a more copious rate than the less luminous ones. A mean injection velocity $v_\infty \cong 10^7$ cm s^{-1} would appear a reasonable estimate.

Condensation of Grains in Supernovae and in Galactic Nuclei. The grain injection rates from cool stars may be augmented to some extent from a contribution by supernovae and from the explosions of massive objects such as may occur in the nuclei of galaxies. We have argued that supernova explosions could lead to the formation of graphite, silica and iron grains (Hoyle and Wickramasinghe, 1970). Particles of radii $10^{-6} - 10^{-5}$ cm would form in the expanding ejecta about a year after the explosion, and such particles are then injected at high speeds into the interstellar medium. According to the abundances calculated by Arnett (1969) about 10% of the mass of the exploded star could be condensible material, e.g. iron, silicon, carbon. The total mass of the exploded matter being about 1 solar mass, we would expect about 0.1 solar mass of dust production per supernova explosion. With a frequency of occurrence of $(30\ \mathrm{yr})^{-1}$ the rate of increase of grain density throughout the volume of the galactic disk $\sim 10^{66}$ cm^3 is

$$\frac{d\rho_{grains}}{dt} \simeq 7 \times 10^{-36}\ \mathrm{g\ cm^{-3}\ yr^{-1}}\ . \tag{6.43}$$

If (6.33) persists for 10^9 years (the average turn–over time of the ISM) we obtain a steady–state grain density

$$\rho_{\text{grains}} \simeq 7 \times 10^{-27} \text{ g cm}^{-3} , \tag{6.44}$$

close to the required value.

We have also discussed the possibility that condensates in novae could augment the supply of refractory grains, particularly carbon grains into the interstellar medium (Clayton and Wickramsinghe, 1976). But the rate of supply is unlikely to compete with that from other sources such as the carbon stars.

6.4. CORE–MANTLE GRAINS

Core–mantle grains were first proposed with a view to alleviating a problem that arose in the context of the graphite grain theory (Wickramasinghe, 1963). If graphite particles with radii $< 0.1\mu m$ were to account for most of the interstellar extinction, their albedo was too low to explain the available observational data on the reflection nebulae and the diffuse galactic light that we discussed in Chapter 4. It was argued that once embedded in cloud gas, graphite particles would accrete molecular mantles. The theory then took on an essential aspect of the older ice–grain theory, differing from it mainly in that nuclei for ice condensation were supplied from cool stars. One of us has discussed this theory in considerable detail in an earlier monograph (Wickramasinghe, 1967). The possibility of accreted mantles was readily extended to include other types of seed nuclei besides graphite. However, just as in the ice–grain theory, highly artificial assumptions were needed to defend the mechanism of ice mantle formation in normal diffuse interstellar clouds. Impinging H atoms may be expected to form H_2 molecules and impede any significant H_2O formation at the grain surface. If such difficulties are supposed to be somehow overcome, the growth rate of a mantle can be calculated on the basis that every impinging O atom was converted to H_2O and then deposited on the grain. The rate of increase of grain radius can then be calculated according to equation (5.52). From equation (5.52) we have for the radius at time t

$$r = r_0 + \frac{\alpha n}{s} (kTm/2\pi)^{1/2} t \tag{6.45}$$

where r_0 is the initial radius of the nucleus and α is the sticking coefficient. For an HI cloud, $T \cong 100°$ K, the oxygen to hydrogen ratio $n/n_H = 10^{-3}$ and $n \cong 10$ cm^{-3}, so that $n \cong 10^{-2}$ cm^{-3}. With these values, together with $s \cong 1$ gm cm^{-3}, $m = 16$, equation (6.45) yields:

$$r = r_0 + \alpha 7.10^{-14} t , \tag{6.46}$$

where the time t is expressed in years. For a condensation nucleus of radius small compared to the final grain radius r we can neglect r_0 in (6.46), and the time t_1 for a grain to grow to a radius r_1(cm) is then given by

$$t_1 \cong 1.4 \times 10^{13}\, r_1\, \alpha^{-1} \text{ years} \tag{6.47}$$

It has been customary to assume that the sticking probability α is close to unity. This assumption is based on experiments measuring the effect of physical adsorption of atoms impinging on metal surfaces. The situation for the case of ice-grain growth is likely to be considerably different, however. An impinging O atom has to form an

H_2O molecule, and also fit into a lattice structure, before it can be regarded as having 'stuck on'. An effective value of α several orders of magnitude less than unity may well be appropriate for this case. In the following, however, we shall set $\alpha \cong 1$ and briefly explore the consequences of such an assumption.

With $\alpha = 1$, (6.47) gives a time of growth for a grain of radius 3.10^{-5} cm of a few times 10^8 years. In the timescale of the age of the galaxy, therefore, ice grains of radius $\sim 10^{-4}$ cm are expected to result on the basis of (6.47). Grains of this size make $2\pi a/\lambda >> 1$ at visible wavelengths, so that Q_{ext} is nearly wavelength independent and equal to ~ 2. Such grains therefore would not produce the desired λ^{-1} extinction law. To overcome this difficulty, Oort and van de Hulst(1946) suggested that, simultaneously with the accretion process, there must also operate random destructive processes, which have the effect of reducing the average grain size.

Extensions of the core–mantle grain model, particularly for silicate core-ice mantle grains, have been dicussed in detail by Greenberg (1968, 1969).

References

Bates, D.R. and Spitzer, L., 1951, *Astrophys. J.*, **113**, 441.

Blander, M. and Katz, J.L., 1967, *Geochim, Cosmochim Acta.*, **31**, 1025.

Clayton, D.D. and Wickramasinghe, N.C., 1976, *Astrophys. So. Sc.*, **42**, 463.

Dalgarno, A., 1976, in P.G. Burke and B.L. Moiseiwitsch (eds.) *Atomic Processes and Applications*, Amsterdam.

Degushi, S., 1980, *Astrophys. J.*, **236**, 567.

Deutsch, A.J., 1960, *Stars and Stellar Systems*, **6**, 543.

Donn, B., Wickramasinghe, N.C., Hudson, J.P., Stecher, T.P., 1968, *Astrophys. J.*, **153**, 451.

Draine, B.T. and Salpeter, E.E., 1977, *J. Chem. Phys.*, **67**, 2230.

Draine, B.T., 1979, *Astrophys. Sp. Sci.*, **65**, 313.

Fix, J.D., 1969a, *Mon. Not. Roy. Astr. Soc.*, **146**, 37.

Fix, J.D., 1969b, *Mon. Not. Roy. Astr. Soc.*, **146**, 51.

Friedemann, C. and Schmidt, K.H., 1967, *Astron. Nachr.*, **289**, 233.

Gilman, R.C., 1969, *Astrophys. J. Lett.*, **155**, L185.

Gilman, R.C., 1973, *Mon. Not. Roy. Astr. Soc.*, **161**, 3P.

Greenberg, J.M., 1960, *Astrophys. J.*, **132**, 672.

Greenberg, J.M., 1963, *Ann. Rev. Astron. Astrophys.*, **1**, 267.

Greenberg, J.M., 1968, *Stars and Stellar Systems*, **7**, 221.

Greenberg, J.M., 1969, *Physica*, **41**, 67.

Greenberg, J.M., 1971, *Astron. and Astrophys.*, **12**, 240.

Hoyle, F. and Wickramasinghe, N.C., 1962, *M.N.R.A.S.*, **124**, 417.

Hoyle, F. and Wickramasinghe, N.C., 1968, *Nature*, **218**, 1126.

Kamijo, F., 1963, *Publ. Astr. Soc. Japan*, **15**, 440.

Kamijo, F., 1969, *Physica*, **41**, 163.

Keenan, P.C. and Morgan, W.W., 1941, *Astrophys. J.*, **94**, 501.

Kramers, H.A. and ter Haar, D., 1946, *B.A.N.*, **10**, 137.

Lindblad, B., 1935, *Nature*, **135**, 133.

Mathis, J.S., Rumpl, W. and Nordsieck, K.H., 1977, *Astrophys. J.*, **217**, 425.

Platt, J.R. and Donn, B.D., 1956, *Astron. J.*, **61**, 11.

Platt, J.R., 1956, *Astrophys. J.*, **123**, 486; 1960, *Lowell Obs. Bull.*, **4**, 278.

Pottasch, S.R., 1970, in H.J. Habing (ed.) *Interstellar Gas Dynamics*, Springer-Verlag.

Salpeter, E.E., 1973, *J. Chem. Phys.*, **58**, 4331.

Salpeter, E.E., 1974a, *Astrophys. J.*, **193**, 579.

Salpeter, E.E., 1974b, *Astrophys. J.*, **193**, 585.

Salpeter, E.E., 1977, *Ann. Rev. Astr. Ap.*, **15**, 267.

Spitzer, L., 1968, in B.M. Middlehurst and L.H. Aller (eds.) *Nebulae and Interstellar Matter Stars and Stellar Systems* VII, Chicago.

Tabak, R.G., Hirth, J.P., Meyrick, G.G. and Roark, T., 1975, *Astrophys. J.*, **196**, 457.

Taft, E.A. and Philipp, H.R., 1965, *Phys. Rev.*, **138**, 197.

ter Haar, D., 1943, *B.A.N.*, **10**, 1.

Thorn, R.J. and Winslow, G.M., 1957, *J. Chem. Phys.*, **26**, 186.

van de Hulst, H.C., 1949, *Rech. Astron. Obs. Utrecht*, XI, Part 2.

Weymann, R., 1963, *Ann. Rev. Astron. Astrophys.*, **1**, 97.

Wickramasinghe, N.C., 1963, *Mon. Not. Roy. Astr. Soc.*, **126**, 99.

Wickramasinghe, N.C., 1965, *Mon. Not. Roy. Astr. Soc.*, **131**, 177.

Wickramasinghe, N.C., 1972, *Mon. Not. Roy. Astr. Soc.*, **159**, 269.

Wickramasinghe, N.C., 1967, *Interstellar Grains* (Chapman and Hall, London).

Zettlemoyer, A.C., 1977, (ed.) *Nucleation Phenomena. Adv. Colloid Interface Sci.*, Vol. 7, Elsevier, Amsterdam.

7. The Organic Grain Model

7.1. INTRODUCTORY REMARKS

By 'organic' material in the present context we mean material comprised of complex arrangements of C, N, O atoms with hydrogen in the form of both aliphatic and aromatic molecules. Such material might be synthesised through abiotic, prebiotic or biological processes although, as we shall show, the last of these processes is likely to be more important in the conversion of inorganic material into organics on a galaxy-wide scale.

Our earlier conclusion, that a large fraction of the C, N, O in the interstellar medium is tied up in grains, admits two distinct possibilities:

(1) The C, N, O may be combined as simple saturated inorganic molecules leading essentially to van de Hulst's icy composition.

(2) The C, N, O combined with H could occur in the form of complex organic polymers.

We have discussed possibility (1) in Chapter 6, with the conclusion that such a formation mechanism poses some serious difficulties. We address our attention in the present chapter to possibility (2). Early ideas relevant to the question of condensed organics in interstellar space go back to the speculations of Platt (1956). Platt (1956) and Platt and Donn (1956) suggested that an interstellar condensation process would more likely result in the formation of large, unsaturated molecules measuring less than 10 Å across rather than crystals consisting of chemically saturated molecules. Although, as we saw in Chapters 5 and 6, thermal spikes in the condensing molecules would vitiate such a condensation model, it does carry the strong hint of a preference towards a basically organic grain model. Recent discussions of the evidence for polycyclic aromatic hydrocarbons (PAH's) to which we shall return in a later chapter imply a similarity between these two types of 'grain species'. Polycyclic aromatic molecules as the cause of interstellar extinction were first proposed in a pioneering paper by Donn (1968) (a paper that is scarcely referenced in the present-day euphoria about detections of PAH molecules). Donn and Krishna Swamy (1969) computed extinction curves for an ensemble of aromatic molecules using laboratory data, indicating a good fit to the $1/\lambda$ extinction law, but their data did not extend into the ultraviolet region of the spectrum. It is worth pointing out that this work was done well before any infrared measurements of interstellar or circumstellar regions were made. The subsequent detections of the $3.3\mu m$ emission feature, and other infrared emissions in a number of reflection nebulae (Sellgen et al., 1983) must therefore be regarded as matching predictions of Donn's PAH model of 1968 (Donn, Allen and Khanna, 1989). Another early suggestion that has a bearing on organic dust grains was made by F.M. Johnson (1967). Johnson, analysing the diffuse interstellar features (discussed in Chapter 3), concluded that these features are strongly suggestive of electronic-vibrational transitions in highly complex hydrocarbon molecules. A particular model involving a complex biomolecule (magnesium porphyrin) was considered by Johnson et al. (1973).

Greenberg (1973a) considered a modification of the dirty–ice grain model in which the effects of cosmic radiation were supposed to generate free radicals that could under suitable conditions lead to the production of quantities of complex organic molecules in a solid state. On occasions the grain could 'explode', resulting in the release of molecules such as are observed by radioastronomers. This idea was based on Greenberg's (1973b) laboratory work in which mixtures of ices yielded organics after intense doses of ionizing radiation were delivered to them. A difficulty results from the fact that the rates of production of such organics cannot be simply scaled in order to make comparisons with the real astronomical situation. The first suggestion that the bulk of the normal interstellar grains was comprised of complex organic polymers was made by one of the present authors (Wickramasinghe, 1974).

7.2. POLYMERISATION OF FORMALDEHYDE

We now develop the arguments that were used to support the contention that grains might be largely in the form of formaldehyde polymers and copolymers (Wickramasinghe, 1974, 1975; Cooke and Wickramasinghe, 1977). It was also argued at this time that cometary grains had a similar polymeric composition (Vanysek and Wickramasinghe, 1975). These ideas appear to have received a measure of support with recent detections of formaldehyde polymers in the gas coma of Comet Halley (Huebner, 1987; Mitchel et al., 1987).

Condensation of H_2CO was envisaged to take place in the interiors of dense molecular clouds. Although a wide variety of organic and inorganic interstellar molecules have been discovered so far, CO and H_2O are the most ubiquitous, being in dense clouds as well as in the more tenuous regions of the interstellar medium. In addition to millmetre-wave observations of CO in denser regions, there are now also ultraviolet observations relating to CO in tenuous regions (Snow, 1975). Reactions leading to CO formation have been described by Glassgold and Langer (1975) and by Langer (1976). A substantial fraction of carbon in dense clouds ($n_H \gtrsim 10^4$ cm^{-3}) may be assumed to be in the form CO. Millimetre–wave observations of CO may be interpreted to give $n_{CO}/n_H \simeq 10^{-5}$ in a typical case (Leung and Liszt, 1976). The actual observed abundance of H_2CO in the gas phase gives a ratio n_{H2CO}/n_H, which is three orders of magnitude lower, $\sim 10^{-8}$ (Zuckerman and Turner, 1975). However, it is not inconceivable that a much larger fraction of interstellar C is tied up as H_2CO, in polymeric form, on grains. The gas phase observations may then merely reflect a number density of molecules which is in equilibrium with polymer/co-polymer phases under different interstellar conditions.

The processes that lead to the formation of H_2CO (and also other more complex molecules) in interstellar clouds are yet somewhat obscure. They could involve ion-molecule reactions in the gas phase, as well as recombination reactions on grain surfaces. One possibility for H_2CO formation involves direct additions of H atoms to CO on grain surfaces (Williams, 1974). It is our present point of view that H_2CO, as indeed all the other organic molecules in space, result from the degradation of highly complex organic particles under interstellar conditions. However, to present the sequence of ideas in an historical perspective, we shall assume in this section that H_2CO, whatever the mechanism of its formation, polymerises on the surfaces of 10^{-6} cm silicate grains which have their origin in Mira stars.

Polymerisation, in general, proceeds in three distinct stages (Bevington, 1961; Blackadder, 1975)

(i) *Initiation*: Initiation of polymerisation, involving a bondbreaking event, essentially converts a saturated molecule into a chemically active state with a free valence bond:

$$\begin{array}{c} H \\ | \\ C \\ | \\ H \end{array} = O \rightarrow \begin{array}{c} H \\ | \\ C \\ | \\ H \end{array} - O\bullet \qquad (7.1)$$

This could be effected in a variety of possible ways – including interaction with UV photons, radicals or ions. Since UV photons will largely be excluded from dense clouds, due to dust opacity, reactions with radicals and ions may provide the preferred initiation routes. For example, if R• denotes a radical (radicals being the most abundant molecular species present in interstellar conditions) we may have an initiation reaction of the form

$$R\bullet + \begin{array}{c} H \\ | \\ C \\ | \\ H \end{array} = O \rightarrow R - \begin{array}{c} H \\ | \\ C \\ | \\ H \end{array} - O\bullet \qquad (7.2)$$

(ii) *Propagation and Co–polymerisation*: Polymer–chain propagation could now proceed by simple addition reactions in the gas phase, i.e.:

$$R - \begin{array}{c} H \\ | \\ C \\ | \\ H \end{array} - O\bullet + \begin{array}{c} H \\ | \\ C \\ | \\ H \end{array} = O \rightarrow R - \begin{array}{c} H \\ | \\ C \\ | \\ H \end{array} - O - \begin{array}{c} H \\ | \\ C \\ | \\ H \end{array} - O\bullet \qquad (7.3)$$

The chain thus propagates spontaneously. H_2CO is known to form highly stable co-polymers with many suitable co-monomers. These include acetaldehyde (CH_3CHO), isocyanic acid (HNCO) and cyanoacetylene (HC_3N), all of which are known to co-exist with H_2CO in dense interstellar clouds. The presence of such molecules, with which co-polymerisation can occur, would lead to the formation of highly complex polyoxymethylene co-polymers.

(iii) *Termination*: Closure or termination of a growing polymer chain could also occur in a variety of ways. In the case of interstellar gas phase polymerisation, a likely mechanism is the approach of a radical or atom which could combine with the chain, rendering its growing tip effectively inert. Closure with radicals other than OH, which seems likely, will block depolymerisation and produce polymers endowed with a high degree of thermal stability (Fawcett, 1975; Wickramasinghe and Santhanan, 1975).

Goldanskii (1979) has also discussed the elaboration of organic molecules in the solid state at very low temperatures. The argument is that interstellar gas phase reactions yield a hybrid mix of low-molecular-weight organic molecules which condense on grains. Polymerisation reactions are then supposed to occur in the solid phase by quantum mechanical tunnelling between adjacent molecules. Goldanskii (1979) has argued that at low temperatures appropriate to grains an entropy factor Q + TS becomes unimportant to an extent that slightly endothermic reactions become weakly exothermic, so that polymerisation reactions can proceed spontaneously. Goldanskii substantiates his claims with experimental results on the polymerisation of H_2CO under laboratory conditions.

Rate of Mantle Growth. The deposition of successive layers of H_2CO co-polymers, which in many cases may be cross-linked, is expected to form highly stable refractory grain mantles. The rate of growth of a formaldehyde co–polymer mantle in a dense molecular cloud of temperature T, is given by

$$\frac{dr}{dt} = \frac{\alpha n_{[c]}}{s} \left[kTm_{[c]}/2\pi\right]^{1/2}, \tag{7.4}$$

where r is the mantle radius, α the sticking coefficient, s is the density of the solid polymer, and $n_{[c]}$ is the number density of C atom containing species whose interaction with grains leads directly to the production of H_2CO and $m_{[c]}$ is the corresonding particle mass. We assume here that $n_{[c]} = n_{co}$ so that $n_{[c]}/n_H = 10^{-5}$. H_2CO formation could then occur, by the exothermic further addition of H atoms as proposed by Williams (1974). With $\alpha/s \simeq 1$ we have from (7.4)

$$\frac{dr}{dt} \simeq 10^{-11} \left[\frac{n_{[c]}}{n_H}\right] n_H \left[\frac{T}{100^\circ K}\right]^{1/2} \text{cm yr}^{-1}. \tag{7.5}$$

It has already been noted that dense massive molecular clouds are not in a state of free-fall collapse (Zuckerman and Palmer, 1974). Collapse could be slowed down by several processes, including effects of magnetic pressure, rotation and turbulence. Turbulence might be generated as well as maintained by the effect of continuing star-formation within molecular cloud complexes of the type considered here. We may suppose that typical contraction time scales for an entire cloud complex, as well as for separate fragments within it, are $\sim 10^6$ yr; such a time, together with the estimated total mass of protostellar clouds, gives a star-formation rate which is in good agreement with observational data (Hoyle and Wickramasinghe, 1976).

Setting $t \simeq 10^6$ yr as the appropriate timescale, together with $n_{[c]}/n_H \simeq 10^{-5}$, $T \simeq 100^\circ K$, we obtain a mantle radius $\sim 10^{-5}$ cm in regions of hydrogen density $n_{H2} \approx 10^5$ cm^{-3}. Such a density is typical of regions in many dense molecular clouds in the galaxy. Clumping of co–polymer–coated grains to form even larger particles could occur in subregions of the cloud which are directly associated with protostellar contraction. A significant fraction of polymer coated grains may be expected to be expelled with systematic gas flows into the general interstellar medium. Such dielectric organic grains with typical radii $\sim 10^{-5}$ cm may be the main source of

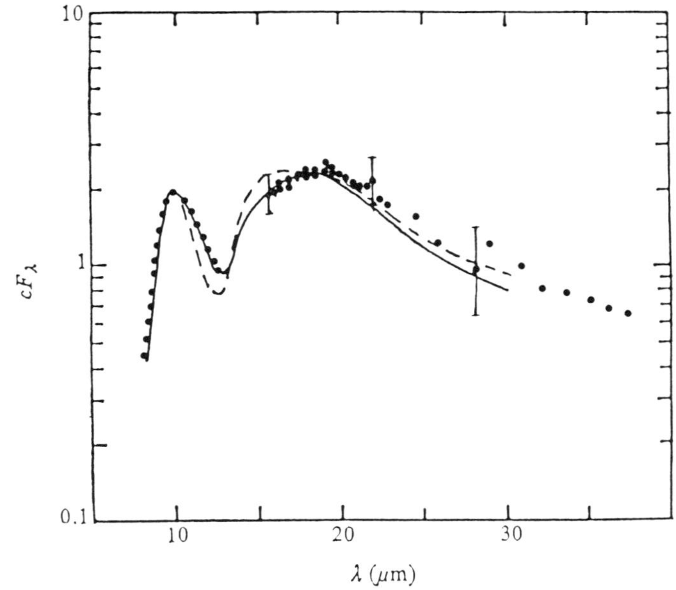

Fig. 7.1 *Normalised flux from the Trapezium Nebula (points) compared with emission from polysaccharide grains of temperature 175° K. The dashed curve is for cellulose, solid curve corresponds to the polysaccharide ensemble characterised by the transmission values given in Table 7.1.*

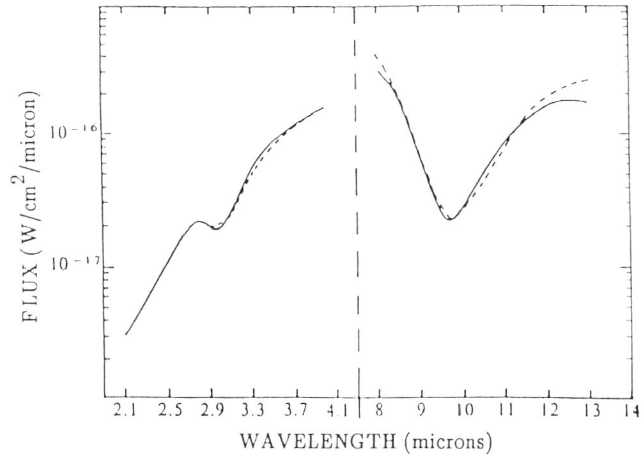

Fig. 7.2 *The solid curve is the calculated normalised flux for an optically thin polysaccharide model with $T = 430°$ K, $\alpha = 2.6$ corresponding to an optical depth $\alpha\tau$ where τ is given in Table 7.1. The dashed curve is the observed flux for the source $H_2$0610+18.*

interstellar extinction at optical wavelengths without violating cosmic abundance constraints (Wickramasinghe, 1976). They could also provide a non-silicate explanation of the 8 – 10μm emission feature in the Trapezium nebula (Cooke and Wickramasinghe, 1977).

These early ideas of polymer formation in interstellar clouds were admittedly naive in their simplicity as were indeed preceding ideas. They had, however, the merit of offering a hitherto unconsidered possibility – that of organic grains with volatilities intermediate between the inorganic ices and the refractory minerals. Data from Comet Kohoutek did indeed suggest a carrier of the 10μm emission feature that became partially volatalised when the temperature rose above 500° K (Vanysek and Wickramasinghe, 1975).

7.3. FROM FORMALDEHYDE TO POLYSACCHARIDES

It was not a big step to proceed from the idea of formaldehyde molecules condensed in long chains to polysaccharides – a group of organic polymers with enormous biogenic significance. The polysaccharide known as cellulose is the structural component in the cell walls of plants, and accounts for a large fraction of the total carbon content on the Earth. Our first inkling of the importance of this macromolecule in astronomy came when we were searching in atlases on infrared spectra of organic molecules to find a material that could give a good fit to the observed infrared emission spectrum of the Trapezium nebula throughout the 8 – 40μm waveband, and also to the absorption spectra of several IR objects including the BN object. It turned out that cellulose gave remarkably close fits as can be seen in Figs. 7.1 and 7.2 (Hoyle and Wickramasinghe, 1977). The fits were indeed so close that we argued for the presence of a material spectroscopically if not chemically identical to cellulose on a galactic scale.

Polysaccharides have substructures built from H_2CO units, substructures with the empirical formula $(H_2CO)_n$ where $n \geq 3$. The commonest polysaccharides, cellulose and starch with $n = 6$, are particularly stable because each $(H_2CO)_6$ is able to form itself into a very stable ring, with the polysaccharide then becoming a chain of hexagonal ring structures. Cellulose can maintain its structure in a vacuum or in an inert atmosphere probably up to a temperature of around 625 – 900° K. Laboratory data for wood cellulose indicate stability up to about 620° K (Shafizadeh, 1971) but in low–pressure interstellar conditions and in the absence of free oxygen there could exist polysaccharides similar to cellulose able to withstand temperatures up to about 900° K.

The condensation and polymerisation of formaldehyde discussed in the previous section was essentially a one-step process. In view of the uncertainties of the relevant reaction kinetics in low-density gas under interstellar conditions, it would seem better to discuss an evolutionary mechanism involving many steps leading sequentially to the production of increasingly stable polymers (Hoyle and Wickramasinghe, 1977). As we mentioned earlier, a combination of ion-molecule reactions and grain surface reactions could lead to H_2CO production. This molecule, however, is relatively fragile with a photodissociation time constant in a diffuse cloud of only $\sim$ 100 yr. If the molecules form in denser conditions they would be broken up on a short timescale when the parent clouds become dispersed to the normal density of diffuse clouds. Thus, since the interstellar gas undergoes alternating phases of compression and evaporation, there must be corresponding alternations of chemical complexity – molecular dissociations taking place during evaporation and association during compression. Such alternations provide a selective process for the emergence of those chemical forms that can best withstand the adverse conditions of the evaporative phases. Starting from H_2CO as the basic unit, it would seem likely that the net outcome of such oscillations would produce $(H_2CO)_n$ ring structures linked into chains becoming polysaccharides in preference to the more fragile long chain polymers comprised of single H_2CO units.

7.4. POLYSACCHARIDE FORMATION IN STELLAR MASS FLOWS

We saw in Chapter 6 that the carbonaceous material emerging from stars must be in the form of graphite, the most stable form of carbon, and that such emergence must be confined to situations where the C/O ratio exceeds unity – e.g. in the atmospheres of carbon stars. We show now that this state of affairs remains valid for mass flows from stars of sufficiently low surface temperatures, but it is not correct for low density flows from stars with colour temperatures > ~ 4000° K or for oscillatory stars with colour temperatures that go above 4000° K over a portion of their cycle. In the latter case we show that carbonaceous material comprised mainly of polysaccharides will be able to condense.

We consider the situation for O stars of high mass where material appears to be flowing out at rates ~ 10^{-2} M⊙ per year (Hoyle et al., 1973). In such a situation, radiation from the star is absorbed and re-emitted by the outflowing material, which produces a shielding of the true surface of the underlying star. The temperature of the material falls with increasing distance from the star, until at a stage where the temperature is in the range ~ 5000° K to ~ 10,000° K, an effective photosphere is formed. The effective photosphere develops at a radial distance

TABLE 7.1

Transmittance exp(−τ) as a function of wavelength λ(μm) for a synthetic polysaccaride ensemble

λ(μm)	2·0	2·5	2·6	2·7	2·8	2·9	3·0	3·1
exp(−τ)	0·70	0·75	0·75	0·70	0·55	0·40	0·34	0·36
λ(μm)	3·2	3·3	3·35	3·5	3·7	4·0	5·0	6·0
exp(−τ)	0·43	0·55	0·5	0·65	0·70	0·75	0·80	0·75
λ(μm)	6·5	7·0	7·5	8·0	8·5	9·0	9·5	10·0
exp(−τ)	0·80	0·50	0·5	0·53	0·40	0·25	0·14	0·15
λ(μm)	10·5	11·0	11·5	12·0	13·0	14·0	15·0	16·0
exp(−τ)	0·23	0·30	0·40	0·55	0·61	0·49	0·43	0·36
λ(μm)	18·0	20·0	22·0	25·0	30·0			
exp(−τ)	0·34	0·43	0·47	0·50	0·50			

$r_0 \sim 10^{14}$ cm from the star, where the hydrogen density $(n_H)_0$ is of order 10^{11} cm^{-3} (for a mass flow of $\sim 10^{-2}$ M⊙ yr^{-1} at a speed of $\sim$ 300 km s^{-1}). Further out in the flow molecular condensations occur which lead to the eventual absorption of radiation emitted by the effective photosphere of the star, and to the re-radiation of the absorbed radiation in the infrared. Would the resulting infrared emission from such a model behave like the simpler polysaccharide models we considered earlier (Hoyle and Wickramasinghe, 1977)? This question is answered affirmatively by Fig. 7.3, which shows the expected emission from the model compared to observations of the BN infrared source (Gillett and Forrest, 1973). The expected emission was calculated for a polysaccharide formation temperature T_m = 850° K and for the relative transmittance values of a synthetic polysaccharide, as given in Table 7.1. The radial opacity through the source was taken to be 4 times the relative values given in Table 7.1.

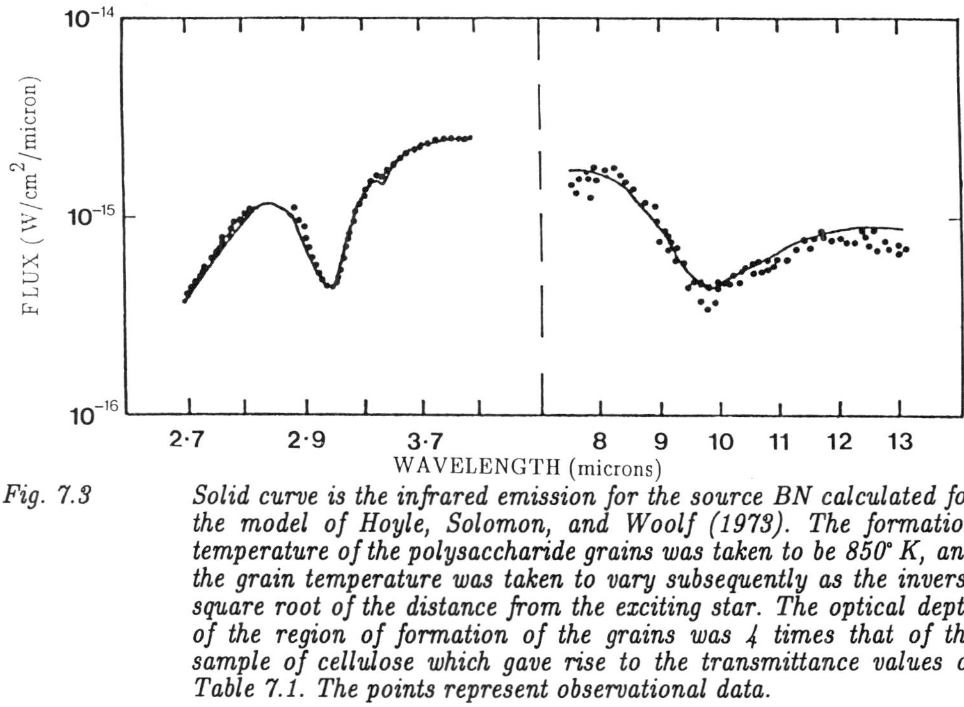

Fig. 7.3 *Solid curve is the infrared emission for the source BN calculated for the model of Hoyle, Solomon, and Woolf (1973). The formation temperature of the polysaccharide grains was taken to be 850° K, and the grain temperature was taken to vary subsequently as the inverse square root of the distance from the exciting star. The optical depth of the region of formation of the grains was 4 times that of the sample of cellulose which gave rise to the transmittance values of Table 7.1. The points represent observational data.*

The agreement between theory and observation shown in Fig. 7.3 gives a strong empirical indication that polysaccharides can indeed form in the mass flows from stars where the C/O ratio is probably less than unity. To give plausibility to such a view, three questions need answering:

(1) Why is the C not almost entirely consumed in the formation of CO?

(2) Why is the C not built into graphite rather than into polysaccharides?

(3) In view of the inevitably small concentrations of the molecules C_2, C_3, how

can the bulk of the carbon manage to condense at all?

The answer to the first of these questions turns on three points:

(i) The ratio CO/C at the effective photosphere is essentially thermodynamic, less than $\sim 10^{-6}$.

(ii) For a constant outflow speed, the hydrogen density n_H falls off according to

$$n_H = (n_H)_0 \, (r_0/r)^2 \tag{7.6}$$

with $(n_H)_0 \simeq 10^{11}$ cm^{-3}, and for a timescale of but a few years available for the recombination of atomic species into molecules, triple collisions of the kind A+B+H → AB+H play no significant role in the formation of diatomic molecules, which must accordingly proceed through two–body radiative recombinations of the type:

$$A + B \rightarrow AB + h\nu \tag{7.7}$$

(iii) With a low cross–section $\sim 10^{-22}$ cm^2 for C+O → CO+hν and for a gas thermal velocity $\sim 10^5$ cm s^{-1} the fraction CO/C built in the available timescale of $\sim 10^7$ s is less than $\sim 10^{-2}$.

To answer the second question, we note that the internal energy of a large polyatomic molecule distributes itself statistically among the many states of the molecule, and it is appropriate to characterise the distribution by an internal temperature, T_m, say. The average value of T_m for a particular molecule is determined by the equation of energy balance between absorption of radiation from the star and the re-emission of radiation by the particular molecule, an equation of the form*

$$\left[\frac{r_0}{r}\right]^2 \int_0^\infty Q_{abs}(\lambda)\, B_\lambda(T)\, d\lambda = 4 \int_0^\infty Q_{em}(\lambda)\, B_\lambda(T_{em})\, d\lambda \tag{7.8}$$

where B_λ is the Planck function. Both the C_6 ring which appears in graphite and the C_5O pyran ring which appears in the commonest polysaccharides absorb mainly in the ultraviolet with comparable values of Q_{abs}. But the pyran ring has a much higher value of Q_{em} in the infrared than the C_6 ring, because the latter is a symmetric molecule with no dipole moment. Hence equation (7.8) leads to a significantly higher value of T_{em} for C_6 than it does for C_5O, and this higher value of T_{em} more than offsets the greater binding energy of the C_6 ring. (the binding energy difference is not large. The binding of the C_6 ring from its constituent atoms is 30.5 eV and that for the C_5O ring is $\sim$ 29 eV (Cox, 1963; Dewar and Harget, 1970)).

---

*The factor 4 is for spherical particles, and it assumes no radiation field in the infrared. The presence of an infrared field reduces the factor 4, in a typical case to 2. This reduction has no significant effect on the above argument since the reduction would apply to both the pyran ring and the graphite ring.

We pass now to the third of the above questions, and for this we begin by assuming the initial existence of a single polymer chain which we take to have been built through C_2, C_3. The logic of our argument will be that this first polymer chain, through rapid building interspersed by repeated fragmentation, can generate a vast cascade of further polymer chains which have nothing to do with the low concentrations of C_2, C_3. The logic is similar to that of the explosion of a nuclear weapon, where a first neutron is enormously amplified by the fission cycles which it provokes, and where the flood of subsequent neutrons have nothing to do with the source of the first neutron.

For a flat chain of width D, length ℓ, the number of carbon atoms arriving in a time dt which could lead to chain growth is

$$dN \simeq 2\, n_c <v_{th}> \ell\, D\, dt\ , \tag{7.9}$$

where v_{th} is the gas thermal speed.
[With $n_O \gtrsim n_C$, the oxygen rate is moderately larger than (7.9)]. Further, assuming a mean length interval between two successive polymer rings to be about 2D and with 6 atoms making up a ring the increment of length corresponding to dN is

$$d\ell \simeq f \frac{dN}{6}\, 2D = \frac{D\ dN}{3} f\ , \tag{7.10}$$

where f is the fraction of impinging carbon atoms that diffuse and attach themselves to the ends of the polymer. Equations (7.9) and (7.10) give

$$\ell = \ell_0\, e^{\tau/\tau_1} \tag{7.11}$$

with

$$\tau_1 = [\tfrac{2}{3}\, n_c <v_{th}> D^2 f]^{-1}\ . \tag{7.12}$$

The phenomenon of exponential growth which we are describing here is the same as that which occurs in the building of 'whiskers', which has been studied in the laboratory (Meyer, 1959; Donn and Sears, 1963). For $n_c \simeq 10^6\ cm^{-3}$, $D \simeq 5 \times 10^{-8}\ cm\ s^{-1}$, we obtain $\tau_1 \simeq 6 \times 10^3\ f^{-1}$ sec, and this is small compared to the available timescale of the order of a year, even for fractions f as low as 10^{-2}. Ample time is therefore available for many exponential 'cycles', again in analogy to the many fission cycles of a nuclear weapon.

We expect polymers to appear in the mass flow from a star when the temperature T_m first becomes low enough for the bond linkages in the polymer to assume stability. For C–O–C linkages of a polysaccharide, the bond strength is $\sim$ 4 eV and the largest values of T_m for which such a bond will be stable lies in the range 800 – 900° K, just the polysaccharide formation temperature used in the calculations leading to Figure 6.3.

The first polymers built through C_2, C_3, will be of short lengths and for them the value of T_m determined by equation (7.8) is less than for longer polymers with lengths $\ell \simeq \lambda_{star}/2\pi$. This is because a tuning effect appears as the polymers grow to a length that is resonant with the main optical radiation of the star, $\lambda_{star} \simeq 5 \times 10^{-5}$ cm. The polymers then experience a strong radiation pressure force due to scattering, which gives them an appreciable drift velocity with respect to the ambient gas. Hoyle et al. (1973) estimated a drift velocity between 10 and 100 km s^{-1} for their model. At 100 km s^{-1} the viscous drag would raise the temperature of the polymer chain by about 100° K. We think even stronger heating than this could well occur. Yet even 100° K is sufficient in a marginally stable situation to begin breaking the C–O–C linkages of the polymer. The reduced lengths of the fragments destroys the resonance, so that the fragments quickly assume lower values of T_m, thereby returning to a stable condition. And because of the explosive exponential growth implied by (7.11), the fragments almost immediately go through the same sequence as the original polymer. Not only do the lengths of the polymers grow exponentially, but the numbers of the polymers grow like the grains of wheat in the old story – one for the first square of the chessboard, two for the second, ..., ... "and not all the granaries in the world can hold enough wheat for the sixty–fourth square".

There is, of course, a limit to the number of particles that can be produced from a single first polymer chain, but the limit turns out to be very large. It can be estimated in the following way.

The zone of influence of any starting polymer chain will be defined by a carbon atom diffusion distance in a transverse direction and by a radiation pressure induced grain–gas drift in the radial direction. Again using the drift velocity obtained by Hoyle et al. (1973) the radial grain diffusion distance in the available timescale $\sim 10^7$ s is $10^{13} - 10^{14}$ cm. The carbon atom diffusion distance in time t in the transverse direction is

$$\sim [\langle v_{th}\rangle \, t/\sigma_c \, n_H]^{\frac{1}{2}}$$

where σ_c is the carbon atom cross–section for collisions with H. With $t \simeq 10^7$ sec, $\langle v_{th}\rangle \simeq 10^5$ cm s^{-1}, $\sigma = 10^{-16}$ cm^2, $n_H \simeq 10^{10}$ cm^{-3},this gives $\sim 10^9$ cm. The zone of influence of an initial polymer is thus $\sim 10^{13} \times (10^9)^2 = 10^{31}$ cm^3. The required formation range through C_3, C_3 of initial short polymer chains therefore takes the exceedingly low value of $\sim 10^{-38}$ cm^{-3} s^{-1}. The rates of formation of C_3, C_4 by $C_2+C \rightarrow C_3$, $C_3+C \rightarrow C_4$ are uncertain, but even with low cross–sections for both reactions, and even with ample allowance for the destruction of C_2 through $C_2+O \rightarrow CO+C$, a fraction $\sim 10^{-12}$ of all the carbon will become C_4 in the available timescale. The molecule C_4 already has a sufficient number of internal states for further additions to occur with comparatively large cross–sections, $\sim 10^{-16}$ cm^2. Thus beyond C_4 the apparent difficulty of small formation rates disappears, and the number of initial polymer chains that could be achieved (if necessary) could be comparable to the number of C_4 nuclei, which is very many orders of magnitude greater than the minimum required number of initial short polymer chains. The first such chains to flow through C_2,C_3,C_4... therefore go on to take the whole of the carbon. This completes the answer to the third question.

If mass loss from a highly luminous O star proceeds for long enough, the oxygen–rich envelope will be replaced by material that has been processed in the star by the CN cycle. The oxygen will now be almost totally depleted, the carbon abundance will be accompanied by a large excess of nitrogen. Under these conditions C_4 rings would evolve into heterocyclic C_4N, C_5N, and C_4N_2 rings. It may well be important in this connection that the porphyrins have a strong absorption band near 4430 Å, the wavelength of a well-known interstellar absorption feature (Johnson, 1971), and that quinazoline and its derivatives have a strong absorption at 2200 Å (Albert and Armarego, 1965) close to another interstellar feature.

Photospheric temperatures of evolved supergiant stars are uncertain. If mass flows in M and N-type stars originate in layers of the atmosphere where the colour temperature $> \sim 4000^\circ$ K, arguments similar to those discussed above would apply.

In this context, we note that the spectra of several carbon stars (including BM Gem V778 Cyg and C1003) show broad emission features at $10\mu m$ which are characteristic of O-rich Mira variables (Little-Marenin, 1986; Williams and de Jong, 1986). The latter have been widely attributed to siliceous material, but because it seems unlikely that the C stars could have produced a shell of siliceous particles, it would appear that cellulose-like polymer chains are present in most cases.

We conclude this section by noting that the infrared sources exhibiting polysaccharide absorption features may be associated with massive stars of the kind discussed by Hoyle et al. (1973). Mass flows from such stars can lead to the production of polysaccharides in the first instance, followed by the condensation of nitrogenated heterocyclic carbon compounds.

7.5. HAC, PAH AND QCC MODELS

These are acronyms for the following: HAC = Hydrogenated Amorphous Carbon; PAH = Polyaromatic Hydrocarbons; QCC = Quenched Carbonaceous Compounds.

Once infrared observations of several astronomical objects revealed both absorption features and emission features that corresponded to the wavelengths of vibrational bands in organic solids, several other models came to be proposed. Duley and Williams (1981) argued in favour of CH bonds and other organic functional groups in amorphous carbon grains, the rationale being that H atoms could easily diffuse into such grains and become chemically attached at suitable sites. Duley and Williams argued that hydrogenated amorphous carbon (HAC) is readily formed when carbon and hydrogen atoms are cooled over a substrate. Later observations of reflection and planetary nebulae indicated continuum emission near $2.5\mu m$ and a set of diffuse bands in the mid-infrared spectral region that pointed to very small organic fragments which have been subject to spike heating by the stellar radiation (Sellgren, 1984). An example of such an observation and the identification proposed by Léger and Puget (1984) is reproduced in Fig. 7.4a (Léger and Hendecourt, 1986). The main identifying features are at 3.28, 6.2, 7.7, 8.6 and $11.3\mu m$ and it would seem that these bands are characteristic of some form of polyaromatic molecule. The structures of molecules proposed by Léger and Puget are shown in Fig. 7.4b. It is clear that these molecules have underlying graphitic structures. In view of the destructive effects of thermal spikes in the diffuse clouds the build up from single carbon atoms to PAH structures would be ruled out. The only viable mechanism for their formation is one involving degradation of larger structures. The breakdown of

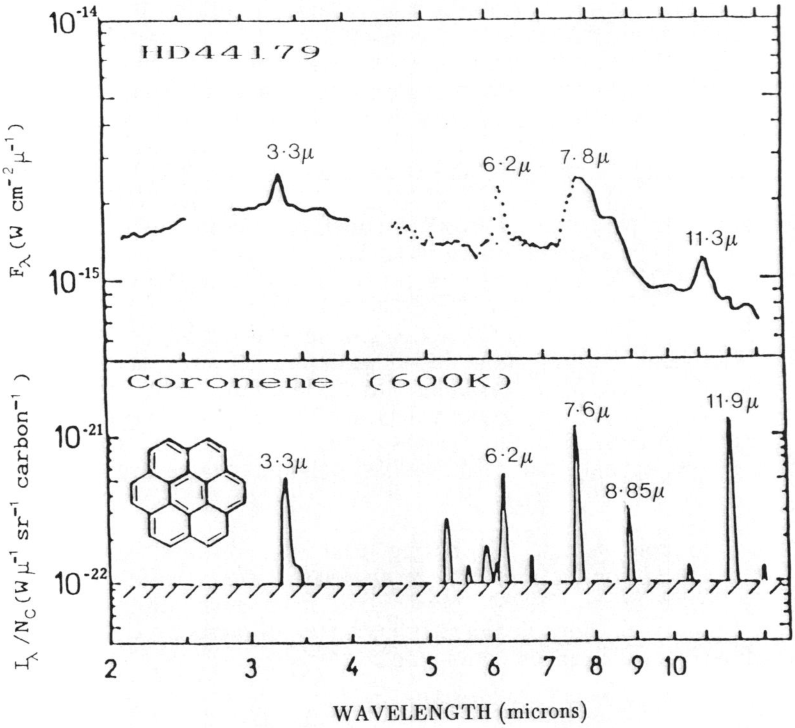

Fig. 7.4a *Observed infrared emission from HD 44179 (adapted from Léger and d'Hendecourt, 1978) compared with the expected emission from coronene heated to 600° K.*

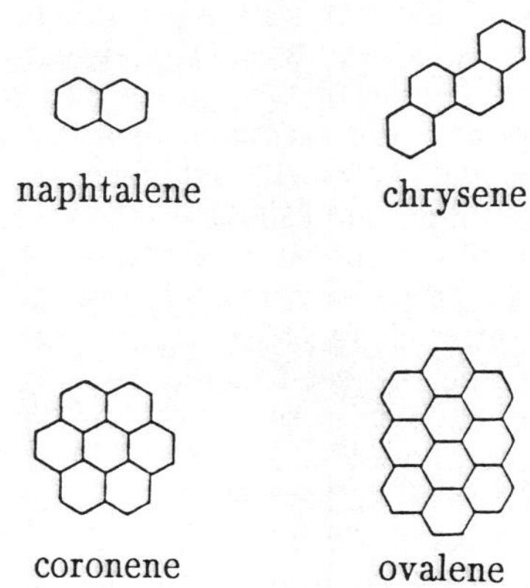

Fig. 7.4b *Structures of graphitic sequence of aromatic molecules.*

graphite in the presence of a hydrogen excess is one possibility, although fragmentation of a highly refractory material may pose a problem. Alternatively, one could imagine a degradation process involving more fragile organic solids to produce such structures as appear to be required by the astronomical data. We shall return to possible alternative identifications of these mid–infrared features in a later chapter.

Duley and Williams (1988) have argued that PAH–like structures may occur as loosely connected 'islands' within larger HAC coated grains. If the HAC coatings are thin enough (~ 100 Å in thickness), they argue that UV photons absorbed within 'PAH islands' would not heat the entire grain. The islands and their immediate environs experience temperature spikes and the infrared radiation corresponding to the higher temperature of the vibrationally excited islands escape, so explaining the observations as in Fig. 7.4a.

A variant of the HAC model is that involving a 'quenched carbonaceous composite' (QCC) as suggested by Sakata et al. (1983). This material is synthesised by quenching (suddenly cooling) a jet of plasma comprised of methane gas. It is yet unclear how an analogous situation could arise in an astronomical context, but an

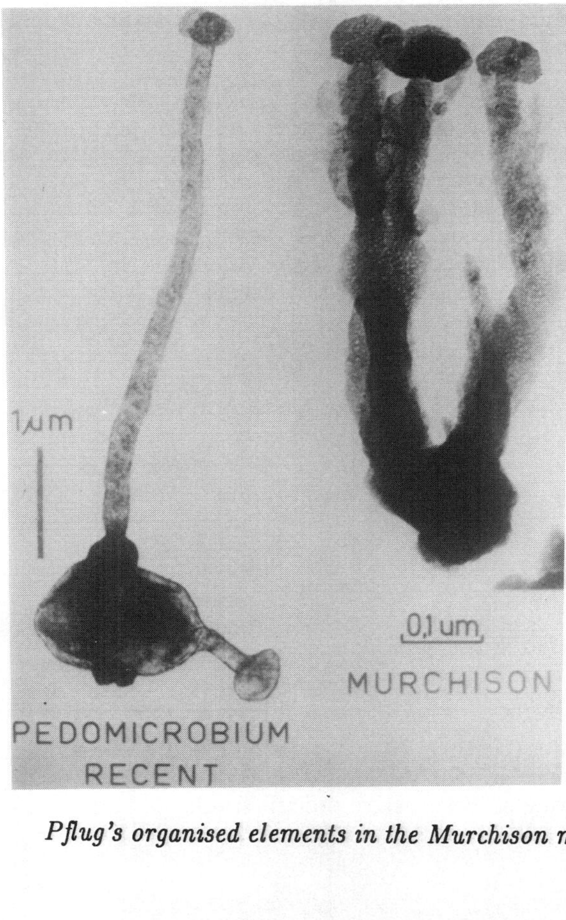

Fig. 7.5 *Pflug's organised elements in the Murchison meteorite.*

interesting complex organic solid has been found to arise from laboratory experiments. The spectra of QCC are found to have infrared and ultraviolet features of potential astronomical interest. In particular the λ 2200 Å ultraviolet absorption is found in such material, as are some of the mid–infrared emission features of reflection and planetary nebulae.

7.6. FISCHER TROPSCH REACTIONS IN THE GAS PHASE

Ever since complex organic compounds and polymers were discovered in carbonaceous meteorites the general belief has grown that these can only be formed in pre-solar nebula conditions by means of catalytic Fischer–Tropsch–type (FTT) reactions (see, for example,. Hayes, 1967). A difficulty arises, however, because the carbon, oxygen and hydrogen in the early solar nebula would occur mainly as CO and H_2. Although the most stable combination of these atoms at room temperature is CH_4 and H_2O, a mixture of carbon monoxide and molecular hydrogen would not, as the solar nebula eventually cooled, be converted into molecules that were thermodynamically most stable without the intervention of suitable catalysts, which are necessary to speed-up reactions that are otherwise far too slow. While suitable inorganic catalysts have to be carefully chosen and prepared, biology proceeds to catalyse reactions such as $CO+3H_2 \rightarrow CH_4+H_2O$ at far higher speeds than inorganic catalysts. The most efficient way by far to effect the conversion of a $CO–H_2$ mixture to thermodynamically more stable combinations would therefore be through the intervention of microbiology.

Organic molecules found in carbonaceous chondrites are exceedingly complex, including biological monomers such as purines, pyrimidines, porphyrins, amino acids and a kerogen-like organic polymer. A biogenic origin of meteoritic amino acids is traditionally denied on the grounds that they are racemic, that is to say they have equal concentrations of D– and L–forms contrasting with biomaterial that has predominantly L–forms. This conclusion has, however, been challenged by the work of Engels and Nagy (1982). Table 7.2 sets out the values for the D/L ratios of amino acids in the Murchison meteorite as obtained by Engels and Nagy:

TABLE 7.2

Murchison Meteorite Amino Acid D/L values

Extract	Glu	Asp	Pro	Leu	Ala
$H_2O(1)$	0·322	0·202	0·342	0·166	0·682
$H_2O(2)$	0·300	0·300	0·300		0·600
HC1	0·176	0·126	0·105	0·029	0·307

These values are consistent with a situation in which the stereochemistry of the original amino acids was of the biological L-forms, with the D-forms being subsequently formed during a degenerative process similar to that which is known to occur in the terrestrial fossilisation of biomaterial (Hare, 1969).

Not only are the organic molecules in carbonaceous chondrites uncannily bio-like in their general character, but it has recently been discovered that much of this material is arranged into structures that bear a striking resemblance to bacterial cells. Figures 7.6 and 7.7 show two examples of such structures recently published by Pflug (1984). It is now supposed by Fischer-Tropsch proponents that entire cellular morphologies of organic material may be formed by such catalytic processes at an early stage in the history of the solar nebula.

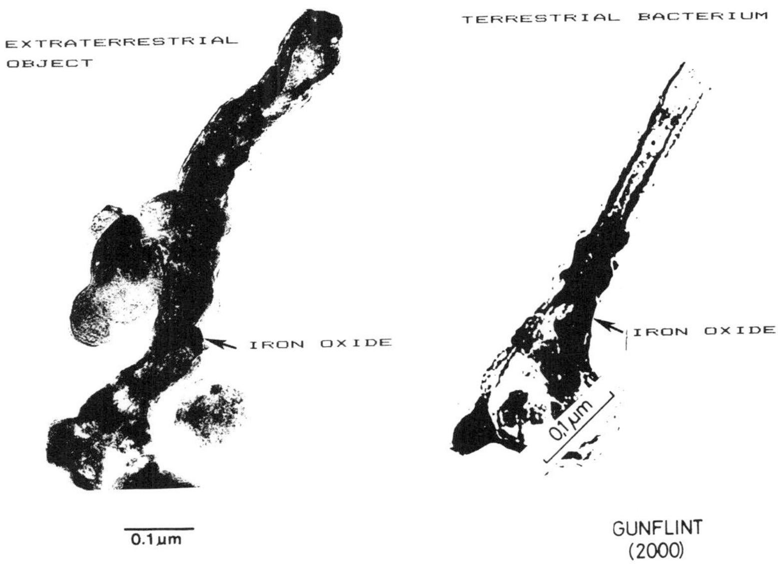

Fig. 7.6 *Comparison of a terrestrial 'fossil' bacterium with a structure recovered in cometary debris (Hoyle et al., 1985).*

Fischer–Tropsch reactions came to the fore in the early 1940's in connection with the German war effort directed at producing artificial petroleum. From a purely economic standpoint it is well-known that FTT reactions were a failure, at any rate for the production commerically of hydrocarbon fuels. It is significant in our view that nowadays, with generally rising oil prices, nobody has thought seriously of attempting the production of petroleum in such an inefficient way, despite the far greater measure of control over catalytic surfaces that can be achieved industrially. In an astronomical situation, catalytic surfaces would readily be poisoned by reactive sulphur gases. Even though a semblance of a case might be made for a catalytic conversion of some CO to CH_4, FTT reactions would be hard put to it to account for the observed quantities of complex, bio–like material in meteorites.

The basic experimental data on which a case can be made relate to the production of hydrocarbons by reactions of the general type:

$$n\,CO + (2n+1)H_2 + \text{catalyst} \rightarrow C_nH_{2n+2} + nH_2O + \text{catalyst} \qquad (7.13)$$

The laboratory experiments are usually carried out at comparatively high pressures and with carefully chosen catalytic surfaces.

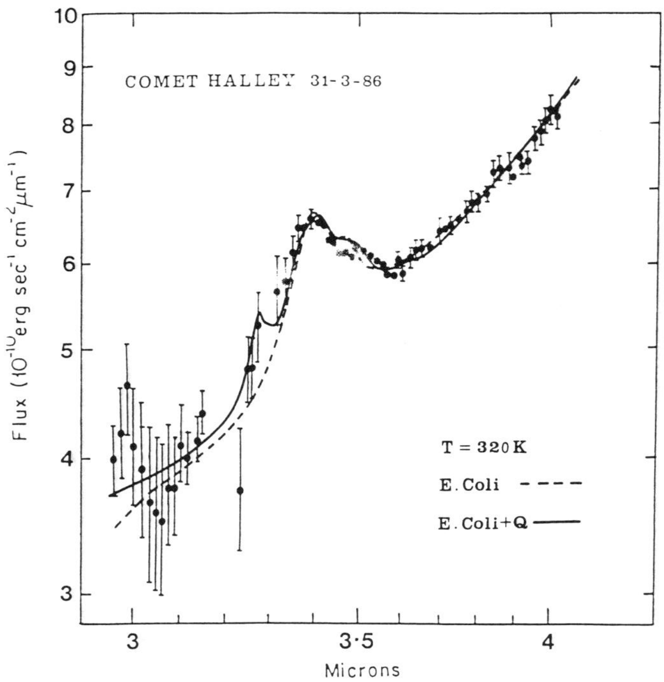

Fig. 7.7 *The observed spectrum of Comet Halley compared with bacterial grain models. The dashed curve includes a contribution from free aromatic structures resembling quinozaline. (Observations are from D.T. Wickramasinghe and Allen, 1986).*

The most commonly cited laboratory experiment in an astrophysical context is one by Lancet and Anders (1970). The experiment described by these authors is said to have been carried out with an equimolar mixture of CO and H_2 at a temperature of 400° K and total pressure 1 atm, using a cobalt catalyst. After the experiment was allowed to run for 320 hr, a significant conversion of CO to CH_4 (87.9%) was claimed, together with a formation of a waxy material to the extent of 1.5%.

The molecules that are synthesised in any FTT scheme and the yields that are actually obtained must depend sensitively on the effectiveness of the catalytic surfaces that are deployed. Lancet and Anders (1970) state that the principal meteoritic mineral phases stable above 350 – 400° K (e.g. olivine, pyroxene, FeS) are not found to be effective catalysts for this reaction. Furthermore, crushed carbonaceous chondritic material that has actually been tried out as a potential catalyst has also been found ineffective. On the other hand, it is claimed that some mineral phases that are stable below 400° K, e.g. magnetite, have been found to be suitable catalysts.

Considerable uncertainties are involved in extrapolating the limited kinetic data available for surface catalysed reactions such as (7.13) to conditions appropriate to the early solar nebula. Firstly, let us note that no laboratory data has been obtained for a gaseous composition that resembles even in broad outline the molecular composition of the solar nebula at 800° K. In particular, no data is available for a mixture containing H_2, H_2O and CO with an excess of H_2O over CO as would be expected to occur. The presence of H_2O might be expected to affect the kinetics of reduction reactions in an unfavourable way.

The only laboratory data available at the present time relate to experimental systems with equal concentrations of CO and H_2 and with a total pressure of $\sim$ 1 atmosphere. The early solar nebula conditions, on the other hand, give a CO/H_2 ratio of 6.5×10^{-4} and a total pressure of 10^{-5} atm. Uncertain extrapolations are needed to obtain characteristic conversion times from CO to CH_4.

Yet another problem for FTT synthesis is the inevitable tendency of catalysts in the solar nebula to become poisoned. Although it is true that a diversity of types of catalytic surface might be initially available, the addition of even a couple of monolayers on these surfaces would lead to an effective cessation of the catalytic process. Apart from sulphur poisoning a surface condensation of organic polymers is bound to occur at the low values of T = 360° K postulated by Hayatsu and Anders (1981), if the FTT process gets started at all.

7.7. THE BIOLOGICAL GRAIN MODEL

Although the possibility exists for the formation of dust by each of the mechanisms that have been discussed, their individual efficiencies remain in doubt. The requirement that $\sim$ 1/3 of the available interstellar carbon is to be processed into highly complex polymers, including substances that are spectroscopically similar to the biopolymer cellulose, made us turn quite naturally to consider the possibility of a biogenic origin for much of the organics in space. The advantage here is that biology offers scope for the operation of a formation mechanism with a positive feedback on an enormous scale. The doubling of the grain of wheat analogy in section 7.4 would now apply on a galactic – even cosmological – scale.

The earliest beginnings of panspermia are buried in the mists of antiquity. Anaxagoras, the Greek philosopher who lived around 500 BC, and who discovered the true nature of eclipses, is credited with being the first person to state clearly the principle of panspermia – that the seeds of plant and animal life are inherent in the cosmos, and that they take root whenever the conditions become favourable. A resurgence of this idea occurred a little more than a century ago. It is not generally remembered that this resurgence took place largely due to the work of Louis Pasteur. Panspermia was in fact a natural corollary of Pasteur's demonstration that life would seem always to be derived from life. Thus the physicist Hermann von Helmholtz wrote in 1874:

> "It appears to me to be a fully correct scientific procedure, if all our attempts fail to cause the production of organisms from non–living matter, to raise the question whether life has ever arisen, whether it is not just as old as matter itself, and whether seeds have not been carried from one planet to another and have developed everywhere where they have fallen on fertile soil ..."

If we look at the geological record, there is no possible case for a denial of the principle that life can only be derived from life all the way back to 3.6 b.y. before the present time. About 3.83 b.y. ago, the Isua sediments were deposited, and these sediments contain less secure evidence for photosynthetic life (Pflug, 1979). At an earlier epoch, however, the Earth was most probably sterile. From lunar data we know that both the Moon and the Earth received intense meteoritic bombardment, so there could not have been a stable crust or an atmosphere on the Earth until the cessation of impacts about 3.9 b.y. ago. Thus the first 600 m.y. of the Earth's history has it seems to be written off as regards life.

This situation leads to two distinct logical possibilities:

(1) There is a requirement for chemical evolution leading to the spontaneous generation of life on the Earth between 3.6 and 3.9 b.y. ago, or

(2) There was no spontaneous generation on the Earth and the principle that life could only be derived from life was maintained throughout by means of panspermia. The seeds of life took root at the first moment that physical conditions became favourable.

The overwhelming majority of present–day scientists have opted for (1), although there is no *a priori* reason for preferring (1) to (2). We shall argue in this section that (2) is indeed more probable and that the vehicle for the transference of panspermia was most likely to have been the comets.

In assessing the possibility (1) referred to above much attention has been focussed on the formation of individual biological monomers. Many ingenious experiments, carried out over the years in a number of different laboratories, have shown that the formation of relevant monomers by inorganic processes may not be too difficult. Nor is it indeed difficult to form non–biological polymers such as, for instance, polypeptides and polynucleotides. But the big question that remains unanswered concerns the origin of the information content of life. The information content of living matter is highly specific in quality and astronomical in quantity. How was this information content acquired from a situation that was initially chaotic? By attempting to tackle this question we believe that a simple argument emerges against the possibility (1).

It is well known that there are some 1000 – 2000 enzymes that are crucial over a wide spectrum of life ranging from simple microorganisms all the way up to man. The variation of amino acid sequences in these enzymes from one species to another are, on the whole, rather minor. A number of key positions on these chains are occupied by almost invariable amino acids.

Let us consider now how these enzyme sequences could have been arrived at in a primaeval soup. Consider a soup of twenty biologically important amino acids in equal concentrations. At a conservative estimate say ten sites per enzyme are crucial for proper biological function. The number of trial assemblies that are needed to produce a single working enzyme is in excess of $(20)^{10}$, and the probability of finding N such enzymes by random assembly is $1:(20)^{10N}$. It is easily seen that we obtain a number of trials exceeding the number of all the atoms in all the stars in the whole visible universe even before we come to N = 10. From this numerical assessment one of three deductions is possible:

(a) Life is a cosmic phenomenon, and we are forced to accept panspermia.

(b) Life is terrestrial, but its information content contains enormous redundancy by a factor $\sim (20)^{2000}$ for the case of the enzymes.

(c) Life is terrestrial. It occurs with such a miniscule probability that it is unique to the Earth.

The consensus view at the moment is that (a) is unacceptable, and that either (b) or (c) has to be true. In our view, however, there is no evidence for (b) while (c) is distinctly pre-Copernican. And so we are left with (a).

Let us now see how (a) connects with the theory of organic or biological grains. The beginnings of such a connection are to be found in the classic work of Svante Arrhenius', *Worlds in the Making*, first published in Swedish in 1904 (in English, Arrhenius, 1907).

Arrhenius essentially followed the logic of Pasteur and Helmholtz. He considered the possibility that bacterial cells (spores in particular) are lifted out of the gravitational potential wells of their planets by electromagnetic effects, and then are dispersed through space by the action of radiation pressure from stars. For particles of bacterial size (radii of a few tenths of a micron) the force of radiation pressure due to a star like the Sun exceeds gravity by a large factor (see Chapter 6). The bacterial particles are then expelled out of the entire planetary system. In very tenuous gas, such as exists between interstellar gas clouds, such grains can attain speeds $\sim$ 100 km/s, and could thus cross the distance between clouds in less than $\sim$ 100,000 years.

A difficult bottle–neck in the Arrhenius picture was the requirement for expelled grains to gain re-entry into another stellar planetary system. Just the same force of radiation pressure that expelled grains from one system would also serve to repel them as they approached a new system. Arrhenius got over this difficulty by arguing that the entry speed of a grain might be checked reversibly by radiation pressure, thereby permitting a transfer of living cells. But it had to be admitted that this was a rare event, and Arrhenius guessed at a number of cells entering the Earth at the present time as being no more than a few dozen every year.

We ourselves have argued in favour of a process that is far more efficient for amplifying life and dispersing it on a cosmic scale (Hoyle and Wickramasinghe, 1981). We proposed that comets which contain both water and complex organic nutrients provide an ideal culture medium for many types of bacteria. The organic content of comets came to be established after the Halley encounter of 1986. Both from direct sampling of grain fragments, using instruments aboard Giotto, and from ground–based infrared observations we now know that the dust in this comet is spectroscopically indistinguishable from bacterial particles (Wickramasinghe and Allen, 1986). Fig. 7.7 shows the infrared flux from particles ejected by Comet Halley on 31/3/86 compared with the predictions for bacterial grains (Wallis et al., 1989; Hoyle and Wickramasinghe, 1989). We regard the closeness of this fit as providing strong evidence in favour of the hypothesis of bacterial grains in comets.

We have argued for comets having once possessed liquid interiors due to the slow release of chemical and radioactive heat sources (e.g. Al^{26}) that were present in the material at the outset (Hoyle and Wickramasinghe, 1978). Once melted an interior region remained liquid for timescales that are long enough to ensure that much of the material of the comets was converted to bacterial grains. The conditions in melted cometary interiors are well suited to the amplification of autotrophic anaerobic bacteria. The reason is that a single bacterium can take up nutrient from the environment and doubles itself in a couple of hours. Two becomes four, four becomes eight, eight becomes sixteen and so on until all the nutrient is used up. In a short time the whole comet is thus transformed into biomaterial, so far as the nutrient quality of its material will allow. When the heat sources are exhausted the comet re–freezes. And the frozen condition generally prevailing in comets is appropriate for the indefinite preservation of almost all forms of microorganisms known to exist on the Earth today.

Next we turn to the cosmogony of the solar system. Suppose that some population of bacterial cells were present in the parent cloud from which the Sun and planets condensed. In our view the bulk of the material of the inner planets accumulated at relatively high temperatures during the early superluminous phases of the Sun, so any biology present in this material would have been destroyed. The outer planets Uranus and Neptune were accreted from cooler cometary bodies. The final stages of the accumulation of these comet–type bodies involved the mopping up of hard–frozen bacterial cells that were present in the original parent cloud. On the road to forming Uranus and Neptune, liquid water would have been retained in abundance for considerable periods of time on a multitude of planetary–sized objects. In such watery objects in the outer regions of the solar system, living cells could have been explosively amplified in number. A fraction of these cells would have been ejected out of the entire solar system by the effect of radiation pressure, and a fraction retained to be mopped up by comets which must have originated at about the orbital distances of the outermost planets. Such cells would have an almost indefinite persistence within comets.

Cells are now spewed off along with volatile gases when comets become deflected into the inner regions of the solar system. Some of these cells could find their way onto planets within the solar system, but the majority would be expelled back into interstellar space.

Cells expelled from the solar system, either now or in the past, are not easily injected directly onto planets of another distant stellar system. They are very efficiently slowed down and stopped in the first dense gas cloud that is encountered. Bacterial cells are amplified on a cosmic scale by the feedback loop shown in Fig. 7.8. For the solar system an estimated 10% or more of the mass of the Sun was returned to the interstellar gas. The bulk of this returned material was of course H

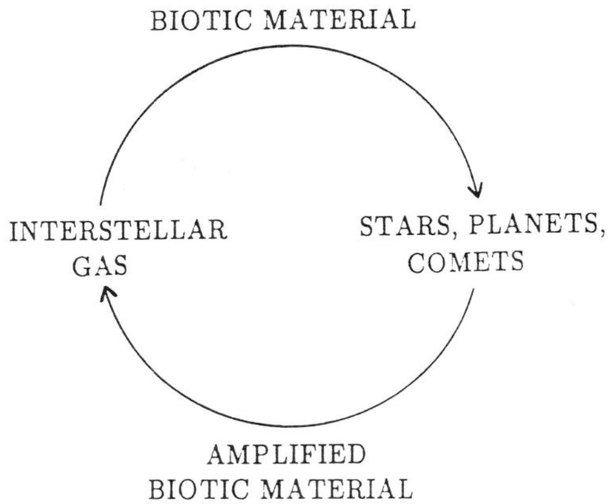

Fig. 7.8 *Cosmic amplification cycle.*

and He, but about 1 percent of it could be in the form of organic/biological material. Each star that is formed in the galaxy is a potential circuit in the feedback loop of Fig. 7.8. With 10^{11} such circuits, one for each sun–like star, $\sim 10^8$ M⊙ of organic/biotic material would be produced throughout the galaxy. Allowing for a grain lifetime of $\sim 10^9$ yr we therefore find that this process can account for the entire mass of dust grains that is known to exist in the spiral arms of the galaxy at the present time.

For the mechanism discussed here to work, we do not require more than a small fraction of the bacterial grains to be in a viable condition when they become incorporated into a new star/comet system. The replicative power of biology (grains of wheat story again) ensures that those cells that remained viable quickly regenerate and colonise the new cometary system. There are several effects that tend to favour retention of viability under interstellar conditions. A thin layer of carbon or graphitic material is known to have strong ultraviolet absorption properties. A layer of thickness 1000 Å coated on a bacterial cell would shield the interior from ultraviolet light. Ultraviolet radiation at wavelengths around 2600 Å is known to be deleterious to biology, although the occurrence of a well-known enzymic process in cells seeks under suitable growth conditions to repair the damage caused by UV radiation. However, accretion of thin mantles or carbonisation of a superficial skin may well go in the direction of ensuring viability to a considerable degree.

Table 7.3, adapted from Vallentyne (1963), shows the range of tolerance for microorganisms subject to various types of environmental stress. The temperature effects are well known. The survival properties of certain types of microorganism after large doses of ionizing radiation would be a mystery to any Earth–bound theory of life. The atmosphere absorbs essentially all X-rays, which are the main source of ionizing radiation in the environs of the Earth. And this situation must have been true from the earliest geological epoch when life was possible. The effect of X–rays is to cause strand breaks in the nuclear DNA of cells. The presence in cells of highly specific enzymes that can put the broken strands together is not easily understood in terms of terrestrial biology. Just as for the case of UV damage, the repair enzymes are redundant for terrestrial biology, but vital of course for panspermia.

TABLE 7.3

Environment Limits for Growth and Reproduction of Microorganisms (adapted from Vallentyne)

Factor	Lower Limit	Upper limit
Temperature	– 18°C (Survival only down to – 270°C)	300°C (deep sea bacteria) 104° (sulfate reducing bacteria at 1000 atmospheres hydrostatic pressure)
Eh	– 450 mv (at pH = 9·5 for sulfate reducing bacteria)	+ 850 mv (at pH = 3 for iron bacteria)
pH	0	>13
Water activity (a_w)	0·65	–
Hydrostatic pressure	~0	1400 atmospheres (deep sea bacteria)
Salinity	Double distilled water	Saturated brines (Dead Sea bacteria)
Ionising Radiation (recovery after)	–	~ 10^6 rad (micrococcus Radiodurans)

References

Albert, A. and Armarego, W.L.F., 1965, *Adv. Heterocyclic. Chem.*, **4**, 1.

Arrhenius, S., 1907, *Worlds in the Making*, Harper and Bros.

Bevington, J.C., 1961, *Radical Polymerisation*, London, Academic Press.

Blackadder, D.A., 1975, *Some Aspects of Basic Polymer Science*, London, The Chemical Society.

Cooke, A. and Wickramasinghe, N.C., 1977, *Astrophys. Sp. Sci.*, **50**, 43.

Cox, J.D., 1963, *Tetrahedron*, **19**, 1175.

Dewar, M.J.S. and Harget, A.J., 1970, *Proc. Roy. Soc. Lond.*, **A315**, 442 and 457.

Donn, B., 1968, *Astrophys. J. Lett.*, **152**, L129.

Donn, B. and Krishna Swamy, K.S., 1969, *Physica*, **41**, 144.

Donn, B.D., Allen, J.E. and Khanna, R.K., 1989, in *IAU Colloquium No. 135: Dust in the Universe*, D. Reidel.

Donn, B.D. and Sears, B.W., 1963, *Science*, **140**, 1208.

Duley, W.W. and Williams, D.A., 1981, *Mon. Not. Roy. Astr. Soc.*, **196**, 269.

Engel, M.E. and Nagy, B., 1982, *Nature*, **196**, 837.

Fawcett, A.H., 1975, *Nature*, **257**, 159.

Gillett, F.C. and Forrest, W.J., 1973, *Astrophys. J.*, **179**, 483.

Glassgold, A.E. and Langer, W.D., 1975, *Astrophys. J.*, **197**, 347.

Goldanskii, V.I., 1979, *Nature*, **246**, 45.

Greenberg, J.M. 1973a, in H. Reeves (ed.) *Symposium on the Origin of the Solar System*, Edition Centre Nat. Recherche Sci., Paris.

Hare, P.E., "Geochemistry of Proteins, Peptides and Amino Acids", in S. Eglington and M.T.J. Murphy (eds.) *Organic Chemistry*, Longman, 1969.

Hayatsu, R. and Anders, E., 1981, in F.L. Borschke (ed.) *Topics in Current Chemistry*, **99**, 1.

Hayes, J.M., 1967, *Geochimica et Cosmochimica Acta*, **31**, 1395.

Helmholtz, 1876, *Populare Wissenschaftliche Vortrage*, Braunschweig, Vol. **iii**, p.101.

Hoyle, F. and Wickramasinghe, N.C., 1976, *Nature*, **246**, 45.

Hoyle, F., Wickramasinghe, N.C. and Pflug, H.D., 1985, *Astrophys. Sp. Sci.*, **113**, 209.

Hoyle, F., Solomon, P. and Woolf, N.J., 1973, *Astrophys. J.*, **185**, L89.

Hoyle, F. and Wickramasinghe, N.C., 1977, *Mon. Not. Roy. Astr. Soc.*, **181**, 51P.

Hoyle, F. and Wickramasinghe, N.C., 1989, *ESA.SP*, **290**, 67.

Hoyle, F. and Wickramasinghe, N.C., 1977, *Nature*, **267**, 610.

Hoyle, F. and Wickramasinghe, N.C., 1981, *Space Travellers: The Bringers of Life*, Univ. Coll. Cardiff Press.

Huebner, W.F., 1987, *Science*, **237**, 628.

Johnson, F.M., Bailey, D.T. and Wegner, P.A., 1973, in H.C. van de Hulst and J.M. Greenberg (eds.) *Interstellar Dust and Related Topics*, D. Reidel.

Johnson, F.M., 1967, in J.M. Greenberg and T.P. Roark (eds.) *Interstellar Grains*, NASA.SP–140.

Johnson, F.M., 1971, *Ann. N.Y. Acad. Sci.*, **194**, 3.

Lancet, M.S. and Anders, E., 1970, *Science*, **170**, 980.

Langer, W.D., 1976, *Astrophys. J.*, **206**, 699.

Léger, A. and Puget, J.L., 1984, *Astr. & Astrophys.*, **137**, L5.

Léger, A. and d'Hendecourt, L., 1986, in F.P. Israel (ed.) *Light on Dark Matter*, D. Reidel.

Leung, C.M. and Liszt, H.S., 1976, *Astrophys. J.*, **208**, 732.

Little–Marenin, I.R. 1986, *Astrophys. J.*, **307**, L15.

Meyer, L., 1959, *Proc. 3rd Conf. Carbon*, **451** (Pergamon).

Mitchel, D.L., Linn, R.P., Anderson, K.A., Carlson, C.W., Curtis, D.W., Korth, A., Reme, H., Sauvaud, J.A., d'Uston, C., Mendis, D.A., 1987, *Science*, **237**, 626.

Pflug, H.D., 1979, *Nature*, **280**, 483.

Pflug, H.D., 1984, in C. Wickramasinghe (ed.) *Fundamental Studies and the Future of Science*, University College Cardiff Press.

Platt, J.R., 1956, *Astrophys. J.*, **123**, 486.

Platt, J.R. and Donn, B.D., 1956, *Astron. J.*, **61**, 11.

Puetter, R.C., Russell, R.W., Soifer, R.W. and Willner, S.P., 1979, *Astrophys. J.*, **225**, 118.

Sakata, A., Wada, S., Okutsu, Y., Shintani, H. and Nakada, Y., 1983, *Nature*, **301**, 493.

Sellgren, K., 1984, *Astrophys. J.*, **277**, 623.

Sellgren, K., Werner, M.W. and Dinerstein, H.L., 1983, *Astrophys. J.*, **271**, L13.

Shafizadeh, F., 1971, *J. Polymer Sci.*, **36**, 21.

Snow, T.P., 1975, *Astrophys. J.*, **201**, L21.

Vallentyne, J.R., 1963, *Ann. N.Y. Acad. Sci.*, **108**, Part 2, 342.

Vanysek, V. and Wickramasinghe, N.C.,1975, *Astrophys. Sp. Sci.*, **33**, L19.

Wallis, M.K., Wickramasinghe, N.C., Hoyle, F. and Rabilizirov, R., 1989, *Mon. Not. Roy. Astr. Soc.*, **238**, 1165.

Wickramasinghe, D.T. and Allen, D.A., 1986, Nature, **323**, 44.

Wickramasinghe, N.C., 1974, *Nature*, **252**, 462.

Wickramasinghe, N.C., 1975, *Mon. Not. Roy. Astr. Soc.*, **170**, 11P.

Wickramasinghe, N.C., 1976, in N.C Wickramasinghe and D.J. Morgan (eds.) *Solid State Astrophysics*, D. Reidel, p, 87.

Wickramasinghe, N.C. and Santhanan, K.S.V., 1975, *Nature*, **257**, 159.

Willems, F.J. and de Jong, T., 1986, *Astrophys. J.*, **309**, L39.

Williams, D.A., 1974, *Observatory*, **94**, 66.

Zuckerman, B. and Palmer, P., 1974, *Ann. Rev. Astron. Astrophys.*, **12**, 279.

Zuckerman, B. and Turner, B.E., 1975, *Astrophys. J.*, **197**, 123.

8. Models of the Extinction and Polarisation of Starlight

8.1. INTRODUCTION

The extinction and polarisation of starlight are perhaps the most important observational properties of interstellar grains. Models of interstellar grain must satisfy the primary requirement that they match to within the observational errors the measured wavelength dependence of extinction that we discussed in Chapter 3. Likewise, interstellar polarisation, the difference in extinction for two orthogonal directions of the electric vector must be explained by the same model. It has been customary to discuss the extinction using computed properties of spherical grains, while for modelling the polarisation data recourse is made to calculations based on infinite cylinder models. The justification for this dual approach is based on the near coincidence between the extinction behaviour of randomly oriented cylinders and spheres that we saw in Chapter 2. Polarisation of starlight results from the alignment effectively of only a few percent of elongated (or flattened) grains. In view of the predominance of whisker growth for many types of crystals and polymers condensing from a vapour, and because bacterial particles of the type discussed in Chapter 7 are mostly elongated rather than flattened, we confine our

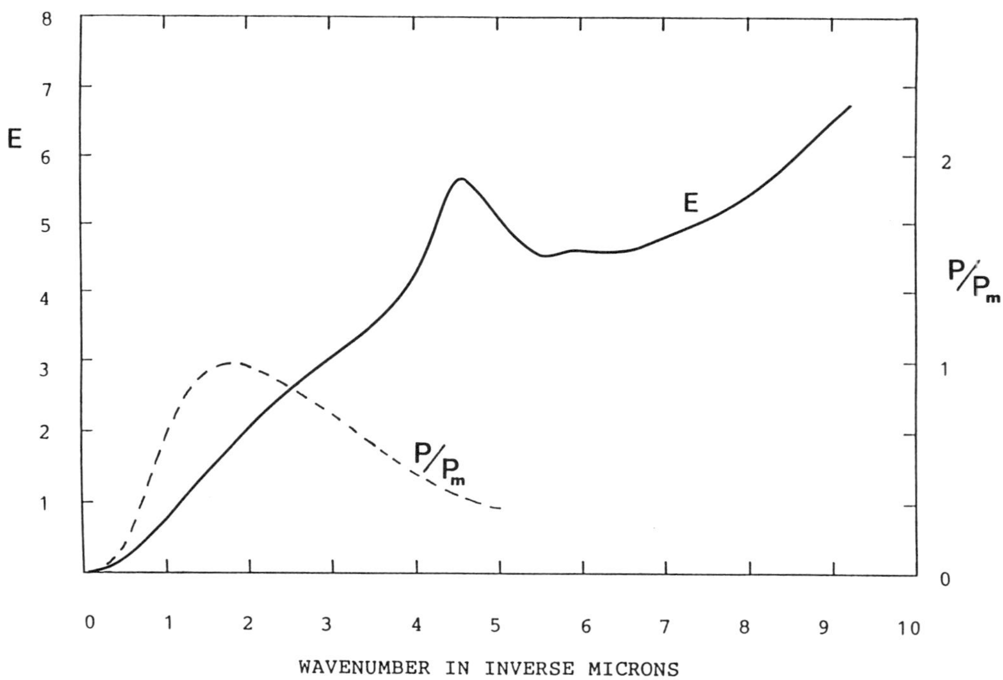

Fig. 8.1 *The mean polarisation and extinction laws in the galaxy. (Note that $E = X/1.744$, where X is given in Table 8.1).*

attention here to long slender grains which will be modelled using the rigorous formulae for infinite cylinders. We also note that laboratory work by Greenberg on microwave analogues has shown that prolate spheroids with an axial ratio $a/b \sim 2.5$ have extinction and scattering properties hardly distinguishable from those for cylinders of radius b (Greenberg, 1968).

Although the observed interstellar extinction curve is known to be somewhat variable in its detailed shape at ultraviolet wavelengths $\lambda < 2800$ Å, it is remarkably constant in shape in the visible region of the spectrum. There are of course exceptional cases such as the Orion stars and stars near the Taurus molecular clouds where the dust properties appear to be somewhat perturbed. In such instances values of $R > 6$ as well as polarisation maxima longer than 6000 Å occur, accompanied also by a broadening of the interstellar polarisation curve (Wilkins et al., 1980, Vrba et al., 1981). We shall mainly be concerned here with the 'normal' extinction curve (the average for thousands of stars) and with the average behaviour in regard to the interstellar polarisation, both linear and circular, as described in Chapter 3. The data we propose to model is summarised in Fig. 8.1.

The average extinction curve expressed in the form $X(x) = A_\lambda/E_{B-V}$ has been shown by Seaton (1979) to be well represented by the analytical expressions and numbers in Table 8.1. Here x stands for the inverse wavelength in μm^{-1}. (The curve plotted in Fig. 8.1 is $E=X(x)/1.744$ giving 1.8 mag/kpc as the average extinction at $\lambda^{-1} = 1.8\mu m^{-1}$). The tabulated columns are from observations of Nandy et al. (1975).

The average linear polarisation data as we saw in Chapter 3, is well represented by the equation

TABLE 8.1

Seaton's (1979) representation of the average interstellar extinction $X(x)$.

Range of x	Expression for X(x)
$2{\cdot}70 \leq x \leq 3{\cdot}65$	$1{\cdot}56+1{\cdot}048x+1{\cdot}01/\{(x-4{\cdot}40)^2+0{\cdot}280\}$
$3{\cdot}65 \leq x \leq 7{\cdot}14$	$2{\cdot}29+0{\cdot}848x+1{\cdot}01/\{(x-4{\cdot}60)^2+0{\cdot}280\}$
$7{\cdot}14 \leq x \leq 10$	$16{\cdot}17-3{\cdot}20x+0{\cdot}2975x^2$

Values of $X(x) = A_\lambda/E_{B-V}$ for $1{\cdot}0 \leq x \leq 2{\cdot}7$, from Nandy et al. (1975), re–normalised to $R = 3{\cdot}2$

x	X(x)	x	X(x)	x	X(x)
1·0	1·36	1·6	2·66	2·2	3·96
1·1	1·44	1·7	2·88	2·3	4·15
1·2	1·84	1·8	3·14	2·4	4·26
1·3	2·04	1·9	3·36	2·5	4·40
1·4	2·24	2·0	3·56	2·6	4·52
1·5	2·44	2·1	3·77	2·7	4·64

$$\frac{P(\lambda)}{P(\lambda_{max})} = \exp[-K \ln^2 (\frac{\lambda}{\lambda_{max}})] \qquad (8.1)$$

with K = 1.15 and the circular polarisation follows an approximately linear law in the range $1.25 < \lambda^{-1} < 2.5\mu m^{-1}$ given by

$$\frac{V}{I} \propto -14 \left(\frac{1}{\lambda} - 1.8\right) \qquad (8.2)$$

The important feature about the circular polarisation is that it changes sign at $\lambda^{-1} \cong 1.8\mu m^{-1}$.

8.2. THE VISUAL EXTINCTION CURVE

As we saw in Chapter 3, early measurements at visual wavelengths of the interstellar extinction curve established a broad result that has survived, namely that the amount of the extinction expressed on a logarithmic scale (or magnitude scale) is approximately proportional to the reciprocal of the wavelength, $A_\lambda \propto 1/\lambda$. This result was refined by Nandy (1964, 1965, 1966) who showed that to a second order of approximation, A_λ could be represented in a plot against $1/\lambda$ by two straight line segments, the two segments intersecting at about $1/\lambda = 2.4\mu m^{-1}$, with the segment appropriate to blue wavelengths being somewhat shallower in slope than the segment appropriate to red wavelengths.

The absolute amount of the extinction is remarkably high, about 2 mag at $1/\lambda = 1.8\mu m^{-1}$ for a star at a distance of 1 kpc, a circumstance which forces the bulk of the grains to be composed of the commonest elements. If abundances in the interstellar medium are taken to be approximately solar, silicate grains would be inadequate in abundance by a factor of about 3, even if their sizes were optimally chosen. Grains based on the C, N, O elements could suffice, however, partly because the CNO group is cosmically more abundant than Mg, Si, Fe, and partly because it can form solids of lower density. Yet with solar abundance the C, N, O elements formed into solid grains of optimal sizes can meet the extinction requirement with only a factor of about 2 in hand.

We do not know the abundance of the elements in the interstellar medium to be solar of course, and arguments by Clegg and Bell (1973) and by Pagel (1974) suggest that present-day abundances might be higher than solar by a factor of about 1.7. By stretching this enhancement a little more, the silicate hypothesis could be saved on abundance grounds alone. Duley (1984) criticizes this result on the assumption that interstellar abundances are in fact close to solar. When abundances are raised even modestly, say by a factor of 1.5, the situation is changed and Duley's criticisms based on abundances do not seem to be valid.

The observation of the near constancy of the wavelength dependence of the extinction curve in the range $1 < \lambda^{-1} < 3\mu m^{-1}$ imposes severe constraints on permissible grain models. One solution is to suppose that this extinction behaviour follows directly from the choice of a conducting grain for which the wavelength dependence of n and k over the visual region is such that

$$Q_{ext}(\lambda) = \frac{8\pi a}{\lambda} \, \mathrm{Im}\left[\frac{1-m^2}{2+m^2}\right] \tag{8.3}$$

has closely the right functional dependence to reproduce the column of extinction values in Table 8.1 (Hoyle and Wickramasinghe, 1962). For this possibility the observed visual extinction law is essentially independent of particle radius as long as the radius is small enough to make (8.3) a good approximation to the extinction coefficient. A difficulty with this approach is that since we are dealing with small conducting grains the albedo is necessarily low and the phase function nearly isotropic, thus leading to a conflict with the data on reflection nebulae and the diffuse galactic light that we discussed in Chapter 4. The problem with the low albedos of small conducting grains forces one to turn to dielectric models. In all such models particle radii have to be chosen so as to optimise the agreement with the extinction data in Table 8.1.

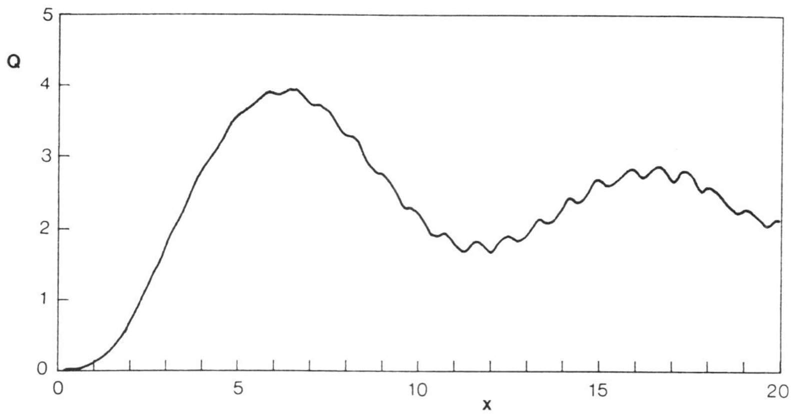

Fig. 8.2 *Extinction efficiency for spheres with m = 1.33.*

Interstellar extinction curves for ice grains were first calculated by van de Hulst (1949). The first step in such a calculation is to compute $Q_{ext}(x)$ for ice spheres (where $x = 2\pi a/\lambda$) with $m = 1.33$, the results of which are shown in Fig. 8.2. It is seen that the curve is approximately linear in the range $2 < 2\pi a/\lambda < 4$. With a suitable choice of radius a, this linear segement of the Q_{ext} curve could be brought into near coincidence with the optical region of the interstellar extinction curve. However, the extinction must then reach a broad maximum in the near ultraviolet. From van de Hulst's (1949) computations it appeared that the properties of ice spheres were ideal with respect to matching the available optical data. The situation reported by van de Hulst was that a single ice particle of radius $a \simeq 0.3\mu m$ comes close to fitting the extinction curve for $\lambda > 3300$ Å, while a suitable size distribution could produce a near perfect fit to the data.

The normalised extinction curve resulting from the distribution of ice particle sizes considered by Oort and van de Hulst (1946) is shown as the dashed curve of Fig. 8.3. The solid curve is the normalised extinction curve for a size distribution given by

$$n(r) \propto \exp[-r/r_0] , \qquad r_0 = 0.075\mu m , \tag{8.4}$$

the points in Fig. 8.3 being the observations set out in Table 8.1*. We note that the theoretical curves fit the optical observations satisfactorily, but deviate markedly from the data in the ultraviolet.

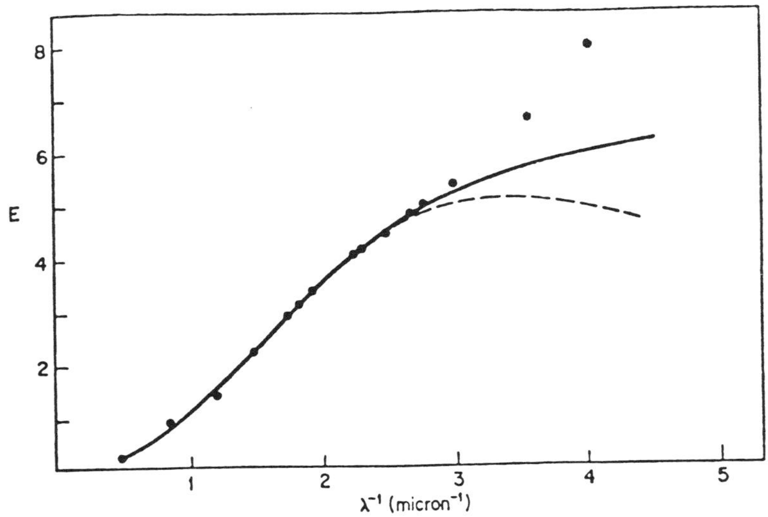

Fig. 8.3 *Normalised extinction curves for ice grains. The solid curve corresponds to an exponential distribution of ice grain radii $n(a) = exp(-a/a_0)$ where $a_0 = 0.075\mu m$. The broken curve is extinction for a size distribution of ice grains adopted by van de Hulst (1946). The points are the observations.*

This disagreement with the ice model first became clear in 1963. Many modifications were attempted with a view to reinstating the model in some form. Greenberg (1968) and Greenberg and Shah (1969) introduced refinements such as bimodal size distributions, the use of measured optical constants of ice in the extinction computations, and the assumption of cylinders rather than spheres. None of these refinements offered a satisfactory explanation for the detailed shape of the extinction curve, particularly the conspicuous extinction hump centred on $\lambda = 2175$ Å. This feature was clearly an absorption profile of some hitherto unidentified component of the grains.

In the visual spectral region, $1 < \lambda^{-1} < 3\mu m^{-1}$, extinction models based on materials with refractive indices $n \gtrsim 1.3$ suffered from the defect that particle sizes

* It should be noted that for the size distribution function given by equation (8.4) the main contribution to extinction arises from grains of radius $0.28\mu m$.

and/or size distributions had to be fixed to an almost unrealistic degree. To account for the visual extinction data with ice grains distributed in size according to equation (8.4) we find that the value of r_0 has to be fixed to within a few percent.

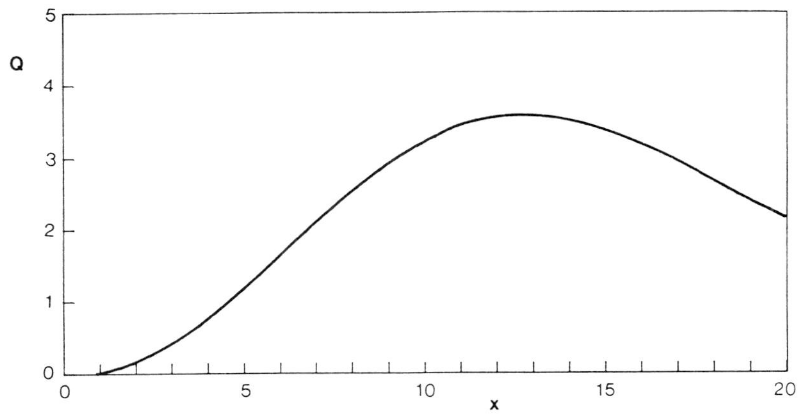

Fig. 8.4 *Extinction efficiency for spheres with m = 1.167.*

This is a difficulty endemic to all high n models of dielectric grains. Only when n becomes somewhat less than 1.2 can significant improvements ensue. Fig. 8.4 shows the Q_{sca} values for a cylindrical particle m = 1.167 in random orientation. We see that Q_{sca} is linear over a substantial range of x, from ~ 2–9, offering scope for fitting the visual extinction curve with a wide spread of sizes or size parameters. The same type of curve is obtained for a hollow cylinder where 1.167 refers to the 'average' refractive index for the entire particle.

8.3. THE ULTRAVIOLET EXTINCTION CURVE : EXTINCTION CURVES FOR GRAPHITE GRAINS

Following our suggestion that graphite grains can be formed in carbon stars and be expelled into the interstellar medium, we proceeded to compute extinction efficiency factors for small graphite particles (Wickramasinghe and Guillaume, 1965). These computations made it amply clear that spherical graphite particles with radii ~ 0.02μm could account for the hitherto unexplained interstellar feature at $\lambda \simeq 2175$ Å. Extinction and scattering efficiencies for a graphite grain of this radius are shown in Fig. 8.5. An assumption inherent in these calculations is that the bulk material of the grain has isotropic optical properties characterised by a complex refractive index $m = m_a$, where m_a are the measured values for light with electric vector parallel to the basal planes of graphite. Since real graphite flakes, which are anisotropic, behave like a dielectric for light with electric vector perpendicular to the planes, and like a metal for electric vector parallel to the planes, it could be argued that a sphere comprised of microscopic flakes in random orientation would behave like the calculated effect in Fig. 8.4.

The situation for interstellar graphite grains remains somewhat uneasy for other reasons, however. On the basis of our calculations for spherical particles, we find

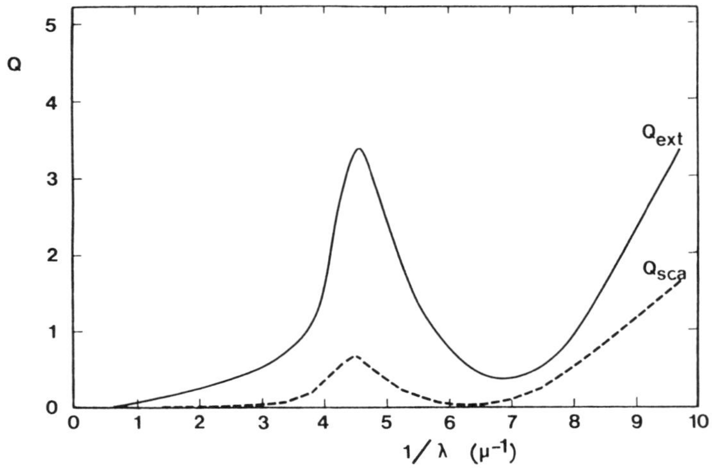

Fig. 8.5 *Extinction and scattering efficiencies for a graphite sphere of radius $a = 0.02\mu m$.*

that grain radii must be fairly narrowly confined close to a $= 0.02\mu$m. For smaller grains the wavelength of peak extinction shifts to a shorter wavelength; for larger particles the shift is to longer wavelength. The question arises as to why the grain size is so sharply defined. Further, any departures from the spherical shape leads to shifts of the resonance wavelength. This is seen from Rayleigh scattering calculations for spheroids shown in Fig. 8.6 (Wickramasinghe and Nandy, 1974). If we are to understand the observed interstellar extinction hump at $\lambda = 2175$ Å in terms of some form of condensed graphite, it would be necessary to discover a good physical reason for the emergence of a spherical shape. One possible way to produce spherical carbon particles may be seen in the analogy of burning a cotton fibre. The fibre clearly 'balls up' to assume a spherical shape from a condition of minimising surface energy. A similar process could be imagined for the ultraviolet degradation of cellulose–type polymers that were discussed in Chapter 7.

C_{60}: Buckminsterfullerene. An interesting variant of the graphite explanation for the λ 2175 Å feature follows from the work of Kroto et al. (1985), who used a laser to vaporise carbon from a graphite disk, finding the vapour to be dominated by clusters of carbon atoms, mostly with upwards of 40 atoms to a cluster. They report that under certain conditions the cluster C_{60} was heavily dominant, which situation they attributed to the formation of a Buckminster–Fuller–type closed surface made up of 12 pentagonal faces and 20 hexagonal faces, a pattern followed in the manufacture of modern soccer balls.

If C_{60} with this structure is considered to be present in great quantity in the interstellar medium, the two previous difficulties of the graphite hypothesis are removed. The particles are indeed nearly spherical in shape, and the sizes would all be the same, about 7 Å in diameter. Moreover the material would derive its electrical conductivity in the same way as graphite, from π–electrons, and so would be expected to possess similar optical constants, thereby giving rise to pronounced

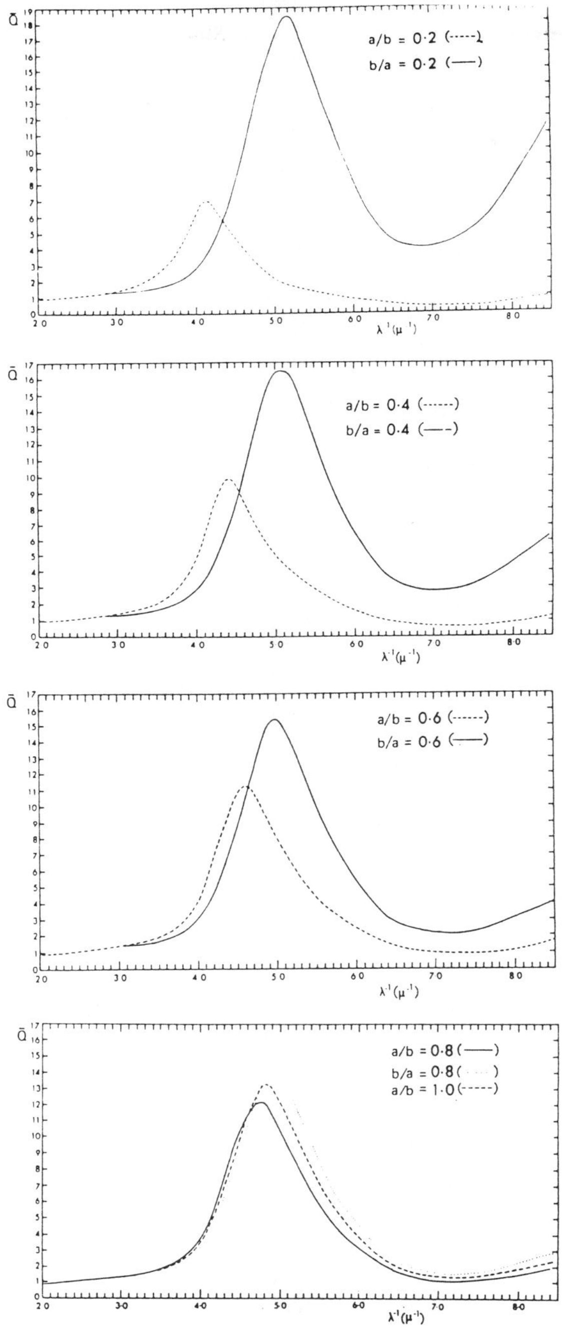

Fig. 8.6 *Normalised extinction curves for graphite spheroids, a denotes the semi–axis of symmetry and b the transverse semi–axis. The normalisation is to unity at $\lambda^{-1} = 2.22\mu m^{-1}$.*

absorption in the ultraviolet. It makes no difference for absorption that the particles are much smaller than was formerly contemplated, diameter 7 Å instead of 200 Å. All the former quantitative aspects of the extinction would remain essentially as before, which is to say an average spatial mass density of ~ 1.8×10^{-27} g cm^{-3} in the form of C_{60} would be required (Hoyle and Wickramasinghe, 1982). Only the question of the wavelength of maximum absorbance requires additional consideration. We consider this question on the basis that the optical constants of C_{60} are the same as graphite.

It was remarked above that for the optical constants of graphite the wavelength of maximum absorbance (WMA) depends on the radius of a particle. For a very small *solid* sphere of radius 7 Å the WMA is about 2105 Å, significantly too short to agree with the observational value for the extinction of starlight, viz., 2175 Å. However, the situation is different when the particles are hollow, as is the case for C_{60}. Considering C_{60} to be a spherical graphite shell with α the ratio of the radius of the inner surface to the radius of the outer surface, calculations of the Güttler formulae for the optical cross–sections of small particles* yield the values of the WMA given in the following table.

TABLE 8.2

Values of the wavelength of maximum absorption for small spherical shells as a function of α

α	Wavelength of Maximum Absorption (Å)
0	2105
0·10	2110
0·20	2120
0·30	2135
0·35	2150
0·40	2165
0·45	2180
0·50	2205
0·60	2260

Since the structure of C_{60} determines a unique value of α, there must be a unique WMA associated with such particles. For the optical constants of graphite the value of α required to match the IUE data is $\alpha = 0.45$. To the extent that the optical constants of C_{60} turn out to be somewhat different from graphite, the value of α required to match the data would be changed. With the optical constants measured, and with α determined from the precise structures of C_{60}, the requirement that the WMA must turn out to be 2180 Å evidently will provide an almost precise test of this new hypothesis.

---

* It is assumed in these calculations that the hollow spheres are sufficiently small, with a radius/wavelength ratio much less than $1/2\pi$, so that asymptotic expressions for the optical cross–sections become valid.

Although extinction calculations based upon the C_{60} idea is essentially as for graphite spheres of radius $0.02\mu m$, there is one interesting difference of detail. Particles of diameter only 7 Å would have essentially zero scattering coefficient, and so would lead to a smaller albedo for the interstellar grains near 2180 Å than was calculated on the older hypothesis. This is to the good, because the formerly calculated ultraviolet albedo has tended to be somewhat higher than reported values, (Chapter 4).

Such graphite as has been found in carbonaceous meteorites has been plausibly attributed to the degradation of organic material. The failure to find graphite particles of appreciable size (diameter ~ 400 Å) in such meteorites has been emphasised as an objection to the earlier graphite hypothesis (for example, Nuth, 1985). If the relevant interstellar particles are C_{60}, however, it becomes more understandable that, with their very small size, C_{60} particles should hitherto have escaped detection. It remains to be seen whether C_{60} will eventually be found among the so–called kerigenous material of the carbonaceous meteorites.

It would obviously be interesting to discover other spectral features of C_{60} in the laboratory and then to see if they are detectable in astronomical sources. Snow and Seab (1989) have done precisely this. They have looked for a feature at 3860 Å found for the neutral molecule C_{60} in the spectra of several reddened stars. From a negative result they infer that the column density of C_{60} must be less than 10^{14} cm^{-2}. Unionized C_{60} has recently been shown to possess an absorption band at about 2700 Å. Since no such absorption at 2700 Å is observed this militates against C_{60} being the source of the absorption at 2200 Å (Krätschmer et al., 1990). However, C_{60} is most likely to be ionized under interstellar conditions and, unfortunately, laboratory measurements of absorption bands do not exist for C^{+}_{60}

The Amount of Graphite and Competing Models. The amount of graphite (or a graphitic equivalent) may be estimated by noting that the observations require an extinction at λ 2200 Å, above a smooth scattering background, of ~ 2.25 mag/kpc. For graphite particles of $a = 0.02\mu m$, which have a mass extinction coefficient of $\kappa \sim 600{,}000$ cm^2 g^{-1}, we obtain an interstellar mass density of ~ 1.25×10^{-27} g cm^{-3}, some 10% or so of the available carbon. Candidates for this interstellar feature that are not carbon-based suffer the handicap of demanding excessively high κ values to compete with a graphite model. This is true for the model proposed by Maclean, Duley and Miller (1982) where it was proposed that the 2175 Å feature could be explained using the transition $2p^6 \rightarrow 2p^5 3s$ of O_2^-–ions at low coordination sites on the surfaces of very small MgO grains. We have shown that this process fails to explain the data by a factor of *at least* 10 (Hoyle et al., 1983).

Another contributor to interstellar absorption at 2175 Å could be in the form of organic chromophores. Absorptions close to this wavelength are known to occur in a wide class of organic molecules arising from $\pi \rightarrow \pi^*$ electronic transitions in conjugated doubles bonds. An important subclass are aromatic molecules which involve nitrogenated heterocyclic groups. We first pointed out that quinazoline isomers ($C_8H_6N_2$) which are typical of such molecules could provide a satisfactory explanation of the λ 2175 Å absorption feature as shown in Fig. 8.7 (Hoyle and Wickramasinghe, 1977). We shall show later in this chapter that an integrated spectrum of 115 naturally occurring aromatic molecules has an ultraviolet

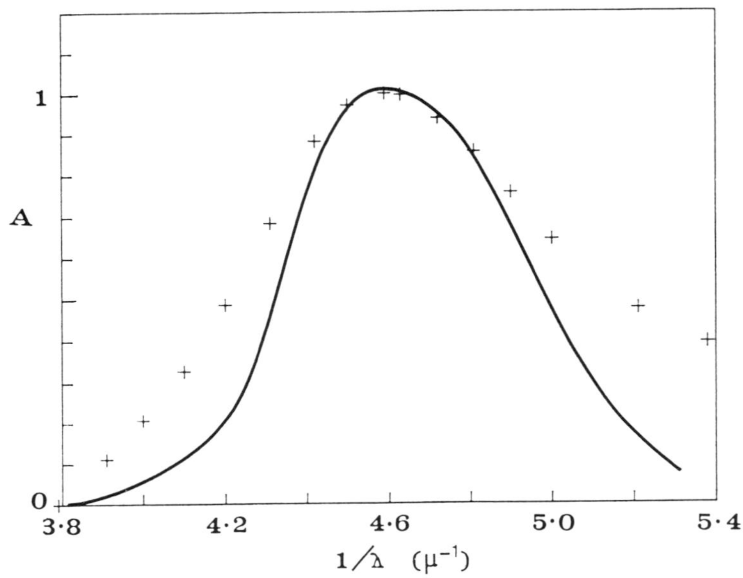

Fig. 8.7 *Normalised average molar absoptivity for $C_8H_6N_2$ isomers (solid curve) compared with interstellar extinction data normalised to 0 at $\lambda^{-1} = 3.8\mu m^{-1}$ and 1 at $\lambda^{-1} = 4.55\mu m^{-1}$. (Data is from compilation by Sapar and Kuusik (1978)).*

absorption profile that matches the astronomical data to a remarkable degree of precision.

Another UV Component. In addition to graphite (or an equivalent particle) it was evident for some time that a further dielectric grain component is required to account for data from $\lambda \sim 2000$ Å – 1100 Å. Grain mixtures including combinations of graphite with iron and mineral grains were considered by ourselves and other authors (Hoyle and Wickramasinghe, 1969; Gilra, 1971; Huffman and Stapp, 1971; Mathis et al., 1977). Although the much quoted Mathis et al. fit to the data on the basis of a graphite–silicate grain mixture would at first sight seem impressive (see Fig 8.8) several reservations might be noted:

(1) the mass ratio of graphite particles to silicate particles appears as a free parameter in the work of Mathis et al which has to be finely tuned,

(2) the size distribution of both kinds of particle is taken without either observational or physical reason to be of the form

$$n(a)\,da = (\text{constant}) \cdot \frac{da}{a^{3.5}},$$

both being spheres with a the radial parameter,

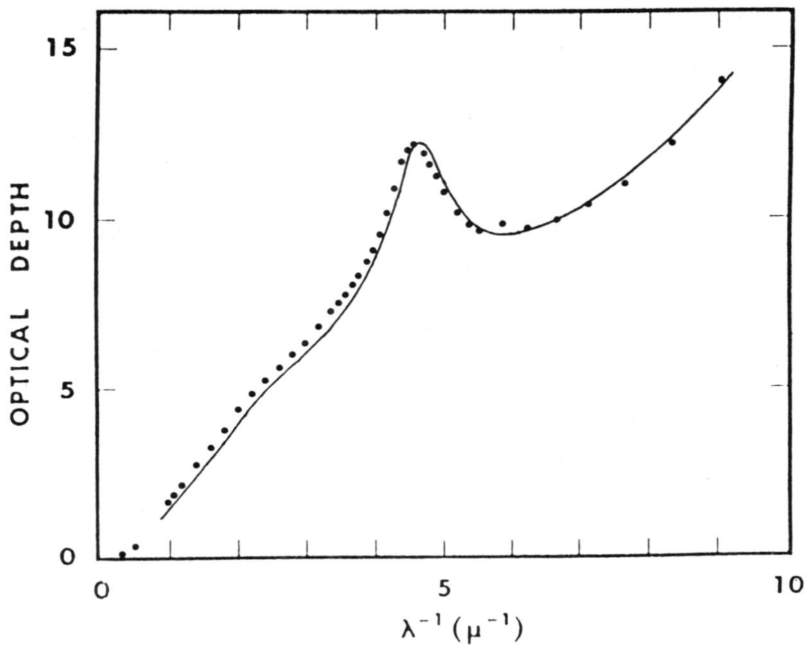

Fig. 8.8 *Best fit to the extinction curve calculated in a model of Mathis et al. (1979). The points are the average extinction data from a compilation by Sapar and Kuusik (1978).*

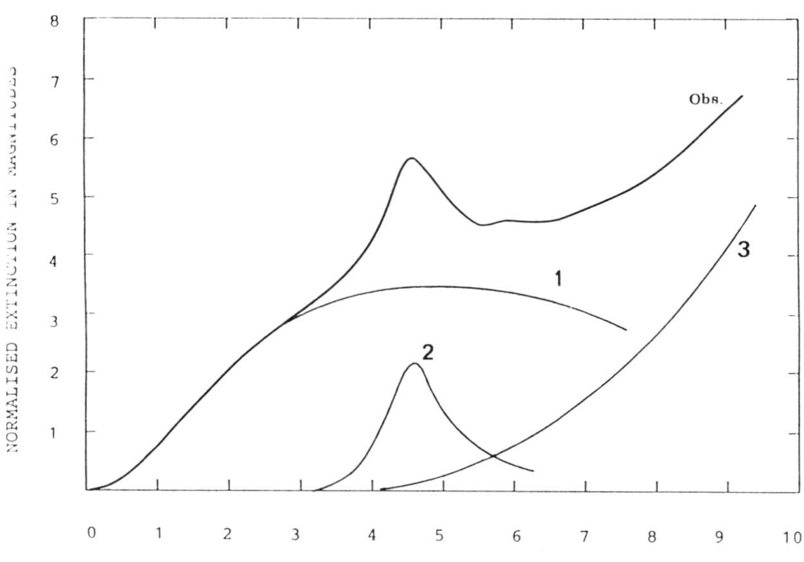

Fig. 8.9 *The observed interstellar extinction curve schematically decomposed into 3 components.*

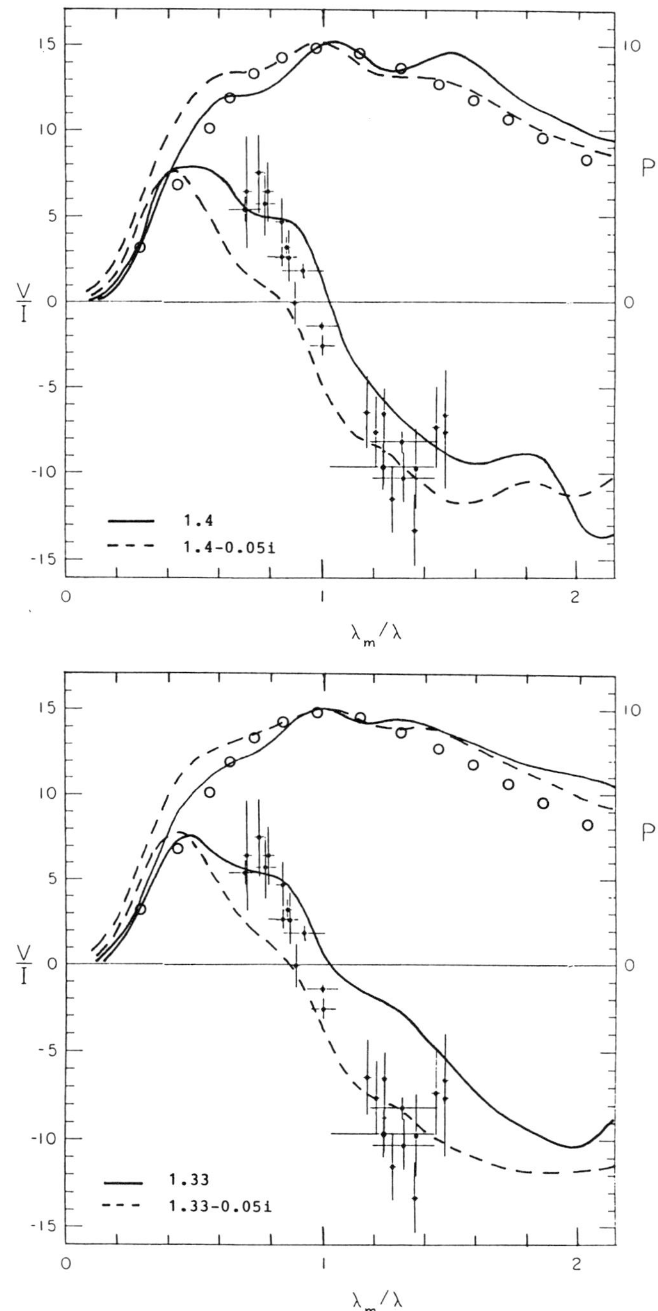

Fig. 8.10a,b *Calculations of $P(\lambda)/P(\lambda_{\max})$ and V/I for infinite cylinders in picket fence alignment with various values of m (= 1.4, 1.4 − 0.05i, 1.33, 1.33 − 0.05i).*

(3) the end points of this distribution are also free parameters, the result being sensitive to the chosen upper limit of $a = 0.25\mu m$, especially for graphite,

(4) the best available data on the optical constants for graphite (Taft and Phillip, 1965) are adjusted for optimising the fit.

Even with such extensive assumptions, our own calculations of the same model have shown that the 'best fitting' solution of the graphite–silicate theory fails to fit the details of the extinction curve to the quite serious extent shown in Fig. 8.8 (Hoyle and Wickramasinghe, 1986).

In general terms the property of the extinction curve rising into the far ultraviolet combined with a high grain albedo would indicate a contribution from small dielectric grains. A schematic distribution of the contributions from 'classical grains' (dielectric cylinders, say with $m = 1.167 - 0.01i$), graphite grains of radius $\sim 0.02\mu m$ and small dielectric spheres with refractive index $m = 1.5$ radius $0.05\mu m$ is shown in Fig. 8.9, together with a curve representing the average values of the observations.

8.4. POLARISATION CONSTRAINTS

A further requirement to be satisfied by a grain model is that it should be able to account for the general features of interstellar polarisation. We require that some fraction of the grains causing visual extinction must be elongated, systematically aligned and give rise to $P(\lambda)$ and V/I curves in accord with the astronomical data (see equations 8.1 and 8.2). Clearly a model such as the one discussed by Mathis et al., which includes only spherical grains, cannot produce polarisation effects.

The general requirements of an elongated grain (in the form of an infinite cylinder) may be examined by calculations for a range of m values as were done by Martin (1973). The computations are made using the formulae set out in Chapter 3. For a given value of refractive index m we assume that the radius a is chosen to give the best fit possible to the data on $P(\lambda)/P(\lambda_{max})$. For the same value of a we plot V/I curves. Calculations for representative values of m are shown in Figs. 8.10 and 8.11. We note the general trend that particles with large values of k are ruled out by the circular polarisation data. Dielectric particles with values of n that are as low as 1.1 also appear to be uncomfortably placed to account for the shape of the $P(\lambda)$ curve. Low values of n lead to linear polarisation curves that are too broad.

8.5. AN ORGANIC/BIOLOGIC GRAIN MODEL

We consider a model of interstellar dust in which individual grains are comprised of clumps of smaller organic units. The primary component within such a clump is taken to be a hollow or porous bacterial particle, closely similar to the model we have discussed before, a particle that we shall define more carefully in Section 8.6. Consistent with the properties of bacteria recovered in freezing nuclei from terrestrial clouds (Jayaweera and Flanagan, 1982), and the organic particles found in carbonaceous meteorites (Pflug, 1982), we take an average radius of a primary unit to be $\sim 0.1\mu m$, about 3 times lower than the average radius of vegetative terrestrial bacteria. A clump of outer radius $\sim 0.5\mu m$ may thus include rather less than $(0.5/0.1)^3$ such units, the precise number depending on the overall degree of

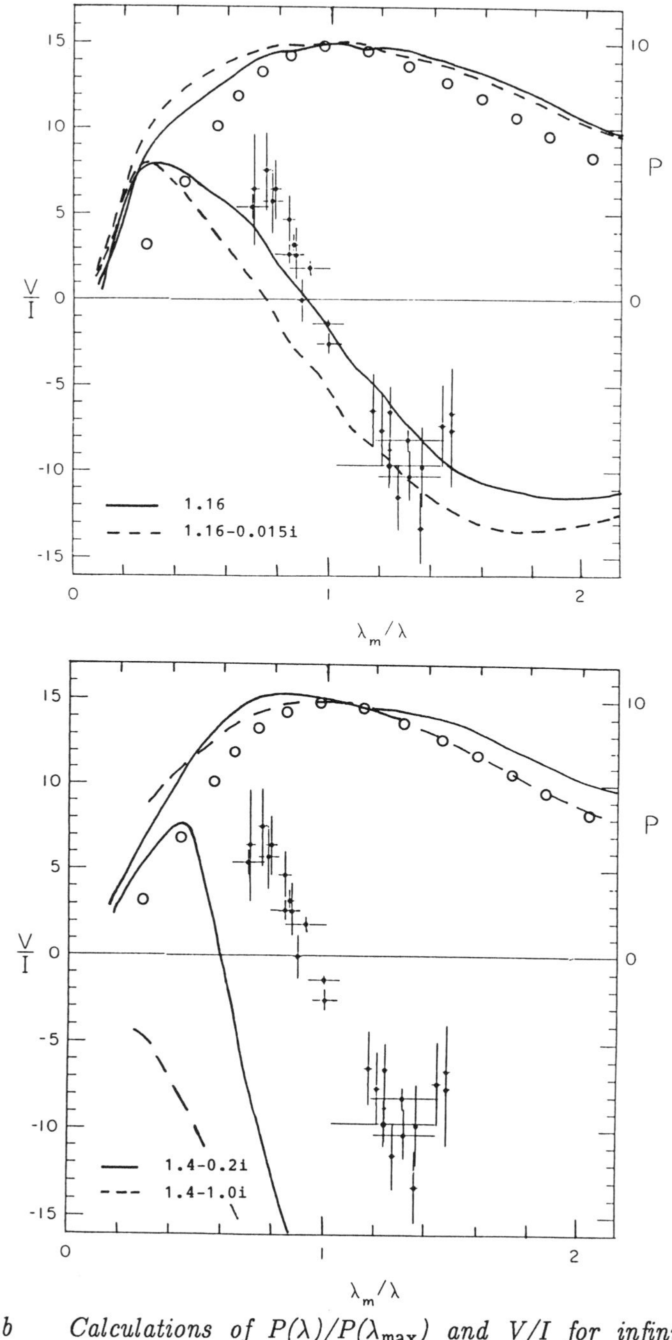

Fig. 8.11a,b *Calculations of $P(\lambda)/P(\lambda_{\max})$ and V/I for infinite cylinders in picket fence alignment for various values of m (= 1.16, 1.16 − 0.015i, 1.4, 1.4 − 1.0i).*

compaction within the clump.

A volume filling factor ~ 1/3 for these particles would lead to ~ 40 particles in a spherical clump of radius 0.5μm and ~ 40 $(a/0.5)^3$ particles in a clump of radius a. If an individual unit (freeze–dried bacterium) has itself a volume filling factor of 1/3, the overall fraction of volume filled in the clumps will be ~ 0.1. An organic clump (or aggregate) of this description would fit well with the properties observed for dust in the coma of Comet Halley. For an average density of condensed polymeric biomaterial of ~ 1.5 g cm^{-3}, the mean density of an aggregate will be ~ 0.15 g cm^{-3}. This value is consistent with reported values for the bulk density of dust in Comet Halley.

The distribution function of grain radii in our model is also linked to data obtained through space–probe sampling of Halley dust (McDonnell et al., 1986). In the best determined part of the mass spectrum of cometary dust (involving relatively large grain sizes) the distribution function for radii was found to correspond to

$$N(a)\,da \propto da/a^{p},\ p \simeq 3.9 \qquad (8.4)$$

Although it is claimed that values of $p < 3.9$ are appropriate for the smallest particles in the distribution, these claims are probably influenced by difficulties of calibrating data from experiments that were not intended to search for small–sized grains. In the present work we assume a cluster distribution

$$N(a)\,da \propto da/a^{3.9},\ a \geq a_c \qquad (8.5)$$

where a_c is some critical cut–off radius. The distribution (8.5) agrees with the observed cometary size distribution for sufficiently large radii.

In addition to the primary aggregates defined above we admit the possibility of two populations of smaller–sized units being associated with the larger grains, and which have fixed proportions by mass. We consider graphite spheres of radius a_1 and condensed polymeric spheres of radius a_2, where a_1 and a_2 are $<<a_c$. The condensed polymeric spheres may be regarded as break-away structures from bacterial grains, and the graphite grains could be organic particles that have been degraded and reduced to free carbon. If the polymeric spheres and graphite spheres have volume fractions f_1, f_2 of the total volume occupied by the primary clumps, the numbers of these units associated with a clump of radius a will be $f_1(a/a_1)^3$ and $f_2(a/a_2)^3$ respectively.

For any specified set of values for the parameters a_c, a_1, a_2, f_1, f_2 we can now proceed to compute the extinction arising from a 'cometary' distribution of sizes defined by equation (8.5), provided we know the optical constants for the various materials that are involved.

Optical Constants. The optical constants for graphite are as before taken from the data of Taft and Phillipp (1965) for electric vector parallel to the basal planes and are denoted by $n_1(\lambda)$, $k_1(\lambda)$. (Here $m = n-ik$ is the complex refractive index, with suffixes assigned to denote various materials.) The optical constants of condensed

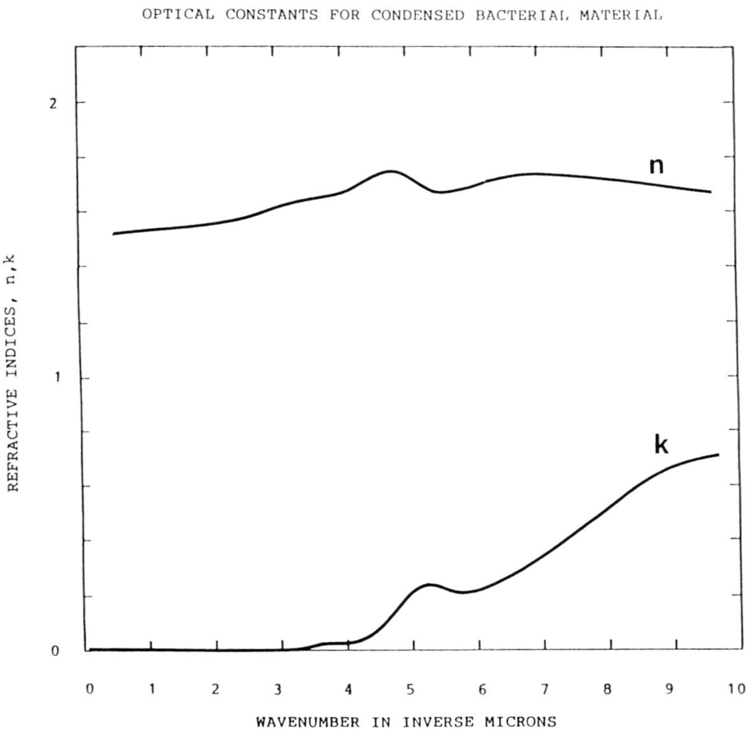

Fig 8.12 *The optical constants of condensed bacterial material from the data of Yabushita et al. (1987) (extrapolated for $\lambda^{-1} < 2.5\mu m^{-1}$).*

bacterial material, designated $n_2(\lambda)$, $k_2(\lambda)$ are taken from a recent publication by Yabushita et al. (1987). Their data was obtained using techniques of vacuum ultraviolet spectroscopy at wavelengths in the range 1030 Å $< \lambda <$ 4000 Å. For wavelengths $2\mu m > \lambda >$ 4000 Å we use an extrapolation according to the empirical formula

$$n_2(\lambda) = 1.5112 + \frac{0.00878}{(\lambda/\mu m)^2}; \quad k_2 = k_2(0.4\mu m) = 0 \tag{8.6}$$

which fits the data adequately over the interval 3000 – 4000 Å and which accords well with the form of the functional dependence of $n(\lambda)$ determined for condensed polyformaldehyde at optical wavelengths (Whittett et al., 1976). The data used for $n_2(\lambda)$, $k_2(\lambda)$ are plotted in Fig. 8.12.

The average optical constants $n_0(\lambda)$ for a fluffy aggregate comprised of bacterial material is given to a good approximation by

$$n_0(\lambda) = 1+[n_2(\lambda)-1]f; \quad k_0(\lambda) = f\,k_2(\lambda) \tag{8.7}$$

where f is the overall volume fraction occupied by material in the condensed polymeric phase. As discussed earlier and consistent with the all available data on the Halley dust grains we take f = 0.1 as an appropriate value in most of the discussion that follows.

The extinction of starlight that would result from an ensemble of grains as defined above may now be calculated for any prescribed set of the parameters in question. If $Q(n,k,a,\lambda)$ denotes the extinction efficiency of a spherical grain of radius a and refractive index $m = n - ik$ at wavelength λ, the average extinction efficiency of the grain ensemble is given by

$$\overline{Q}(\lambda) \propto \int_{a_c}^{\infty} [\pi a^2 Q(n_0, k_0, a, \lambda) + \pi a_1^2 (a/a_1)^3 f_1 Q(n_1, k_1, a_1, \lambda)$$

$$+ \pi a_2^2 (a/a_2)^3 f_2 Q(n_2, k_2, a_2, \lambda)] N(a)\, da \quad (8.8)$$

where N(a) is given by equation (8.5) and the function $Q = Q_{ext}$ is computed using the Mie formulae.

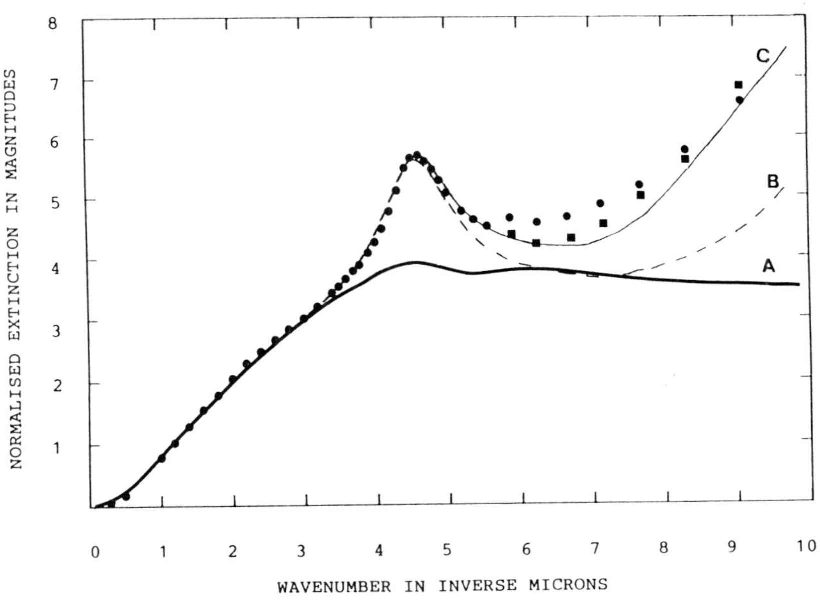

Fig. 8.13 *The average interstellar extinction data (points) compared with calculations for polymeric grain model. Normalisation is to 1.8 mag at $\lambda^{-1} = 1.8\mu m^{-1}$. Curve A: fluffy assembly with 10% of volume filled with organic polymers; curve B: admixture of 0.55% volume fraction of graphite spheres of radius 20 nm; curve C: further admixture of condensed polymeric units radii 12 nm, volume fraction 0.8% (Points represent data from Sapar and Kuusik, 1978).*

The integral was found to converge rapidly to a limit at the upper value of the radius for most cases considered, due to a combination of factors. The approximately a^{-4} dependence of N, and the fact that contributions to Q from porous clumps mainly arose with a dependence a^{-q} with $q < 1$, combined to yield a convergent integral at an upper integration limit of $6\mu m$ in most cases. We accordingly truncated our numerical procedure at $a = 6\mu m$ in the calculations we now report. For the purpose of comparison with relevant astronomical data the numerical values computed from equation (8.8) were normalised according to the relation

$$\Delta m(\lambda) = 1.8\ Q(\lambda)/Q(\lambda_V) \tag{8.9}$$

where $\lambda_V^{-1} = 1.8\mu m^{-1}$. A full range of calculations for this model is discussed elsewhere, and we refer the reader to this work (Hoyle and Wickramasinghe, 1988). Here we present a few representative results.

Curve A in Fig. 8.13 shows the normalised extinction behaviour of a fluffy grain distribution with no additional associated components – i.e. $f_1 = f_2 = 0$. The points are the mean interstellar extinction data. Already we find here that the calculated extinction curve agrees remarkably well with observations over the visual and near infrared wavebands. An admixture of a 0.55% volume fraction of 20 nm radius graphite grains ($f_1 = .0055$, $a_1 = 0.02\mu m$) yields the better agreement of curve B, and a further admixture of ~ 0.8% by volume contribution of condensed polymeric

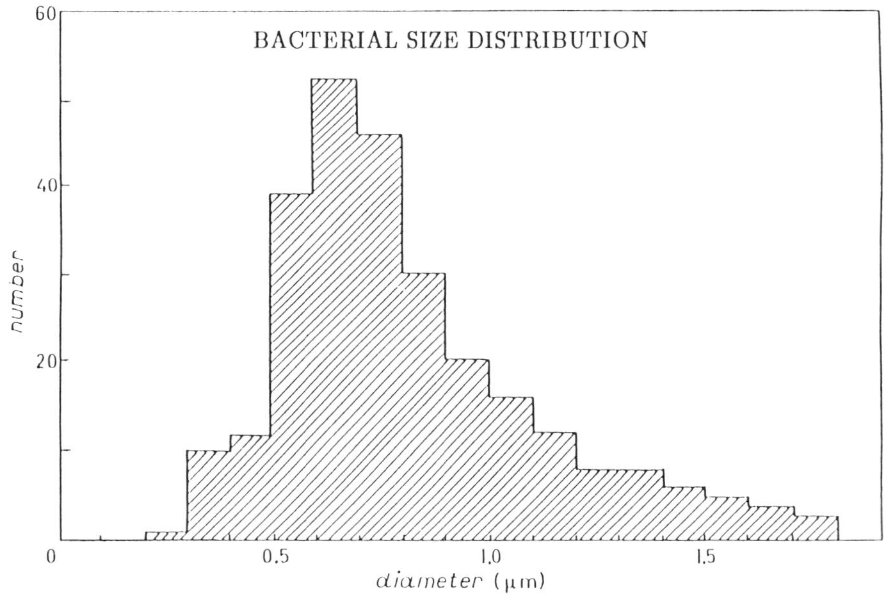

Fig. 8.14 *Size distribution of terrestrial spore–forming bacteria (see Hoyle and Wickramasinghe, 1979).*

particles of radii 12 nm ($f_2 = 0.008$, $a_2 = 0.012\mu m$) yields the overall fit to the data shown in curve C. The correspondences with data seen in Fig. 8.13, particularly for curve C, must be considered to lend strong support to the general class of interstellar grain model comprised of organic polymeric particles supplied from cometary objects.

8.6. ANALYSIS OF A BIOLOGICAL GRAIN MODEL

The extinction curve of a biological/organic grain model described in Section 8.5 involved three basic components: hollow bacterial grains modelled as spheres of low $|n-1|$, graphite particles of radius 0.02μ and condensed organic polymers of radius $0.012\mu m$. The three components were integrated into a single model of grains derived from comets, and the computational treatment had the advantage of using optical constants for all the components that were obtained from laboratory measurements. The defect in our earlier treatment, however, was that our calculations were confined to spherical particles, so that comparisons with polarisation data were not possible.

We now discuss an equivalent three component bacterial model including hollow cylindrical particles to represent dehydrated cells. The treatment we follow is essentially as we have described in an earlier paper (Jabir et al., 1983). We use a size distribution for bacteria as determined in the laboratory, measured optical constants of graphite and a grid of n, k values to explore the extinction properties of hollow bacterial cells. The model, it may be noted, is defined mainly from refractive index criteria and is therefore separable from a specific biological or organic interpretation. We return to the question of more decisive organic/biogenic identifications from infrared spectroscopy in the next chapter.

Fig. 8.14 shows the observed size distribution of a representative sample of spore-forming bacteria (cross–sectional diameters of rods and spheres (see Hoyle and Wickramasinghe, 1979)). Terrestrial bacteria are bound by rigid outer cell walls comprised mainly of polysaccharides and lipoproteins, with an interior that is comprised of a rich variety of biochemicals and water. Water makes up about 80% by volume of a bacterium. We estimate that interstellar cloud conditions lead to evaporation of water to 60% by volume, assuming the bulk of interstellar C is tied up as bacteria. Thus the mean refractive index of a freeze–dried bacterium is made up of 20% organic material with $m = 1.5$, 20% water with $m = 1.3$ and 60% vacuum with $m = 1$, giving a mean value

$$\overline{m} = 1 + 0.2 \times 0.5 + 0.2 \times 0.3 = 1.16$$

The Three Component Model. We now consider a three component grain model defined as follows:

Component 1: Bacterial grains in the form of long hollow needles. The average refractive index n is taken to be $m = 1.16 - ik$ with k varied in the range 0 to 0.025. The distribution function of radii is taken as in Fig. 8.14.

Component 2: Graphite spheres of radius $a = 0.02\mu m$.

Component 3: Dielectric spheres of refractive index m = 1.5, radius 0.04μm.

To specify the relative proportions of the three particle species, we define two parameters w_g, w_d, such that the contributions from the individual species to the extinction at $\lambda^{-1} = 1.8\mu m^{-1}$ are in the ratio:

hollow needles (1) : graphite (2) : dielectric spheres (3)
$= 1.0 : w_g : w_d$

To specify species (1) more fully we note that the calculations are carried out for cylinders of length ℓ long compared with radius, $\ell >> a$. The distribution function of cylinder radii a is taken to be given by the size–distribution function n(a) shown in Fig. 8.14. The axes of the grain are assumed to be randomly oriented in space. We consider also a further refinement that includes the effect of Si, Fe atoms condensed within or on the grains. If a significant fraction of the total interstellar complement of these elements, say 50%, is condensed as SiO_2 and Fe, along with C, N, O, we estimate an average absorptive index k = 0.015 at optical wavelengths. The complex refractive index of the grain material (for the bulk of grains in component (1)) is thus taken as m = 1.16 – 0.015i.

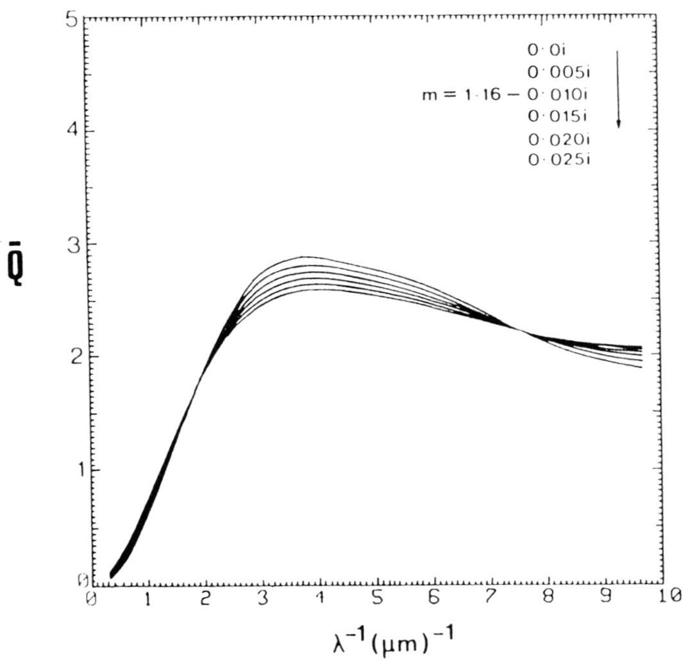

Fig. 8.15 *Average $\overline{Q}(\lambda)$ for bacterial size distribution defined in Fig. 8.14 for m = 1.16–ik, k = 0 (0.005) 0.025.*

Computational Results. We use the Mie formulae for spheres to compute the extinction properties of the spherical grain species (2) and (3), and the corresponding formulae for infinite cylinders (set out in Chapter 2) to compute the properties of the cylindrical grain species (1). For the cylindrical grain ensemble the average extinction coefficient $\overline{Q}^{(1)}_{ext}$ is given by

$$\overline{Q}^{(1)}_{ext}(\lambda,m) \propto \int_0^\infty 2a\ell[Q_{//}(m,a,\lambda)+2Q_{\perp}(m,a,\lambda)]n(a)da \Big/ \int_0^\infty n(a)\,2a\ell da \qquad (8.11)$$

where $Q_{//}$, $Q_{\perp}$ are the extinction cross–sections of a cylindrical grain of refractive index m for electric vector respectively parallel and perpendicular to the cylinder axis, and n(a) is the size distribution function defined by the histogram in Fig. 8.14. Here a is the radius and ℓ is the length of the cylinder. Fig. 8.15 shows the resulting $\overline{Q}$ curves for $m = 1.16 - ik$ with $k = 0(0.005)0.025$. Note that $\overline{Q}$ has a nearly exactly linear shape for $0.8 \lesssim \lambda^{-1} \lesssim 2.4\mu m^{-1}$, particularly for $k < 0.015$. The linear shape of $\overline{Q}(1/\lambda)$ in the visual spectral region turns out to be an asset for accounting for the observed behaviour of interstellar extinction over approximately this range of visual wavelengths.

For computing the extinction coefficient $Q^{(2)}_{ext}$ of the graphite component (2) we use the optical constants for graphite measured in the laboratory by Taft and Phillipp (1965) and the Mie formulae for spheres of radius 0.02μm. The solid curve

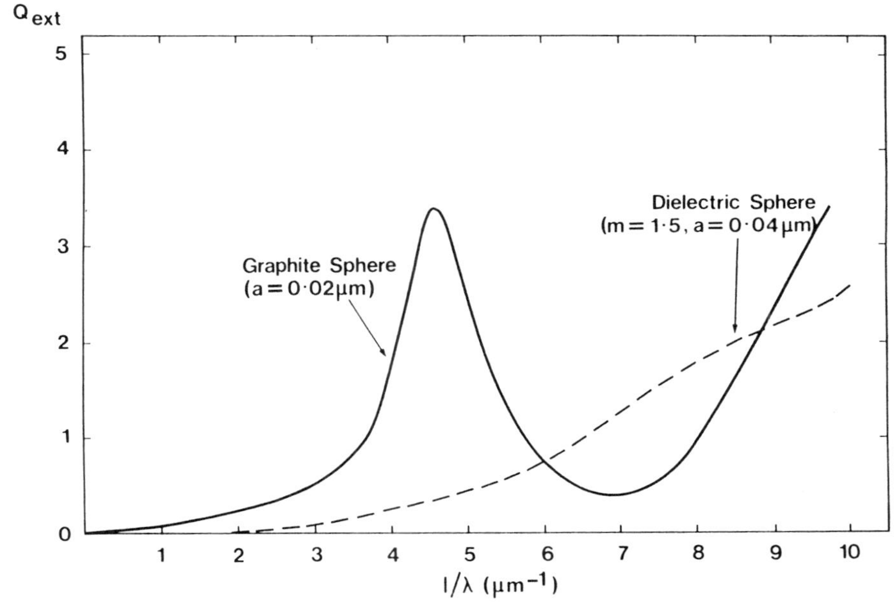

Fig. 8.16 *Extinction curve for graphite spheres of radii 0.02μm (solid curve). Extinction curve for dielectric spheres of radii 0.04μm and refractive index m = 1.5 (dashed curve).*

of Fig. 8.16 shows the results of our computations for this case. We note the conspicuous symmetrical hump centred on ~ 2175 Å which is coincident with the astronomical data as we have earlier noted. The dashed curve of Fig. 8.16 shows the extinction $Q_{ext}^{(3)}(\lambda)$ of the dielectric component (3) with m = 1.5, a = 0.04μm, also calculated using the Mie formulae.

TABLE 8.3

The best fit to the whole interstellar extinction curve

(1) Bacterial size distribution ($\bar{m}$ = 1·16–0·015i)

(2) Graphite, radius 0·02 μm w_g = 0·1

(3) Dielectric particles, a = 0·04 μm, w_d = 0·01, m = 1·5–0·0i
Normalisation to 1·8 mag at λ^{-1} = 1·8 μm^{-1}

$\frac{1}{\lambda}$ (μm^{-1})	Δm	$\frac{1}{\lambda}$ (μm^{-1})	Δm
0·30	.118	4·50	5·486
0·50	.257	4·55	5·506
1·00	.777	4·60	5·504
1·40	1·286	4·65	5·486
1·60	1·546	4·70	5·445
1·80	1·800	4·76	5·380
2·00	2·041	4·83	5·282
2·27	2·340	4·90	5·173
2·40	2·472	5·00	5·016
2·60	2·659	5·10	4·873
2·90	2·906	5·20	4·744
3·00	2·981	5·30	4·628
3·23	3·165	5·40	4·525
3·47	3·372	5·81	4·285
4·00	4·127	6·00	4·246
4·10	4·408	6·29	4·239
4·20	4·766	6·60	4·340
4·30	5·071	7·41	5·043
4·40	5·345	7·69	5·383
		8·85	7·146

The combined extinction behaviour of the three–component model is calculated according to the expression:

$$\bar{Q}(\lambda) = \frac{\bar{Q}_{ext}^{(1)}(\lambda)}{\bar{Q}_{ext}^{(1)}(\lambda_0)} + \frac{w_g Q_{ext}^{(2)}(\lambda)}{Q_{ext}^{(2)}(\lambda_0)} + \frac{w_d Q_{ext}^{(3)}(\lambda)}{Q_{ext}^{(3)}(\lambda_0)}, \qquad (8.12)$$

where $\lambda_0^{-1} = 1.8\mu m^{-1}$ and w_g and w_d are the weighting parameters defined earlier.

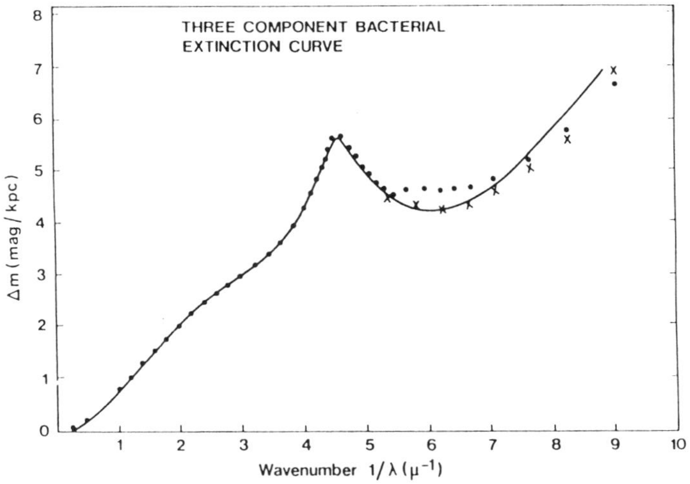

Fig. 8.17a *Comparison of model extinction curve with* $m_1 = 1.16 - 0.015i$, $w_g = .10$, $w_d = .01$, $m_3 = 1.5$ *with data points denoting astronomical observations. Filled circles are average data (Sapar and Kuusik, 1978); crosses denote data in the ultraviolet by Bless and Savage (1972). Normalisation is to* $\Delta m = 1.8$ *mag at* $\lambda^{-1} = 1.8 \mu m^{-1}$.

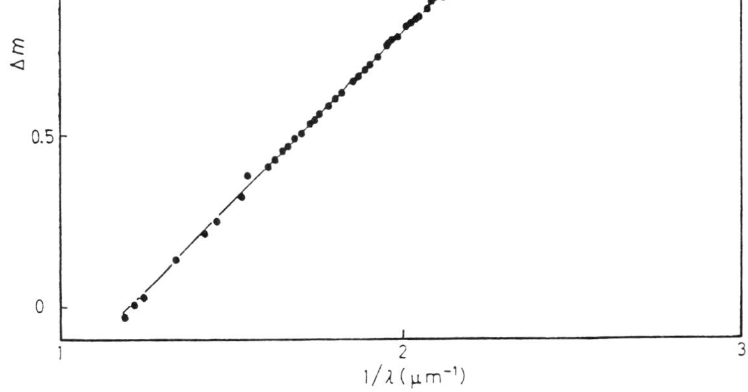

Fig. 8.17b *Details of fit of the model in Fig. 8.17a to Cygnus extinction data (Nandy 1964).*

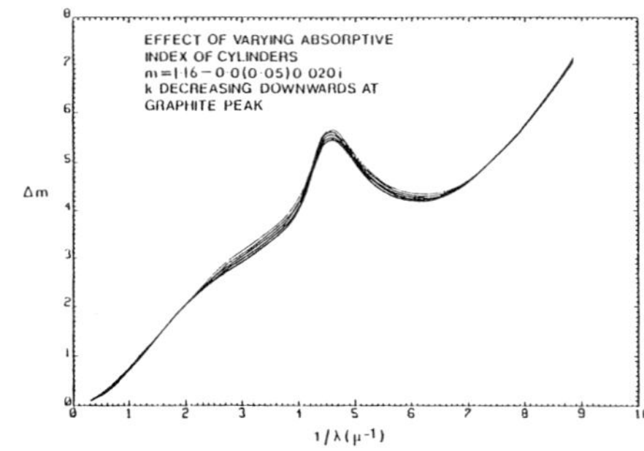

Fig. 8.18 *Effect on the normalised extinction curve as absorptive index of hollow cylinders is varied. Stack of curves is for $k = -Im\ (m_1) = 0.0\ (0.05)\ 0.02$, k decreasing downwards at graphite peak. All other parameters are fixed at best–fit values as defined in Fig. 8.17.*

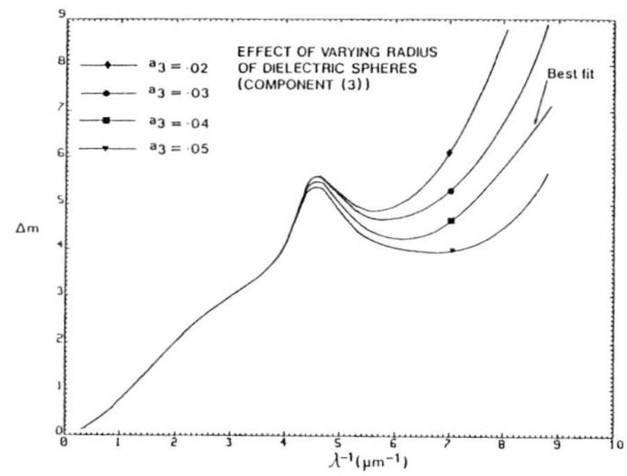

Fig. 8.19 *Effect on the normalised extinction curve of varying the radius of the small dielectric spheres a_3, $a_3 = 0.02\ (0.01)\ 0.05$. All other parameters are fixed at the best–fit values as defined in Fig. 8.17.*

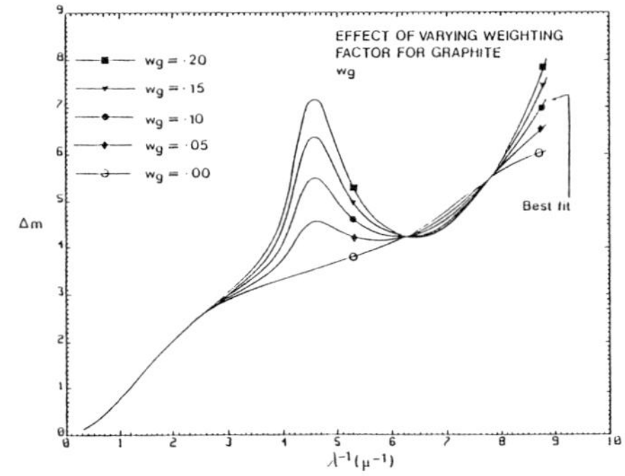

Fig. 8.20 *Effect on the normalised extinction curve of varying weighting factor for graphite spheres, w_g, $w_g = 0.0\ (0.05)\ 0.2$. All other parameters are fixed at the best–fit values as defined in Fig. 8.17.*

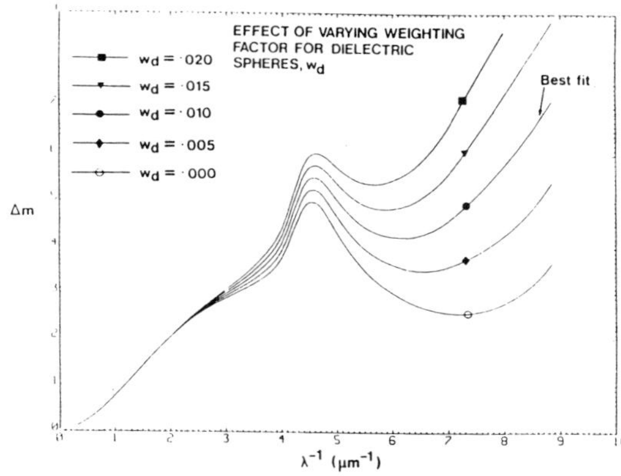

Fig. 8.21 *Effect on the normalised extinction curve of varying weighting factor for small dielectric spheres, w_d, $w_d = 0.0\ (0.005)\ 0.02$. All other parameters are fixed at the best–fit values as defined in Fig. 8.17.*

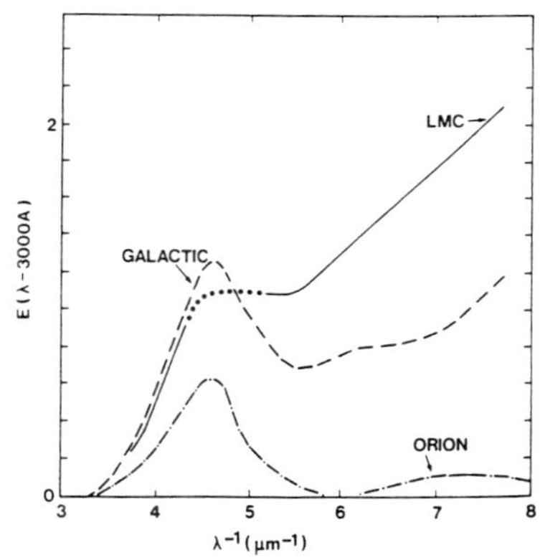

Fig. 8.22 *Colour excesses $E(\lambda - 3000$ Å$)$ for the mean galactic extinction, the Orion stars in the galaxy, and the Large Magellanic Clouds.*

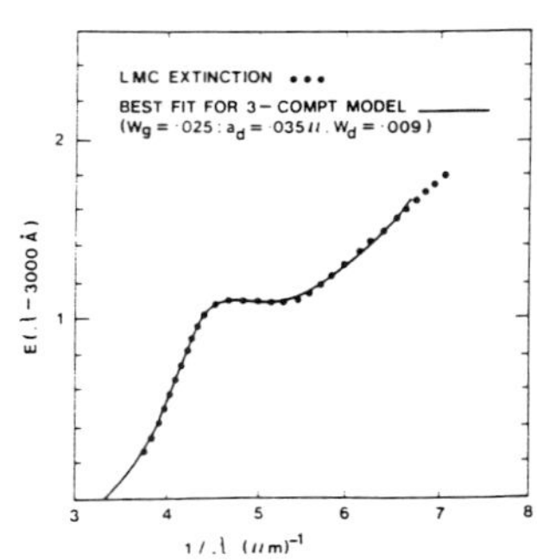

Fig. 8.23 *Best fit to LMC data for 3–component bacterial grain model with the following values of the model parameters: $w_g = 0.25$, $a_g = 0.02\mu m$; $w_d = .009$, $a_d = 0.35\mu m$. Normalisation at 2200 Å peak to agree with observations.*

Considering the observational data over the entire waveband $0 \lesssim \lambda^{-1} \lesssim 10\mu m^{-1}$ the best fit to the mean extinction curve was obtained for $m_1 = 1.16 - 0.051i$, $w_g = .10$, $w_d = .01$ and $m_3 = 1.5$. Our calculation in a tabular form is given in Table 8.3.

The correspondence with the data for this case is shown in Fig. 8.17a,b. (The normalisation is to $\Delta m = 1.8$ at $\lambda^{-1} = 1.8\mu m^{-1}$). The best fitting 'conventional' grain model of Mathis et al. involving a mixture of graphite and silicate grains, produced as its closest agreement, a considerably inferior fit to the one obtained here.

Figs. 8.18 – 8.21 show the effect of variations from our 'best fit' values of each of the several free parameters in our model. We note that significant variability of the extinction curve is more or less confined to wavelengths shortward of 3300 Å, exactly as the observations seem to imply (Fig. 8.22). In particular we note that a weak or smeared out graphite peak as is found, for instance, in the Magellanic Clouds, is obtained by lowering the weight of graphite w_g. The remarkably close fit to this data obtained by such a procedure is shown in Fig. 8.23. A weaker than average far–ultraviolet extinction at ~ 1000 Å (as for σ Sco) arises from lower values of w_d and/or w_g. Our best fit for this case is shown in Fig. 8.24. All the observed properties with respect to deviations from the mean galactic extinction law could be readily obtained by varying the free parameters in the present model.

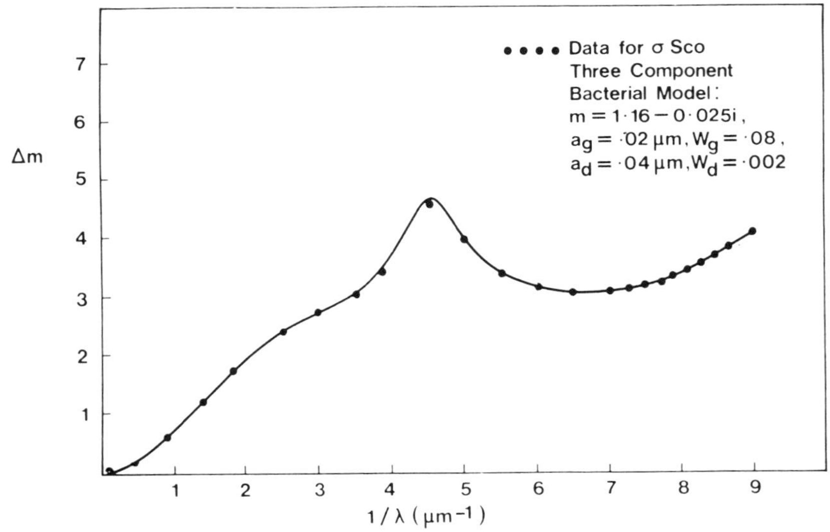

Fig. 8.24 *Best fit to σ Sco data for 3–component bacterial grain model with the following values of the model parameters: $w_g = .08$, $a_g = .02\mu m$, $w_d = .002$, $a_d = .04\mu m$ and $m = 1.16 - 0.025i$ for refractive index of hollow cylinders.*

The Albedo of Bacterial Grains. A further requirement to be satisfied by any model of the interstellar grains is that it should possess an albedo

$$\bar{\gamma} = <Q_{sca}>/<Q_{ext}> \tag{8.13}$$

and a phase parameter g = $<\cos\theta>$ (g = 0 for isotropic scattering; g = 1 for forward scattering), which are *generally* consistent with observational data. The relevant observational data are the measurements of scattered light from reflection nebulae as well as the diffuse galactic light, which were discussed in Chapter 4. Fig. 8.25 shows the $\bar{\gamma}(\lambda)$ curve for our 3–particle model which gives the best fit to the mean galactic extinction law. Our computed g values range from 0.8 to 0.6. The points show the g and γ values implied by the observations according to Lillie and Witt (1976). The observations are seen to be in overall general agreement with our model, except that the albedo value at λ = 2175 Å is a little too high. It should be noted, however, that the inclusion of a contribution from aromatic molecules, which is to be expected, would lead to the required reduction of the minimun γ value at just this wavelength. If all the graphite in this model is replaced by aromatic molecules which are pure absorbers over the 2175 Å band the minimum albedo is ~ 0.35 as indicated by the dashed curve in Fig. 8.25. Such a situation will be relevant to the model we shall discuss at the end of the present chapter.

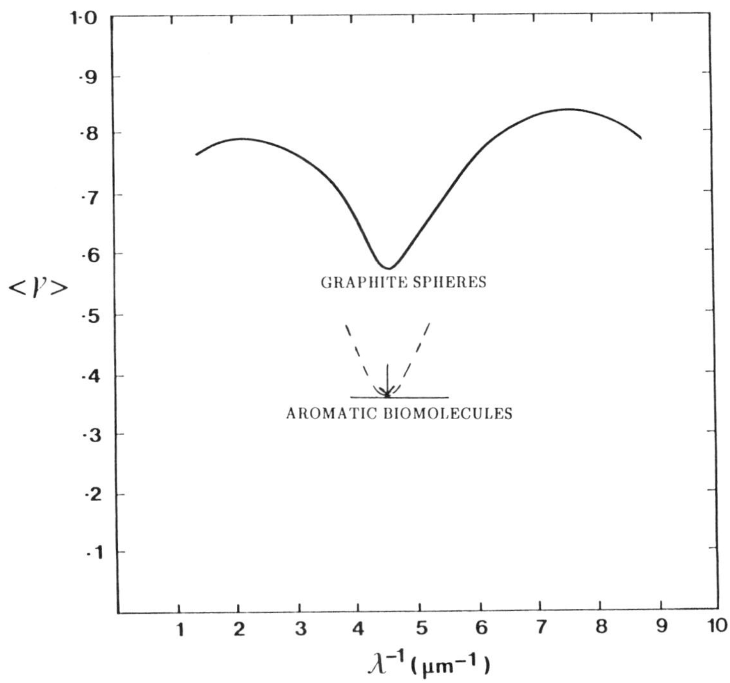

Fig. 8.25 *The albedo of the best fit 3 component bacterial model: Solid curve includes graphite; dashed curve corresponds to 100% replacement of graphite with 2175 Å aromatic absorber.*

Polarisation. In the above 3–component model polarisation can arise only from the cylindrical grains (component (1)). For such grains, if they are *all* systematically aligned, the average polarisation is computed according to

$$P(\lambda) \propto \frac{\overline{Q^{//}_{ext}} - \overline{Q^{\perp}_{ext}}}{\overline{Q^{//}_{ext}} + \overline{Q^{\perp}_{ext}}} \tag{8.14}$$

where the averages of the Mie coefficients are done separately for electric vector parallel ($Q^{//}_{ext}$) and perpendicular ($Q^{\perp}_{ext}$) to the cylinder axis. The normalised plot $P(\lambda)/P(\lambda_{max})$, where λ_{max} is chosen at the maximum of $P(\lambda)$, is given as the solid curve in Fig. 8.26. The points are the average observations of Coyne (1974). The wavelength dependence of the bacterial grain ensemble of low (n–1) (eg. m = 1.16 – 0.015i) is seen to be inconsistent with the average polarisation curve, being too broad in its profile compared with the observations.

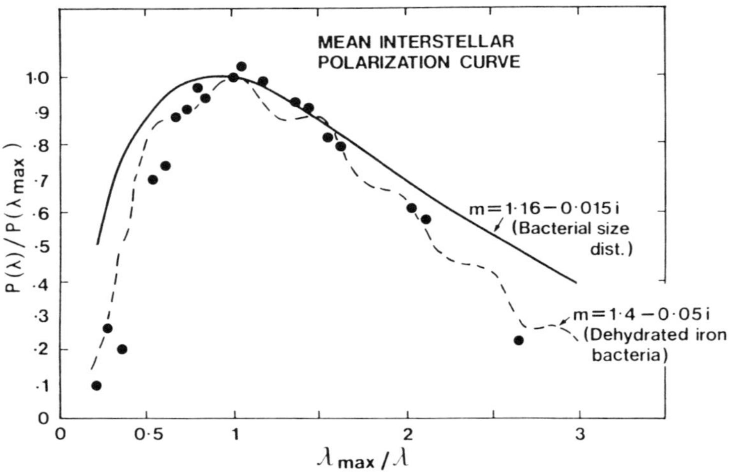

Fig. 8.26 *The observed mean interstellar polarisation curve (dots) compared with model calculations. Dashed curve is polarisation for entire ensemble of dehydrated bacteria, solid curve is for dehydrated iron bacteria with m = 1.4 – 0.05i, a = .33μm.*

We now discuss a further refinement to deal specifically with alignment and polarisation problems. We discuss the possibility that interstellar Fe and Si might condense in the form of skins of more or less uniform thickness over the entire distribution of cylinder radii, the iron condensing mainly in metallic form, the Si mainly as SiO_2. It is this material that could give bacterial grains non–zero k values in the visible spectral region. We assume further that the elemental abundance ratios in the mantles occur in their full cosmic proportions (Cameron, 1970) resulting in a volume ratio of iron to siliceous material in the entire grain population of ~ 0.13 : 0.44. This ratio will be appropriate for the composition of skins on grains of all sizes. The ratio of carbonaceous material to siliceous material for the overall grain mixture would be 5.2 : 0.44 by volume provided all the interstellar C is in the form of grains of average density. (If only a fraction ξ of the carbon relative to Si is in grains this ratio is 5.2 ξ : 0.44.)

Grains that become preferentially aligned by the Davis–Greenstein process in a galactic magnetic field are those that have a higher mass fraction of iron–bearing mantle material and hence would tend to be the smaller particles. To calculate the polarisation behaviour of these grains, we suppose that a long cylindrical grain of radius a has a skin of thickness t comprised of a mixture of iron and SiO_2 in the volume ratio Fe : SiO_2 = 0.13 : 0.44. If n is the average refractive index of the central hollow cylinder, n_1–ik_1 is the refractive index of iron (Lenham and Treherne, 1966) and n_2 that of SiO_2 the average refractive index of the composite cylinder is $\bar{m} = \bar{n} - i\bar{k}$, where

$$\bar{n} = 1 + (1-f)(n_0-1) + f[\frac{13}{57}(n_1-1) + \frac{44}{57}(n_2-1) ,$$

$$\bar{k} = \frac{13f}{57} k_1 , \tag{8.15}$$

with f given by

$$f = \frac{(t+a)^2-a^2}{(t+a)^2} . \tag{8.16}$$

Calculations were carried out for various values of the radius a, thickness of coating t, and the cylinder refractive index n_0. $\Delta m_p(\lambda)$ is calculated and normalised to give Δm_p = 10 at the peak; q = V/I is calculated and multiplied by an arbitrary constant to make for the best overall agreement with the observed points.

With values of n_0 chosen in the range 1.15 – 1.17 we explored the (a–t) parameter space with a view to obtaining reasonable fits to the polarisation data. No agreement was discovered to be possible for a > 0.125μm, as is illustrated in the results plotted in Fig. 8.27. The worst discrepancies were found in the wavelength dependence of V/I. We next fixed the radius at a = 0.1μm, the average refractive index at n_0 = 1.15, and determined the effect of varying the Fe–SiO_2 coating thickness t. From Fig. 8.28 we note that the best agreements occur for the case t = 0.006μm. Fig. 8.29 demonstrates the effect of varying the value of n_0 keeping t at 0.006μm. For a cylinder of radius a = 0.1μm the best agreement occurs for n_0 = 1.15.

In our preferred solutions indicated by the solid curves of Figs. 8.28 and 8.29, we have a thickness of mantle ~ 0.006μm. From arguments relating to the constancy of condensation rates from the gaseous phase, one could infer that this value of mantle thickness must be applicable for grains of all sizes. For particles of typical radius 0.5μm, which are in the size range that makes the major contribution to the visual extinction, the value to t = 0.006μm together with n_0 = 1.15 gives $\bar{n}$ = 1.166, $\bar{k}$ = 0.016 at λ = 5500 Å, using equations (8.15) and (8.16). Such $\bar{n},\bar{k}$ values are remarkably close to the refractive index data used for calculating the extinction curve of Fig. 8.17.

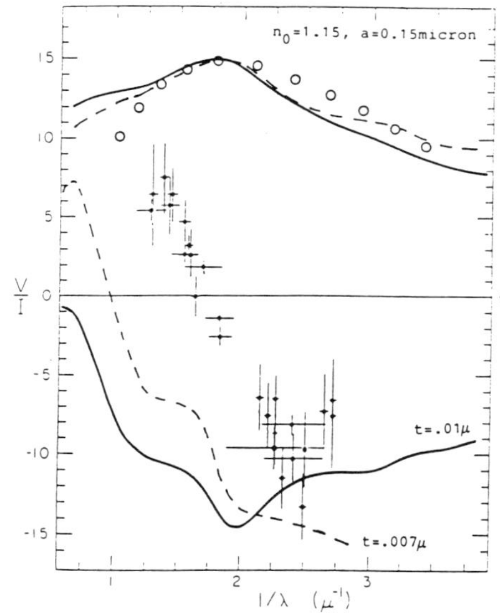

Fig. 8.27 *Normalised curves for linear and circular polarisation by a hollow cylinder of radius 0.15μm and refractive index $n_0 = 1.15$ with Fe–SiO$_2$ skins of thickness 0.007, 0.01μm. The open circles are the mean linear polarisation data (Serkowski, 1973) and the points with error bars are circular polarisation data (Martin, 1974; Kemp and Wolstencroft, 1972; Shapiro, 1975).*

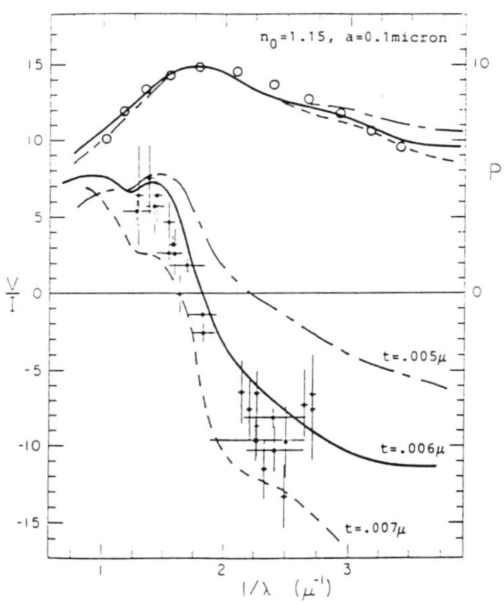

Fig. 8.28 *Normalised curves for linear and circular polarisation by a hollow organic cylinder of radius 0.1μm and refractive index $n_0 = 1.15$ with Fe–SiO$_2$ skins of thickness 0.005, 0.006, 0.007μm. Points are the same as in Fig. 8.27.*

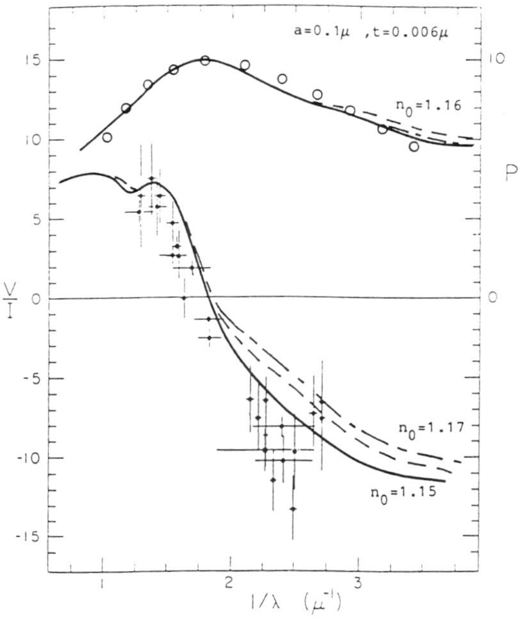

Fig. 8.29 *Normalised curves for linear and circular polarisation by a hollow organic cylinder of radius 0.1μm with a coating of thickness $t = 0.006\mu m$ for values of n_0 = 1.15, 1.16, 1.17μm. Points are the same as in Fig. 8.27.*

The above considerations would seem to imply that our size distribution of bacterial hollow particles must include a substantial fraction of grains that have radii close to ~ 0.1μm. Although our original size distribution of spore–forming bacteria (Jabir et al., 1983) contained only ~ 1% of grains in the radius range 0.1 – 0.15μm, the actual fraction for any realistic distribution of grains is expected to be much higher. Organic structures recovered from the Murchison meteorite, for instance, have been found to include a considerable proportion of grains of radius ~ 0.1μm (Pflug, 1984). Furthermore, nutrient–starved cultures of bacilli are known to possess radii peaking in the same general range ~ 0.08 – 0.12μm. The inclusion of such particles in the size distribution, sufficient to contribute ~ 10% of the total visual extinction, would not alter the possibility of fits to the extinction curve as close as that shown in Fig. 8.17. Indeed for the organic model that led to the agreement in Fig. 8.13, the proportionality $n(a) \propto a^{-3.9}$ led to about the required amount of 0.1 – 0.15μm radius grains.

A final requirement to be satisfied is that particles of radii 0.1 μm with SiO_2–Fe mantles are selectively aligned to a high degree by the galactic magnetic field. Estimates of the magnetic field strength needed to produce adequate alignment of 'standard' elongated paramagnetic grains of any composition have been found to be an order of magnitude or more in excess of ~ 10^{-6} G, the strength of the galactic magnetic field, as we saw in Chapter 5. The same difficulty of alignment will persist for the hollow organic grain model excepting the smaller particles of the type considered here which have a high mass ratio of Fe. We could reasonably argue that such particles are either ferromagnetic or superparamagnetic, in which case they may be aligned to a high degree even with fields as low as 10^{-6} G.

Mass–Density of Grains. The average values of the mass extinction coefficient at $\lambda^{-1} = 1.8\mu m^{-1}$ for the several components of our composite best fitting 3–component grain model are as follows:

(1) $\kappa_{\text{hollow needles}} = 2.81 \times 10^4 \text{ cm}^2 \text{ g}^{-1}$

(2) $\kappa_{\text{graphite spheres}} = 3.47 \times 10^4 \text{ cm}^2 \text{ g}^{-1}$

(3) $\kappa_{\text{dielectric spheres}} = 1143.75 \text{ cm}^2 \text{ g}^{-1}$

These values are calculated from the Mie formulae as described earlier, and with $\rho_s = 0.8, 2.25, 1.6 \text{ g cm}^{-3}$ respectively assumed for the three grain species. The latter two values are direct experimental values for graphite and organics; the former value 0.8 g cm^{-3} for the hollow cylinders takes account of the iron and metal component of the grain, and for the vacuum cavity within. Since astronomical observations give a total extinction of 1.8 mag/kpc at $\lambda^{-1} = 1.8\mu m^{-1}$ in the solar vicinity, the partitioning of this total between the three components for the values of w_g and w_d in Table 8.3 is: 1.62 mag/kpc from hollow needles, 0.162 mag/kpc from graphite spheres, 0.016 mag/kpc from dielectric spheres, with corresponding mass densities given by

$$\begin{aligned} \rho_{\text{hollow needles}} &= 18.66 \times 10^{-27} \text{ g cm}^{-3} \\ \rho_{\text{graphite}} &= 1.51 \times 10^{-27} \text{ g cm}^{3} \\ \rho_{\text{dielectric spheres}} &= 4.53 \times 10^{-27} \text{ g cm}^{-3} \end{aligned} \tag{8.18}$$

The total grain density is then 2.47×10^{-26} g cm^{-3}. With a total hydrogen density in the solar neighbourhood $\rho_{H,H2} = 2 \times 10^{-24} \text{ g cm}^{-3}$ we now obtain

$$\frac{\rho_{\text{dust}}}{\rho_{H,H2}} = 1.24 \times 10^{-2}, \tag{8.19}$$

implying the condensation of more than 50% of the total amount of CNO elements that is present in the interstellar gas, assuming solar abundances (Cameron, 1970).

8.7. REFINEMENTS TO BIOLOGICAL EXTINCTION MODEL

In an attempt to refine the 3–particle extinction models discussed above we recently used the rigorous Kerker–Matijević scattering formulae for hollow infinite cylinders (coaxial double cylinders with vacuum cavity) to recalculate the $\overline{Q}(\lambda)$ curves of Fig. 8.15. We also used laboratory measurements of the optical constants of silica and of iron to compute the extinction efficiencies of iron whiskers of radii 0.01μm and silica spheres of radii 0.03μm (Wickramasinghe et al, 1991).

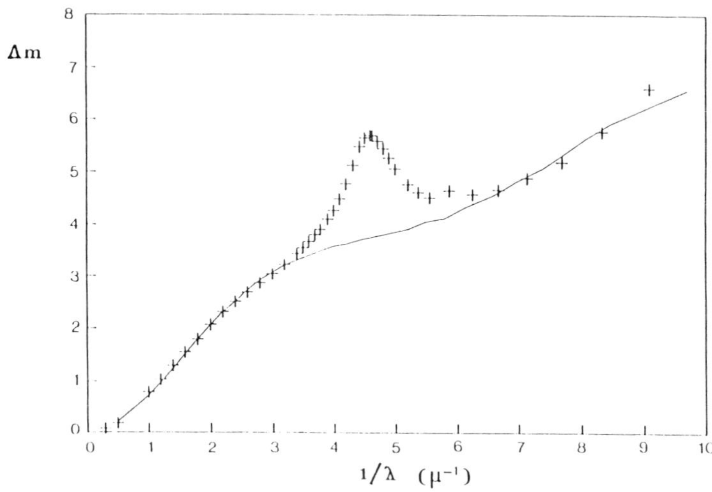

Fig. 8.30 *Normalised extinction curve for a mixture of hollow organic cylinders, iron whiskers (a = 0.01μm) and silica (spheres (a = 0.03μm)) in the mass ratios 2.37 : 0.094 : 1. The points are the average interstellar extinction data compiled by Sapar and Kuusik (1978). Normalisation is to $\Delta m = 1.8$ mag at $\lambda^{-1} = 1.8\mu m^{-1}$.*

The curve in Fig. 8.30 shows the calculated normalised extinction for a mixture of bacterical grains, silica spheres (radius 0.03μm) and randomly oriented iron whiskers (radius 0.01μm) in the mass ratio 2.37 : 1 : 0.094. These values of mass ratios are in good accord with cosmic abundances (Cameron, 1970) provided that silicon is taken to be selectively depleted (condensed) with respect to C, N, O by a factor of 1.8. We note that the correspondence of the calculated extinction in Fig. 8.30 to the observations, excluding the 2175 Å absorption feature, is entirely satisfactory. The latter feature could of course be fitted by introducing a spherical graphite component as discussed earlier, but we now prefer to consider a model involving clusters of aromatic molecules, each cluster being comprised of 50–100 atoms. Such clusters may be regarded as the most stable and long–lasting degradation products of bacterial grains. Further reasons for this model will emerge in Chapter 9.

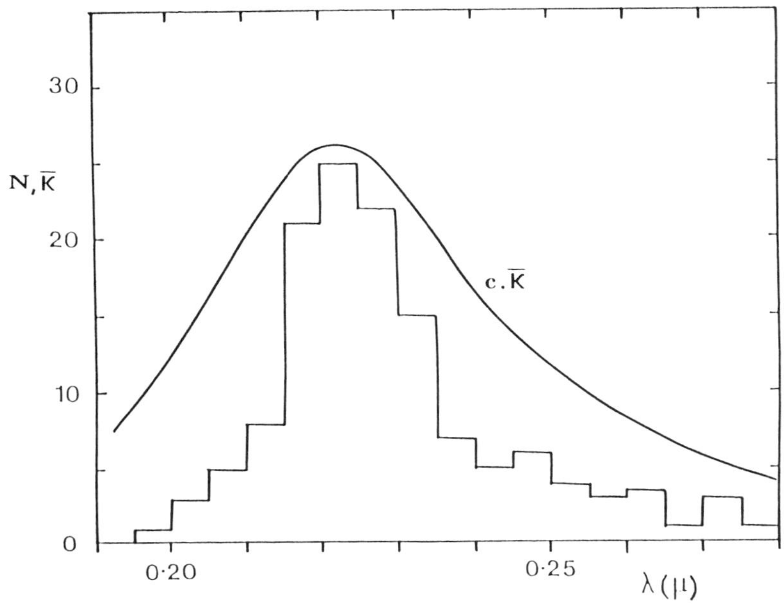

Fig. 8.31 *Histogram shows the distribution of principal ultraviolet absorption peaks for a set of naturally occurring aromatic molecules of types listed in Table 9.5. The curve is the calculated mean absorption profile for the ensemble.*

The distribution of ultraviolet absorption peaks in 115 aromatic molecules found in biomaterial is shown in histogram of Fig. 8.31. The solid curve is the computed average opacity curve for this ensemble assuming that the absorption curve for each individual molecule is similar in shape to that of quinazoline isomers (Fig. 8.7). Fitting the band profile in Fig. 8.31 to the discrepant region of Fig. 8.30 leads to the situation depicted in Fig. 8.32. The heavy curve in Fig. 8.32 shows the result of combining our extinction solution in Fig. 8.30 with an appropriate quantity of aromatic molecules in the form of small clusters. Adopting an average mass extinction coefficient of 500,000 $cm^2\ g^{-1}$ at the centre of the 2175 Å band we require less then 10 percent of the available interstellar carbon to be in this form.

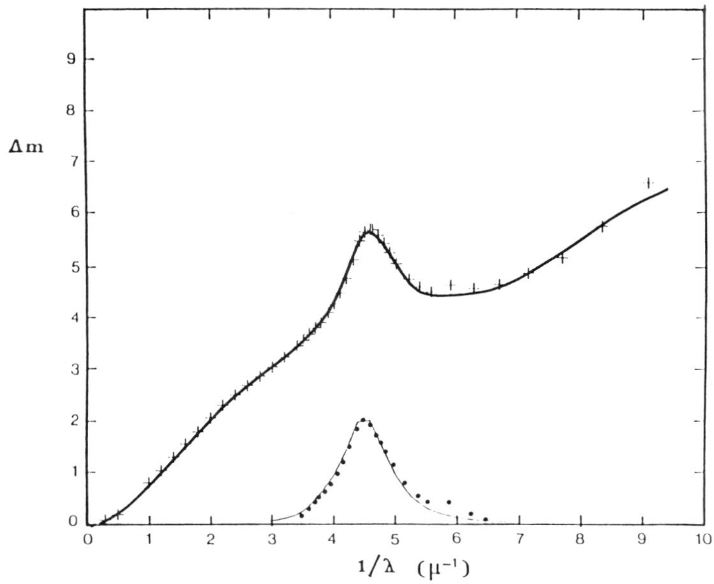

Fig. 8.32 *The filled circles (points) are the excess interstellar extinction values over the curve calculated in Fig. 8.30. The thin solid curve is the absorption due to aromatic molecules occurring in form of free clusters. The heavy solid curve is the total extinction curve including the contribution from aromatic clusters. The observations and the normalisation are the same as in Fig. 8.30.*

References

Cameron, A.G.W., 1970, *Sp. Sci. Rev.*, **15**, 121.

Clegg, R.E.S. and Bell, R.A. 1973, *Mon Not. Roy. Astr. Soc.*, **163**, 13.

Coyne, G.V., 1974, in T. Gehrels (ed.) *Planets, Stars and Nebulae*, University of Arizona Press.

Duley, W.W., 1984, *Q. Jl. Roy. Astr. Soc.*, **25**, 109.

Gilra, D.P., 1971, *Nature*, **229**, 237.

Greenberg, J.M., 1968, in *Stars and Stellar Systems*, **Vol. VII**.

Greenberg, J.M. and Shah, G.A., 1969, *Physica*, **41**, 92.

Hoyle, F. and Wickramasinghe, N.C., 1962, *Mon Not. Roy. Astr. Soc.*, **124**, 417.

Hoyle, F. and Wickramasinghe, N.C., 1969, *Nature*, **223**, 459.

Hoyle, F. and Wickramasinghe, N.C., 1986, *Q. Jl. Roy. Astr. Soc.*, **27**, 21.

Hoyle, F., Wickramasinghe, N.C. and Jabir, N.L., 1983, *Astrophys. Sp. Sci.*, **124**, 417.

Hoyle, F. and Wickramasinghe, N.C., 1977, *Nature*, **270**, 323.

Hoyle, F. and Wickramasinghe, N.C., 1988, *Astrophys. Sp. Sci.*, **140**, 191.

Hoyle, F. and Wickramasinghe, N.C., 1979, *Astrophys. Sp. Sci.*, **66**, 77.

Huffman, D.R. and Stapp, J.L., 1971, *Nature Physical Science*, **229**, 45.

Jabir, N.L., Hoyle, F. and Wickramasinghe, N.C., 1983, *Astrophys. Sp. Sci.*, **91**, 327.

Jayaweera, K. and Flanagan, P., 1982, *Geophys. Res. Lett.*, **8**, 386.

Kemp, J.C. and Wolstencroft, R.D., 1972, *Astrophys. J.*, **176**, L115.

Kissell, J. and Krueger, F.R., 1987, *Nature*, **326**, 755.

Krätschmer, W., Lamb, L.D., Fostiropoulos, K. and Huffman, D.R., 1990, *Nature*, **347**, 354

Kroto, H.W., Heath, J.R., O'Brien, S.C., Curl, R.F. and Smalley, R.E., 1985, *Nature*, **318**, 162.

Lenham, A.P. and Treherne, D.M., 1966, in F. Abeles (ed.) *Optical Properties and Electronic Structure of Metals and Alloys*, Metals and Alloys, North–Holland Publ. Co., Amsterdam.

Lillie, C.F. and Witt, A.N., 1976, *Astrophys. J.*, **208**, 64.

MacDonnell, J.A.M. et al., 1986, *Eur. Space Agency Spec. Publ.*, SP–250, **2**, 25.

Maclean, S., Duley, W.W. and Millar, T.J., 1982, *Astrophys. J.*, **256**, L61.

Martin, P.G., 1973, in J.M. Greenberg and H.C. van de Hulst (eds.) *Interstellar Dust and Related Topics*, D. Reidel.

Mathis, J.S., Rumpl, W. and Nordsieck, K.H., 1977, *Astrophys. J.*, **217**, 425.

Nandy, K., 1964, *Publ. Roy. Obs. Edin.*, **3**, 142.

Nandy, K., 1965, *Publ. Roy. Obs. Edin.*, **5**, 13.

Nandy, K., 1966, *Publ. Roy. Obs. Edin.*, **5**, 233.

Nandy, K., Thompson, G.I., Jamar, C., Monfils, A. and Wilson, R., 1975, *Astron. Astrophys.*, **44**, 195.

Nuth, J.A., 1985, *Nature*, **318**, 166.

Oort, J.H. and van de Hulst, H.C., 1946, *Bull. Astron. Soc. Netherlands*, **10**, 187.

Pagel, B.E.J., 1974, *NATO Conference*, July 12–August 9, Cambridge, UK.

Pflug, H.D., 1984, in C. Wickramasinghe (ed.) *Fundamental Studies and the Future of Science*, University College, Cardiff Press.

Sapar, A. and Kuusik, I., 1978, *Publ. Tartu. Astrophys. Obs.*, **46**, 71.

Seaton, M.J., 1979, *Mon. Not. Roy. Astr. Soc.*, **187**, 73P.

Serkowski, K., 1973, in J.M. Greenberg and H.C. van de Hulst (eds.) IAU Symposium 52, *Interstellar Dust and Related Topics*, Dordrecht: Reidel.

Shapiro, P.R., 1975, *Astrophys. J.*, **201**, 151.

Snow, T.P. and Seab, C.G., 1989, *Astron. Astrophys.*, **213**, 291.

Taft, E.A. and Phillipp, H.R., 1965, *Phys. Rev.*, **138A**, 197.

van de Hulst, H.C., 1946, *Rech. Astron. Obs. Utrecht* **XI**, part I.

van de Hulst, H.C., 1949, *Rech. Astron. Obs. Utrecht* **XI**, part II.

Vrba, F.J., Coyne, G.V. and Tapia, S., 1981, *Ap. J.*, **243**, 489.

Wickramasinghe, N.C. and Guillaume, C., 1965, *Nature*, **207**, 366.

Wickramasinghe, N.C. and Nandy, K., 1974, *Astrophys. Sp. Sci.*, **26**, 123.

Wickramasinghe, N.C., Jazbi, B. and Hoyle, F., 1991, *Astrophys. Sp. Sci.*, in press.

Whittett, D.C.B., Dayawansa, I.J., Dickenson, P.M., Marsden, J.P. and Thomas, B., 1976, *Mon. Not. Roy. Astr. Soc.*, **175**, 197.

Wilking, B.A., Lebofsky, M.J., Martin, P.G., Rieke, G.H. and Kemp, J.C., 1980, *Ap. J.*, **235**, 905.

Yabushita, S., Inagaki, T., Kawabe, T. and Wada, K., 1987, *Astrophys. Sp. Sci.*, **132**, 409.

9. Spectroscopic Identifications

9.1. INTRODUCTION

Spectroscopic identifications in chemistry normally proceed along the following lines. Firstly a wavelength range over which instruments are available is chosen. Next, a spectrum (usually an absorption spectrum) of the material in an appropriately prepared form and under a well–defined set of physical conditions is obtained. Finally, the wavelengths of features in this spectrum are examined in detail and compared with wavelength sets in reference spectra of known materials taken under similar conditions. If there are a sufficient number of correspondences over a wide enough wavelength interval a chemist would assert that an identification is made. The requirement that similar spectroscopic conditions (e.g. solvent) are used for the sample and for the reference material is important, particularly in ultraviolet spectra in which large wavelength shifts occur due to interactions with polar solvents. In the infrared spectral region identifications on the basis of wavelength positions alone are difficult because a wide range of substances could have closely similar wavelengths in the principal absorption bands. For instance, for an organic material calculations place the stretching frequencies of the following bonds in the general wavelength ranges as indicated below:

C–C, C–O, C–N	$7.7 - 12.5\mu m$
C=C, C=O, C=N, N=O	$5.3 - 6.7\mu m$
C≡C, C≡N	$4.4 - 5.0\mu m$
C–H, O–H, N–H	$2.6 - 3.7\mu m$

Even within a single streching mode, e.g. for CH, there is a wide range, $\lambda \sim 3.3 - 3.5\mu m$, depending on the particular type of configuration in which it exists. Apart from simple stretching modes there are CH bending vibrations, out of plane bending in aromatic structures etc., which add to the complications of individual spectra. It is not always possible to make unequivocal assignments of all the minor features that are seen in the infrared spectra of a complex organic material.

A chemist is normally loathe to identify a complex biopolymer on the basis of an infrared spectrum alone. The reason is that such spectra are dauntingly complex due to the numerous interacting groups that exist. All that is attempted then is to assert identifications of basic functional groups, with other techniques (e.g. mass spectroscopy) deployed for a fuller identification. However, there is little doubt that an infrared spectrum can often be characteristic of a particular material. Fig. 9.1a shows the infrared spectrum of cellulose A. Although much hair splitting could be done to decipher the fine details of this spectrum, no polysaccharide chemist seeing this spectrum will doubt its identification with cellulose. Fig. 9.1b shows the same spectrum degraded by averaging over $0.03\mu m$ bins. The smoother spectrum loses all the minute characteristic wiggles but maintains an average functional dependence $\tau(\lambda)$ that is almost unchanged. The question as to whether this smoothed–out spectrum, as one would find in a typical astronomical case, can be used to identify cellulose, is one that we need to address. If astronomical data requires with high accuracy a transmittance curve as given in Fig. 9.1b we would assert that the changes of an alternative substance being available must be reckoned to be small. The issue hinges on attaching importance not only to the placements of the detailed absorption features but to the overall shape of the underlying continuum as well.

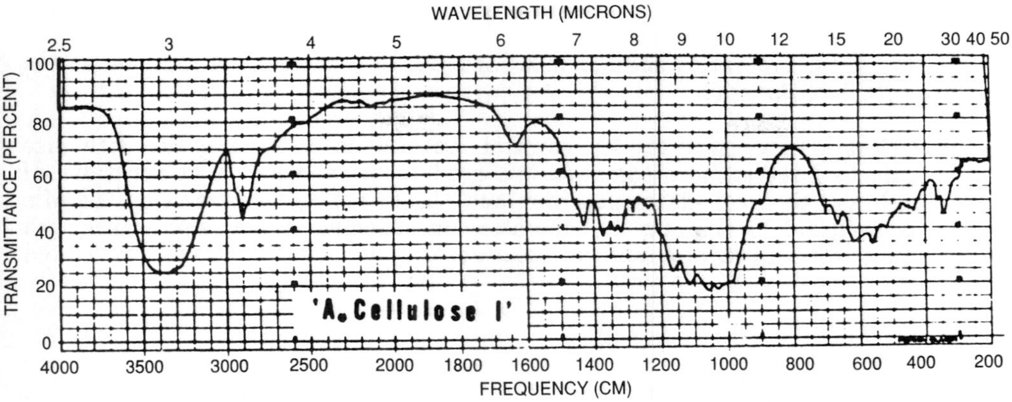

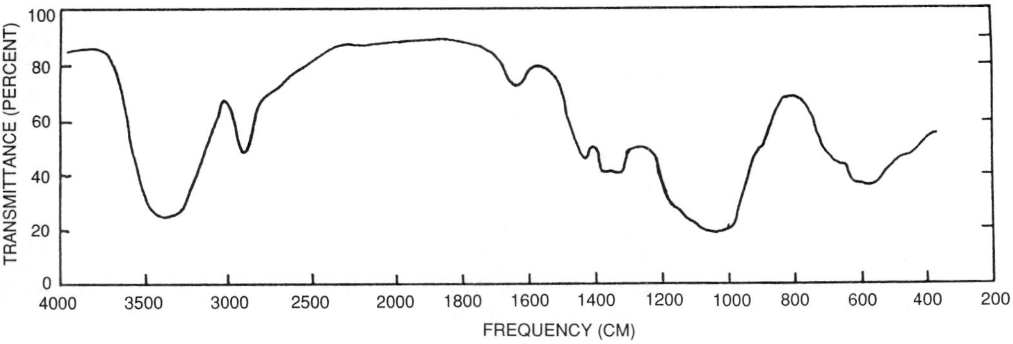

Fig. 9.1a (above) *The infrared spectrum of cellulose A.*

Fig. 9.1b (below) *Spectrum of cellulose A 'degraded' by taking rolling averages over 0.01μm.*

For this we must ensure that the laboratory experiment is suitably designed so as to exclude a scattering contribution and that the transmittance scale is properly calibrated. It is not only the wavelength positions that are important, the ratios of transmittances $\tau(\lambda_i)$, $\tau(\lambda_j)$ for every pair of wavelengths (λ_i, λ_j) in the interval under consideration must be considered to be defined by the distribution of oscillators in the sample.

If interstellar grains are small compared to the wavelength being observed, i.e. $2\pi a/\lambda << 1$, the extinction is due to absorption only and we can safely assume that the extinction optical depths in an astronomical source are proportional to the $\tau(\lambda)$ values derived from a laboratory sample of candidate material. This simplifying assumption is met for all the cases of IR extinction we shall discuss in the present chapter. Similarly, for a small particle emitting infrared radiation we can assume that the flux is

$$F(\lambda) \propto \tau(\lambda)\, B_\lambda(T) \,, \tag{9.1}$$

where $\tau(\lambda)$ is the laboratory measurement and $B_\lambda(T)$ is the Planck function at the temperature T of the grain.

The situation, in general, for spectral features in a solid grain with size comparable to $\lambda/2\pi$ is more complex. We need to know n and k as functions of wavelength and then calculate extinction and absorption efficiencies using the Mie formulae discussed in Chapter 2. Laboratory measurements over any wavelength region of interest lead to values of the linear absorption coefficient α which is related to k by

$$k = \alpha\lambda/4\pi \tag{9.2}$$

The function $k(\lambda)$ leads then to a determination of the refractive index $n(\lambda)$ via the Kramers–Krönig dispersion relations. Strictly one needs to know the behaviour of k over a fairly extended spectral region in order to calculate $n(\lambda)$ in a reliable way. Tabulations of n, k are available in the literature for several materials of astrophysical interest, including ice, graphite and silicates. Resonant absorptions in small spherical particles arise whenever

$$m^2 + 2 \approx 0 \tag{9.3}$$

The precise wavelength of such an absorption varies with particle radius for spherical grains, and also with particle shape. The $\lambda \simeq 2175$ Å absorption due to small graphite spheres is caused by such an effect. Similarly, ice grains, if spherical, have a strong resonant absorption near $\lambda \simeq 3.1\mu m$.

Apart from the central wavelengths of absorption bands astronomical data give an indication of the mass absorption coefficients at the band centre. High values of k give rise to bands that are detectable with only small values of the column density to an astronomical source. Thus ice grains, whether crystalline or amorphous, have mass absorption coefficients at the $3.1\mu m$ band centre of $\sim 30{,}000\ cm^2\ g^{-1}$, so that a density $\rho \approx 10^{-26}\ g\ cm^{-3}$ of ice grains would have an extinction of ~ 1 mag/kpc. This enormous extinction coefficient of ice grains near $\lambda \approx 3\mu$ made the early ice–grain theory susceptible of proof or disproof as soon as it was possible to observe spectra of stars with measurable fluxes at this wavelength. From the early observations of Danielson et al. (1965) and Knacke et al. (1969), which led to non–detections of the $3.1\mu m$ ice band in the spectra of highly reddened supergiants, it was possible to exclude a significant contribution from ice grains in the general interstellar medium. This conclusion has not changed in the intervening years as far as the general interstellar medium is concerned, although there is evidence for water–ice associated with dense clouds.

In strong contrast to the high values of k for ice at the $3.1\mu m$ band centre, OH, NH, CH stretching modes in organic polymers give broad weak features over the $2.5 - 3.9\mu m$ waveband, as for the case of cellulose in Fig. 9.1. A typical band centre value of the mass absorption coefficient in a broadened OH band near $\lambda = 2.9\mu m$ is $1000\ cm^2\ g^{-1}$, and in a broad CH band near $\lambda \sim 3.4\mu m$ it is $\sim 500\ cm^2\ g^{-1}$. These low values imply that even the occurrence of very weak bands at these wavelengths imply large mass fractions of organic material, a point that we had stressed as early as 1980 when astronomers seemed almost unanimously opposed to the idea of any appreciable quantity of condensed organic matter in the galaxy (Hoyle and Wickramasinghe, 1980).

9.2. THE 8 – 13μm FEATURES IN ASTRONOMY

The first positive detection of any spectral feature of grains in the infrared waveband was made in 1969. Woolf and Ney (1969), Knacke et al. (1969), Ney and Allen (1969) and Stein and Gillett (1969) detected a strong infrared excess (above a thermal continuum) in the 8 – 12μm waveband in the spectra of several oxygen–rich Mira–type stars (Fig. 9.2a). In view of broad similarities between observations and the spectra of terrestrial silicates in this waveband the 8 – 12μm feature of Mira stars has widely been attributed to silicate grains that were thought to condense in stellar mass flows. A feature such as this could arise from Si–O bending vibrations, but the identification is not unique.

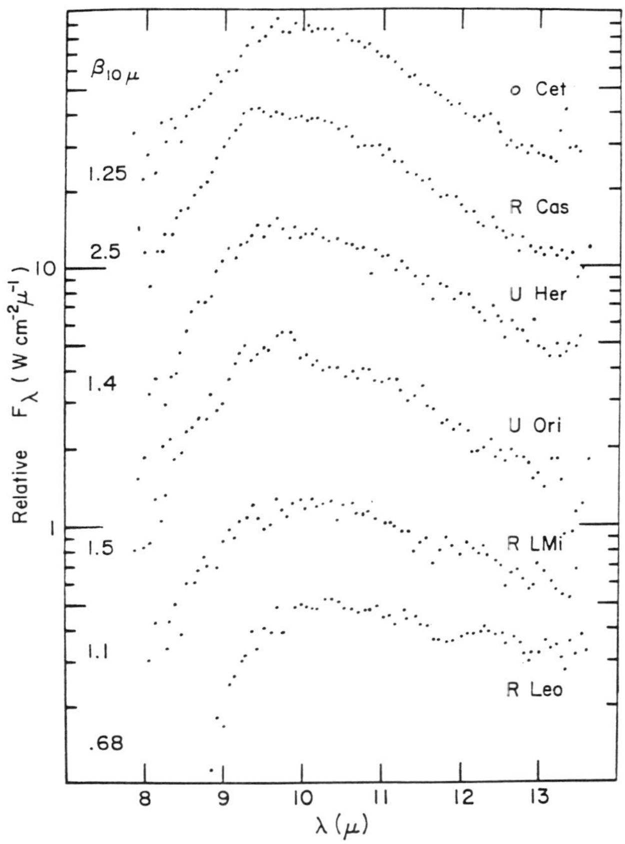

Fig. 9.2a *Spectra of several Mira stars (Forrest et al., 1975).*

On the other hand, cool carbon stars are found mainly to have infrared continua that correspond to emission from spectrosocopically featureless circumstellar dust shells up to $\lambda \simeq 10.5\mu m$; but emission increases sharply from $\lambda = 10.5\mu m$ to $\lambda \simeq 12.7\mu m$, consistent with the behaviour of specified size and shape distributions of silicon carbide particles (Forrest et al., 1975a; Gilra, 1971). An alternative explanation of the 10.5 – 12.7μm excess from carbon stars is that these are due to hydrocarbon molecules (Tarafdar and Wickramasinghe, 1975). The 10.5 – 12.7μm excess from the carbon star V Cyg (Forrest et al., 1975a), together with the spectra of several hydrocarbons, are shown in Figs. 9.2b and 9.2c. In support of the latter explanation of the 10.5 – 13μm feature it is relevant that the central band strength of SiC is 1.4 x 10^4 cm^2 g^{-1} (Dorschner et al., 1977). Such a strong band has not been seen in absorption in the general interstellar medium.

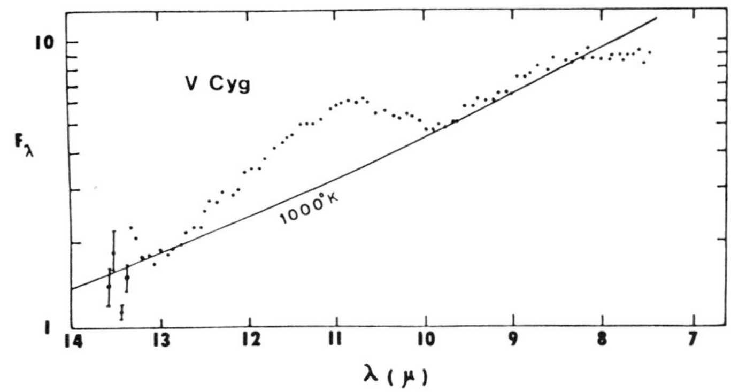

Fig. 9.2b *Spectrum of V Cyg (Forrest et al 1974a).*

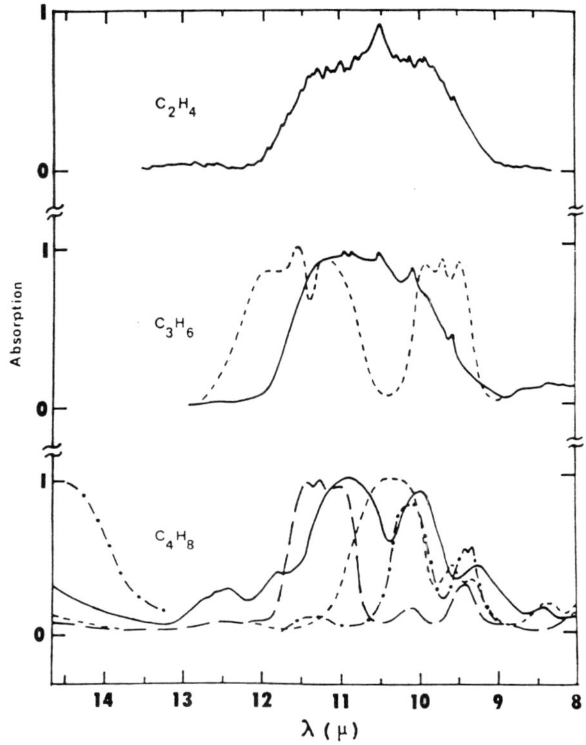

Fig. 9.2c *Spectra of Hydrocarbons Tarafdar and Wickramasinghe, 1975).*

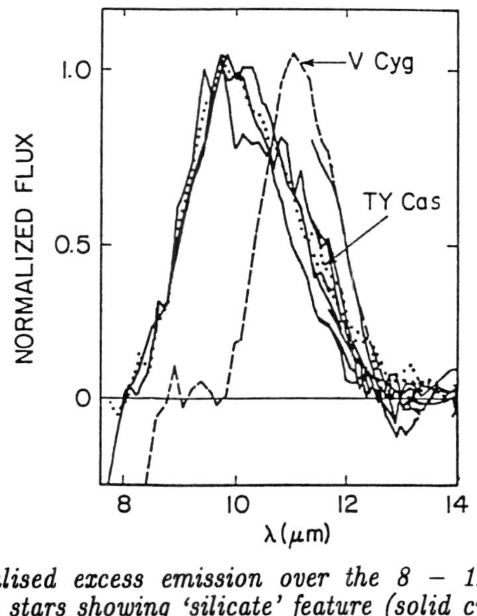

Fig. 9.2d *Normalised excess emission over the 8 – 14μm waveband for 3 carbon stars showing 'silicate' feature (solid curves) compared with a normal carbon star V Cyg (dashed curve)(adapted from Little–Marenin, 1986) and normal Mira star TY Cas (dots).*

9.3. THE 8 – 14μm FLUX FROM THE TRAPEZIUM NEBULA

At the present time the 8 – 12μm feature has been observed both in emission and in absorption in a variety of types of astronomical object – these include the Trapezium nebula, planetary nebulae, compact HII regions, OH/IR sources, the galactic centre and comets (Russell et al., 1975; Woolf, 1973; Cohen, 1980; Capps et al., 1978; Aitken et al., 1979). It is significant that in a small fraction of cases the same feature has also been seen in the spectra of carbon stars (Little–Marenin, 1986). The shapes of the bands in all these instances appear fairly constant, resembling the M star 10μm feature, as can be seen in Fig. 9.2d. It would seem unlikely that this carbon star 10μm band is due to silicates – thus casting doubt on the usual 'silicate only' hypothesis.

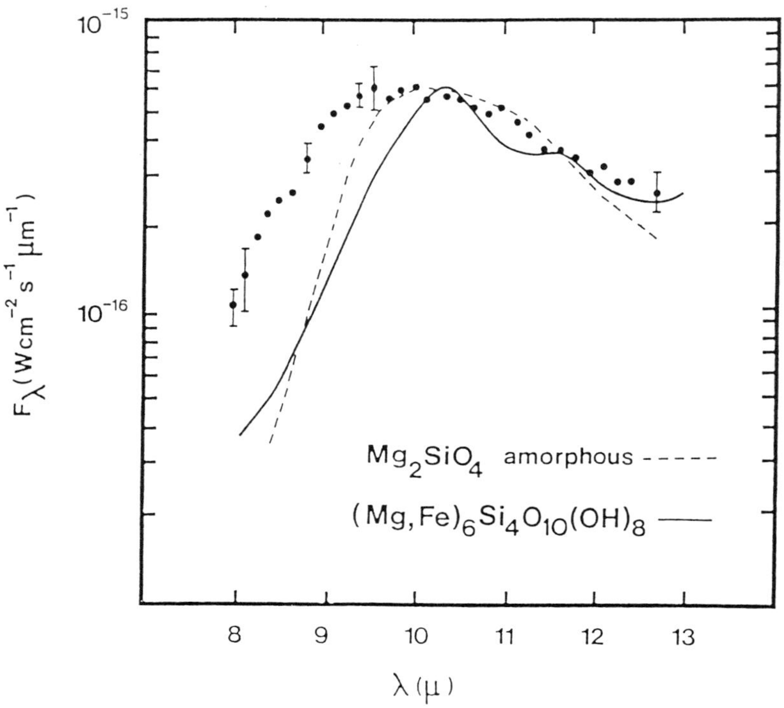

Fig, 9.3 *Flux data for the Trapezium nebula (dots) compared with the theoretical flux for models of magnesium silicate grains at a temperature of 175° K. Normalisation is to $F = 6 \times 10^{-16}$ W cm^{-2} s^{-1} μm^{-1} at the wavelength of maximum flux. The data is taken from Forrest et al., (1975a, b).*

Modelling the 8 – 12μm feature is relatively straightforward for the case of optically thin dust emission as in the case of the Trapezium nebula. The points in Fig. 9.3 show the data for the Trapezium emission obtained by Forrest et al. (1975a,b) compared with calculated flux curves for small particles comprised of amorphous and hydrated silicates (Knacke and Kratschmer, 1980; Day, 1979). The

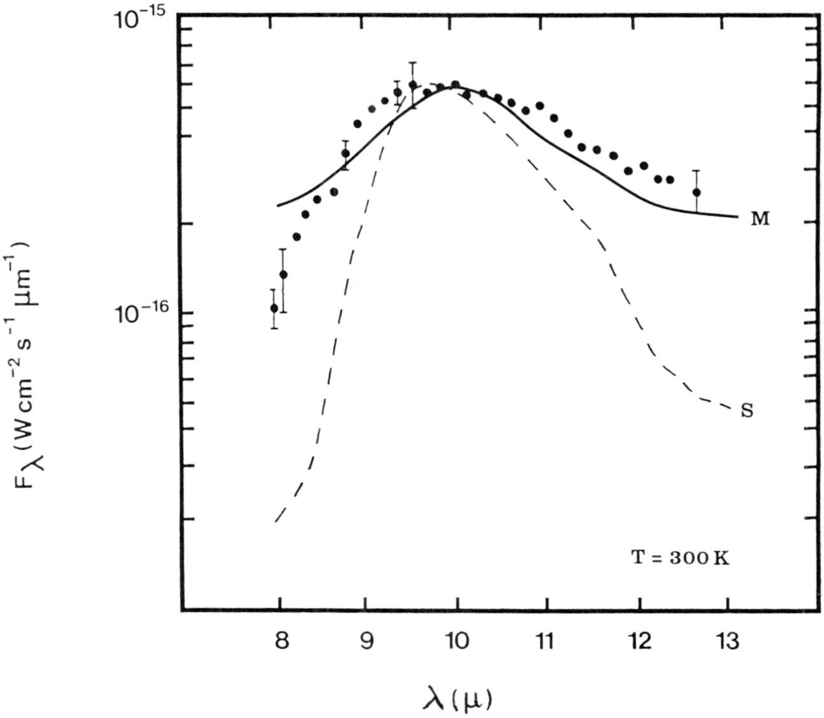

Fig. 9.4 Normalised flux over the waveband 8 – 13μm calculated for amorphous silicate particles and for Murchison material at 300° K compared with data points for the Trapezium nebula from Forrest et al. (1975a, b).

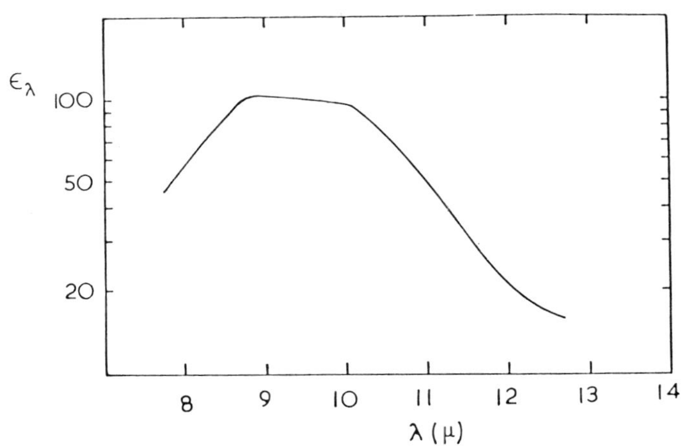

Fig. 9.5 *Normalised emissivity of soot with 98.5 per cent graphite content calculated from the data of Foster and Howarth.*

agreement between the data and the silicate models is seen to be poor, and this evident mismatch has not been rectified even by considering the various ways in which silicates might be modified under astronomical conditions – e.g. irradiation. The best 'mineral' fit to the $10\mu m$ feature is for grains comprised of material derived from carbonaceous meteorites. Fig. 9.4 shows a comparison between the data and a calculation based on optical constants of finely crushed Murchison meteorite material (Majeed et al., 1988). It should be remembered that this material includes quantities of carbonaceous organic matter, which can also contribute to absorptions in the $8-12\mu m$ band through C–O, C–O–C bending vibrations. As early as 1969 we pointed out that carbonaceous grains comprised of soot with 1–2% hydrogen produces a $8-12\mu m$ emissivity profile which is qualitatively similar to that of silicates. Fig. 9.5 shows a 98.5% carbon–soot 1.5% hydrogen emissivity curve (Hoyle and Wickramasinghe, 1969). This is a material none other than hydrogenated amorphous carbon (HAC) – a grain material that was discussed over a decade later by Duley and Williams.

Fig. 9.6 shows an improvement to the earlier silicate models that was obtained by one of us for the polyformaldehyde grain model (Wickramasinghe, 1974), a fit that was further improved using data based on laboratory measurements of the transmittance of cellulose (Hoyle and Wickramasinghe, 1977).

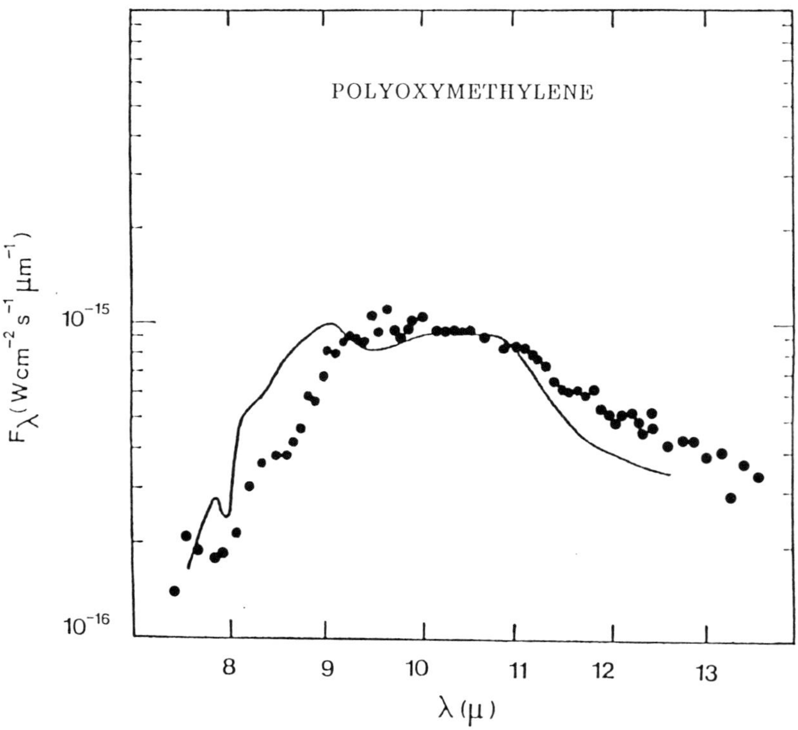

Fig. 9.6 *The infrared spectrum of the Trapezium nebula compared with prediction for a 445° K polyformaldehyde grain model.*

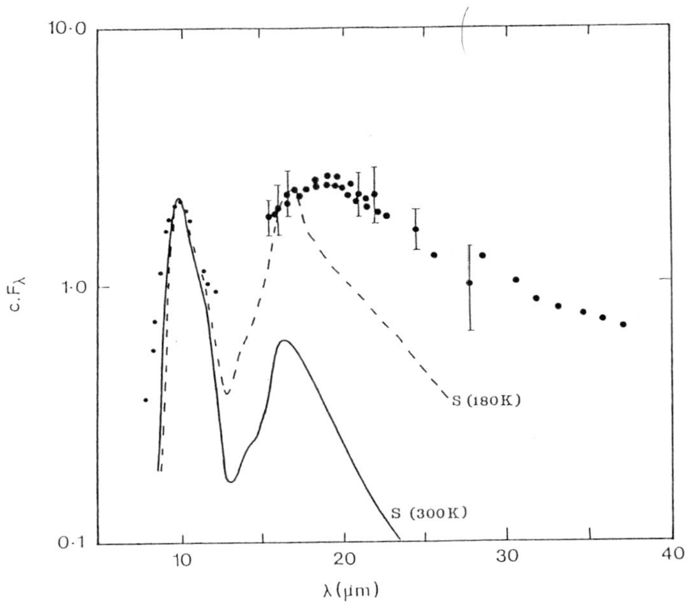

Fig. 9.7 *The infrared spectrum of the Trapezium nebula compared with amorphous silicate models at two temperatures.*

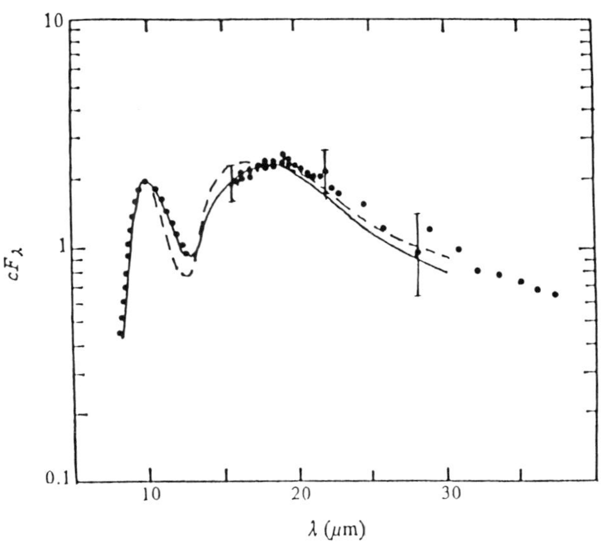

Fig. 9.8 *Normalised flux from the Trapezium nebula (Forrest et al., 1976) (points) compared with emission from polysaccharide grains with temperature 175° K (Hoyle and Wickramasinghe, 1977).*

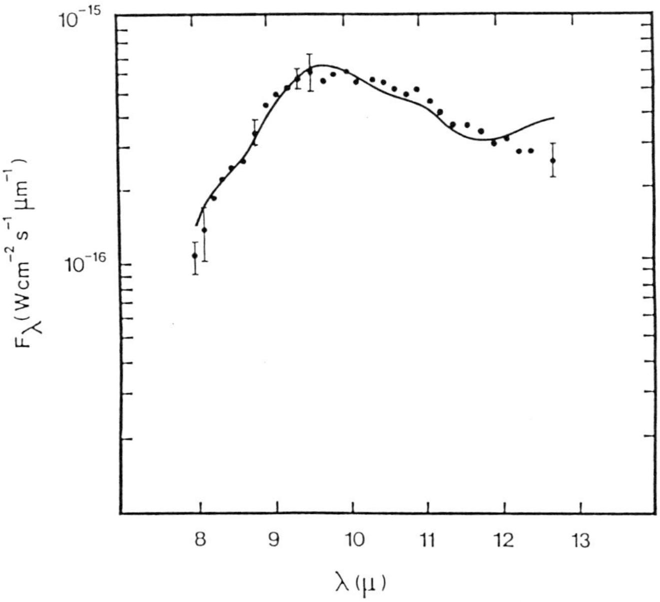

Fig. 9.9a *Normalised flux curve for the 'average' bacterium–diatom silica mixture calculated for a temperature of 170° K. The observational points are the same as in Fig. 9.3.*

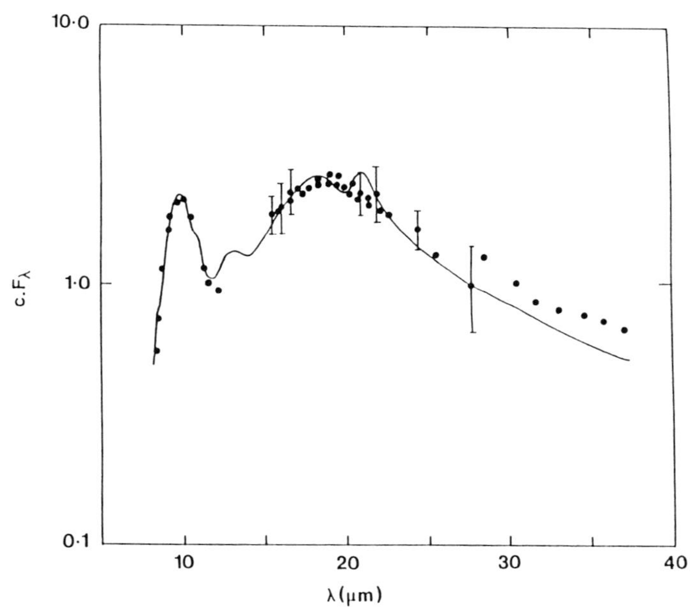

Fig. 9.9b *Best fit over the waveband 8 – 37μm for the same model as in Fig. 9.9a.*

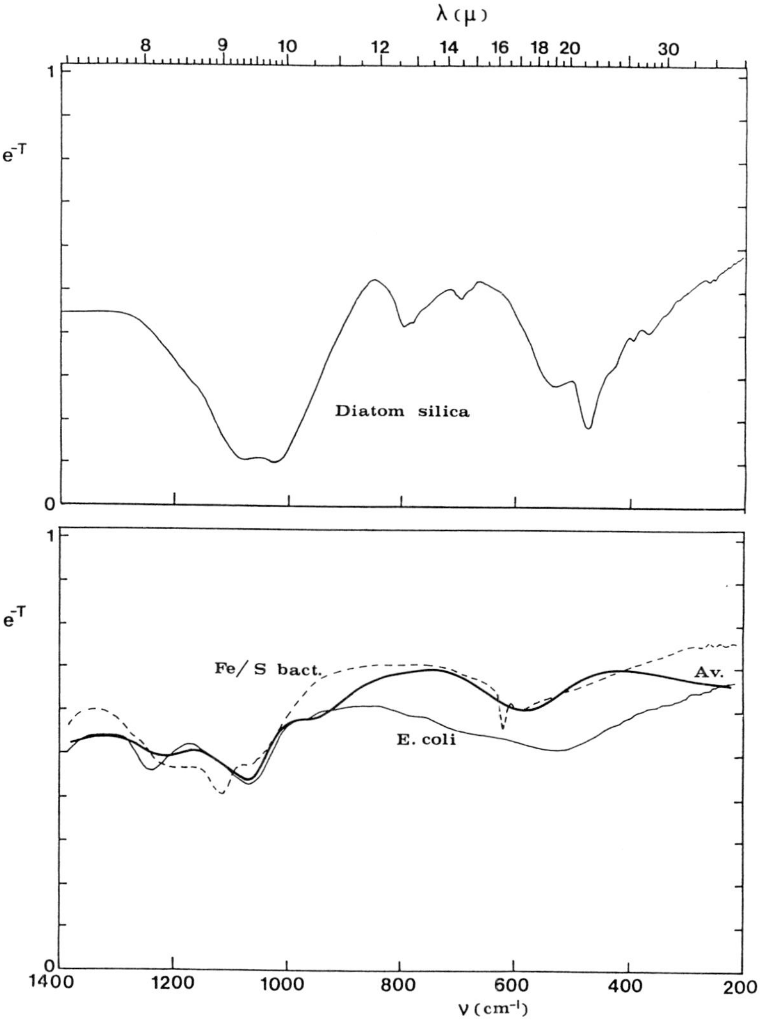

Fig. 9.10 *Upper panel: transmittance curve for diatom silica. Lower panel: transmittance curves for E. coli (Fine solid line); for a mixed culture of iron and sulphur bacteria (dashed curve), for an 'average' interstellar organic particle (heavy solid line).*

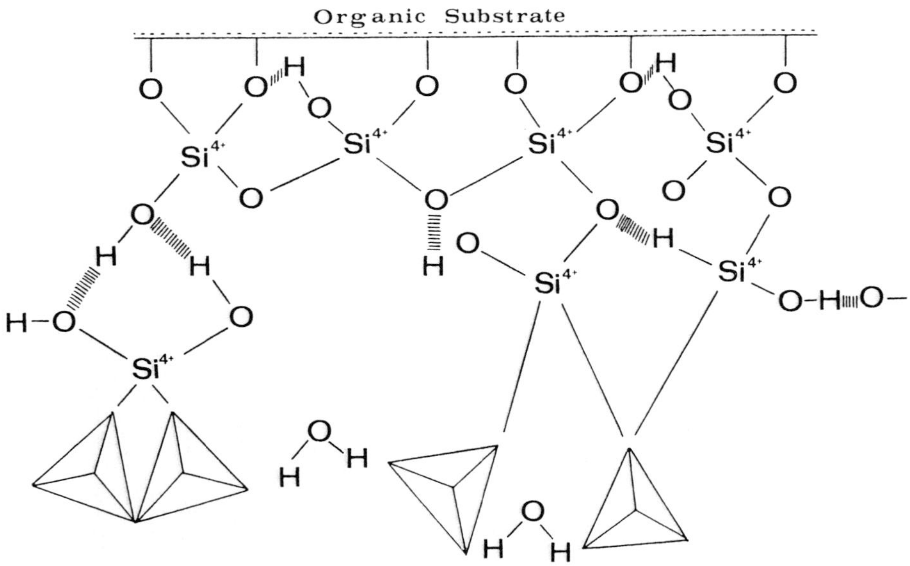

Fig. 9.11 *Schematic depiction of bond arrangements in diatom silica.*

A question now arises as to whether the siliceous material is condensed around organic grains, or, alternatively, is it possible that organic matter (characterised by the infrared spectroscopic properties of *E. coli*) could be condensed as mantles around silica grains? From the absorption and emission data alone it is clearly not possible to distinguish between these options. There is also a further option: namely that the two components coexist in the form of separate particles, occurring quite independently of each other. The last of these options would seem to be inappropriate, however, at any rate for the dust in the Trapezium nebula, for the reason that separate silica and *E. coli*–type particles would have widely separated values of temperature, which would not be consistent with the available infrared spectral data. Gaseous organic material condensing around silica grains would give rise to composite particles with a high average value for (n–1), where n is the refractive index, and this situation also presents difficulties with regard to the visual wavelength dependence of interstellar extinction, as we have discussed in Chapter 8. A monolayer of small $0.01\mu m$ silica–type grains on the surfaces of hollow bacterial grains of radii $0.1 - 1\mu m$ would, on the other hand, provide a consistent model for extinction and polarisation data over a wide range of wavelengths.

Polarisation in the 10μm Absorption Band. Many galactic infrared sources are found to show a broad absorption band in the $8 - 12\mu m$ region. Observations of the infrared source GC–IRS 3 and the BN object both show significant changes in polarisation across this $10\mu m$ absorption feature (Aitken et al., 1988). Such data can be used to discriminate between pure silicate and organic/mineral models because polarisation curves are sensitive to the average $n(\lambda)$, $k(\lambda)$ values near the absorption peak (Martin, 1975). Whereas amorphous silica has a maximum k value of ~ 2.5 in the 8 – 12μm region, the organic/mineral combination proposed here has an average k value at the 10μm peak of ~ 0.5, and a purely organic model has a maximum k value ~ 0.1.

In the calculations that follow we use $k(\lambda)$ curves determined directly in the laboratory to generate corresponding infrared $n(\lambda)$ curves, using a routine that computes the Kramers–Krönig dispersion relations, and assuming the generally accepted values of n for these materials in the optical spectral region. For our mineral–organic combination we use an average $\bar{k}(\lambda)$ and the appropriate average $\bar{n}$ value in the optical wavelength region. The computed $\bar{n}(\lambda)$, $\bar{k}(\lambda)$ functions are then used in the Mie formulae for homogeneous cylinders to compute extinction efficiencies Q_{eE}, Q_{eH} for light with electric vectors respectively parallel and perpendicular to the cylinder axes. Computations for infinite cylinders might be assumed to hold approximately for oblate spheroidal particles with an axial ratio ~ 2.5/1 or more. The ratio of polarisation to optical depth arising from a column of partially aligned grains in the line of sight of a source is given by

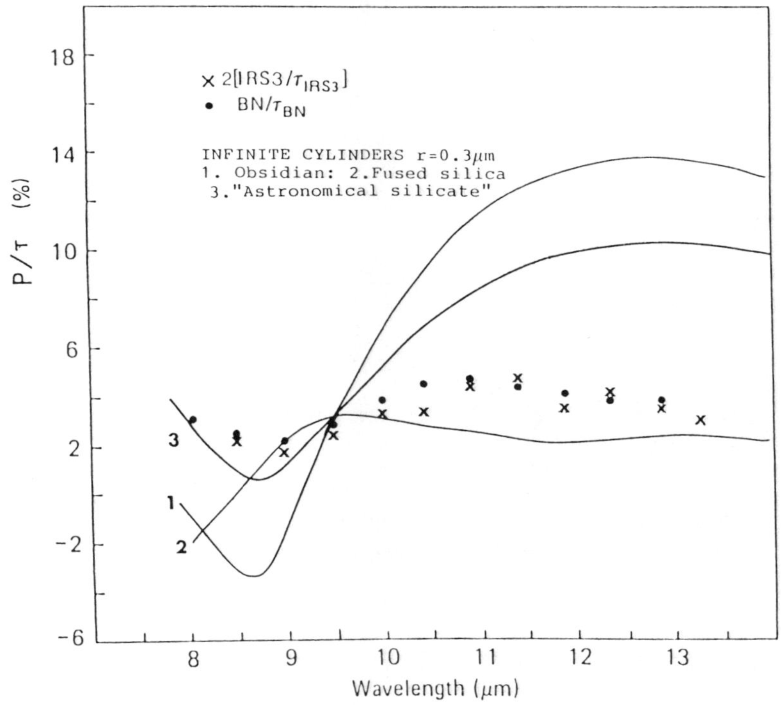

Fig. 9.12 *The P/τ ratio expressed as a percentage for the sources GC–IRS 3 and the BN object compared with the corresponding quantity calculated for partially aligned cylinders comprised of three different types of silicate material. The degree of alignment is an arbitrary parameter in the model (Data points from Aitken et al., 1988).*

$$\frac{P}{\tau} \propto \frac{|Q_{eE} - Q_{eH}|}{Q_{eE} + Q_{eH}}, \tag{9.4}$$

the constant of proportionality depending on the degree of alignment of the cylinders in a particular case.

Fig. 9.12 shows the observational data for the galactic centre source GC–IRS 3, and for the BN object, compared with theoretical calculations for three types of purely mineral grains of radii $r = 0.3\mu m$. The theoretical curves are normalised to agree with the data at a wavelength close to $\lambda = 9.5\mu m$. In all three cases the theoretical curves differ markedly from the trends in the observations. These results are not sensitive to our choice of the radius r. The implication must be that theories involving purely mineral grains are not viable. Fig. 9.13 shows the corresponding calculations for hollow organic grains (characterised by the data for bacteria) and for organic/silica mixtures with a mass ratio 4.3/1. Such particles with typical radii $0.3\mu m$ are seen to give far better agreement with the observational data.

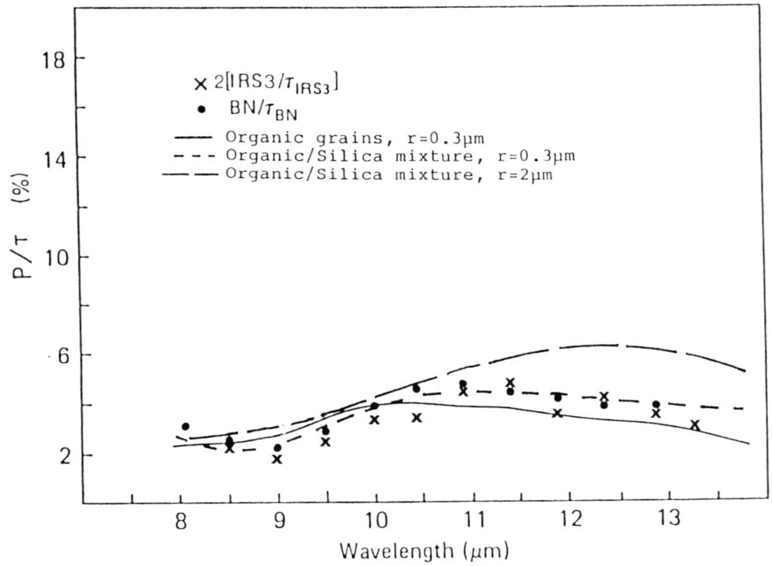

Fig. 9.13 *Same as Fig. 9.12 for organic grains and organic/silica composite grains.*

9.4. THE 3.4μM BAND: PROOF THAT GRAINS ARE MAINLY ORGANIC

The 10 and $20\mu m$ emission features in the Trapezium Nebula provide suggestive evidence of a major organic contribution to the interstellar dust. Clearly the deficiencies of emissivity in the $8 - 9\mu m$ spectra of most silicates has to be augmented. It is natural to seek a contribution in the $8 - 9\mu m$ region from organic solids, which then possess spectral features over a wider infrared waveband. In particular, the observation of a broad CH stretching feature near $\lambda \sim 3.4\mu m$ (in various aromatic/aliphatic configurations) would be more conclusive evidence of organic grains. Many astronomers had asserted that such an absorption band could not possibly exist in the grains to any degree that implied a substantial mass fraction of organic grains (Duley and Williams, 1979; Whittet, 1979). The facts as they eventually emerged turned out quite differently, however.

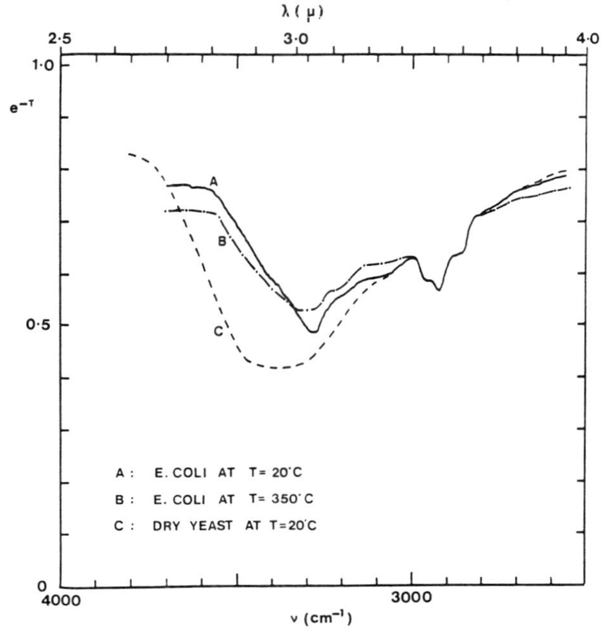

Fig. 9.14 *The measured transmittance curves of microorganisms. For E. coli a dry mass of 1.5 mg was dispersed in a KBr disc of radius 0.65 cm. The transmittance data for yeast was normalised to agree with the E. coli curves at $\lambda = 3.406\mu m$.*

When we ourselves first proposed the organic/biological grain model there were no suitably calibrated transmittance spectra of dessicated bacteria at infrared wavelengths. Together with S. Al–Mufti, we set out to obtain such spectra with a view to testing the bacterial grain hypothesis. Our experimental procedure was designed so that the measured transmittance curves were calibrated to give $\tau(\lambda)$ on an absolute scale, and our results were not contaminated by scattering because of our choice of KBr as the dispersing agent, a material that possesses the same $n(\lambda)$ values in the $2.5 - 15\mu m$ band as bacteria to within a small margin.

Fig. 9.14 shows our measured transmittance curve over the wavelength range $2.6 - 3.9\mu m$ for three systems we studied in the laboratory: (a) *E. coli* at room temperature, (b) *E. coli* at 350°C and (c) dehydrated vegetative yeast cells. For these investigations, micro–organisms obtained in a pure form were dried in a dessicator and sealed in KBr discs of radii 0.65 cm under a pressure of 6.8 tonnes cm^{-2}. A cell containing such a disc was then placed in a furnace, specially designed to be operated in conjunction with a Perkin Elmer 257 infrared spectrometer, where an assigned temperature could be maintained as the furnace itself was continuously flushed with nitrogen gas at a pressure of 28 atmospheres. Infrared spectra over the wavelength range 2.5 – 15μm were thus obtained for various temperatures. Our experiments indicated thermal stability of the biological material to temperatures of about 400° C. By 350° C all traces of free water disappeared, but the chemical integrity of the biomaterial as judged by infrared spectroscopy was otherwise preserved.

The mass of *E. coli* used was 1.50 mg, yielding a mass absorption coefficient at 3.4μm close to 500 $cm^2\ g^{-1}$. The resolution of the laboratory measurements is to $\Delta\lambda/\lambda \sim 1/600$ over the entire experimental waveband. Whilst the transmittance curve is seen to be generally the same for the whole wavelength range shown in Fig. 9.14, we note that there is a surprising invariance of shape between 3.3μm and 3.5μm. Still more strikingly, the same invariance is seen to hold for eukaryotes as the curve for yeast cells shows.

When these spectra were first obtained on 21 May 1981 it was noticed that the 3.3 – 3.9μm opacity behaviour bore a general resemblance to the spectrum of GC–IRS 7 published by D.T. Wickramasinghe and D.A. Allen (1980). The significance of this particular comparison arises because GC–IRS 7 happens to be ideally placed for studying the properties of interstellar dust over a 10 kpc path length from the solar system to the galactic centre. In May 1981 Allen and D.T. Wickramasinghe observed the same object at a higher spectral resolution than before and over a slightly more extended wavelength region. Their spectrum is shown in Fig. 9.15. The source of the primary radiation in which absorption takes place is thought to be a late–type supergiant. After experiencing general interstellar reddening, the radiation over a limited wavelength region centred at 3.4 microns would if specific absorptions were absent approximate to a black–body distribution for a lower temperature than the supergiant itself. The observations actually have an envelope that is close to a black–body distribution at 1100° K.

It is against this background continuum that absorptions by both gas and dust can be seen. The 2.4 micron absorption band seen here is due to CO, and the 3.4 and 2.95 micron absorptions are most probably due to organic grains with OH, CH and NH linkages. To confirm our earlier result, there is again no evidence whatsoever for a 3.1 micron water–ice absorption in this source.

From Fig. 9.15 it is easily seen that the depth of the 3.4μm absorption peak below the continuum drawn here amounts to $\sim$ 0.3 mag. With grains of mass absorption coefficient 500 $cm^2\ g^{-1}$ distributed along the 10 kpc path length to the galactic centre we require a smoothed–out organic grain density of 2×10^{-26} g cm^{-3}, essentially the bulk of the interstellar carbon. This conclusion cannot be relaxed unless one adopts higher values for the mass absorption coefficient. Even $\kappa \simeq 1000$ $cm^2\ g^{-1}$, which is high for a band of this width, leads to $\rho_g = 10^{-26}$ g cm^{-3}.

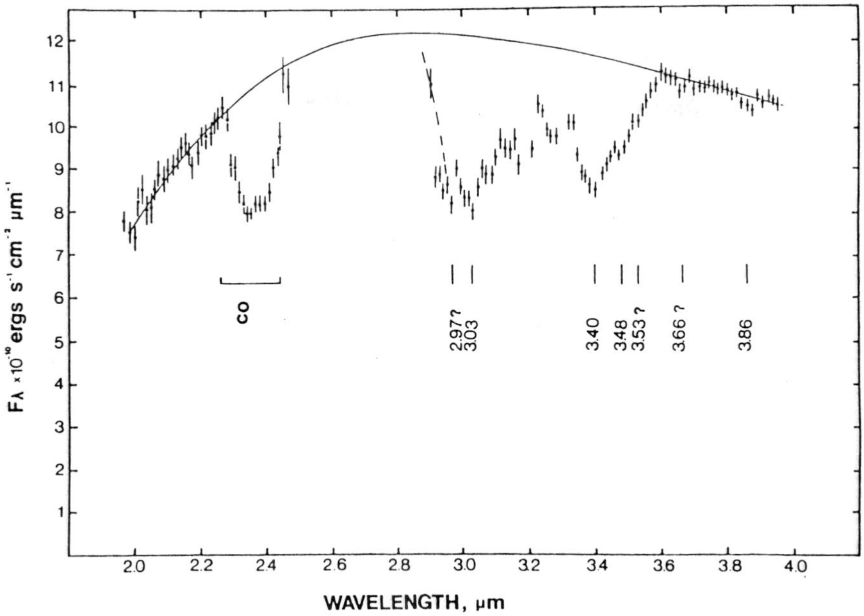

Fig. 9.15 *Observed flux from GC–IRS 7 (From Allen and D. T. Wickramasinghe).*

This conclusion can only be avoided if a large part of the 3.4μm absorption in the GC–IRS 7 spectrum arises within a circumstellar environment close to the source. Recent observations by Okuda et al. (1988) of 6 other obscured supergiants all lying within a few parsecs radius of GC–IRS 7 can be used to discount this possibility, however. Fig. 9.15a shows the relative intensity curves of these sources along with the spectrum of GC–IRS 7. The solid lines are the best–fitting background continua represented by Planck curves with temperatures ranging from 450 – 1100° K. Against such black–body continua we note that absorptions near 3.4μm for each source correspond to optical depths that agree to within about 10% with the value for GC–IRS 7. These results confirm beyond reasonable doubt that the bulk of the 3.4μm absorption does not arise from grains local to the sources, but arises instead from grains in the general interstellar medium.

9.5. MODELLING THE 2.9 – 4μM IR DATA FOR GC–IRS 7

We shall now model this data more carefully with a view to identifying the organic materials of the grains, starting with the calibrated laboratory spectrum of dessicated *E. coli* in the 2.6 – 3.6μm waveband measured in the laboratory by S. Al–Mufti and plotted in Fig. 9.16 (Hoyle et al., 1982). Table 9.1 shows

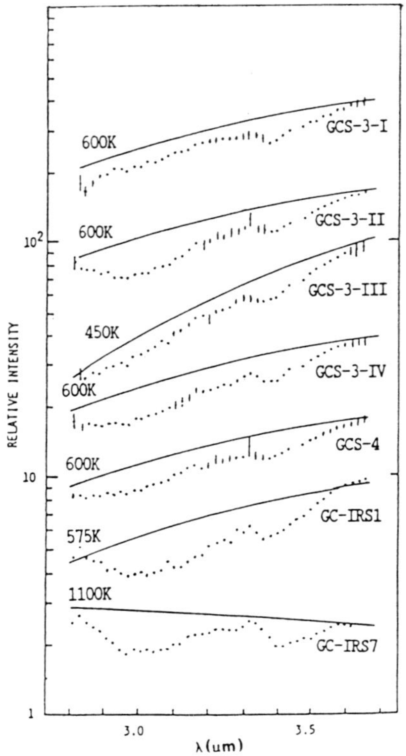

Fig. 9.15a *Spectra of several discrete infrared sources in the galactic centre region distributed within a volume of 3 cu.pc. The solid lines are proposed underlying black–body continua. The data are from Okuda et al., 1989.*

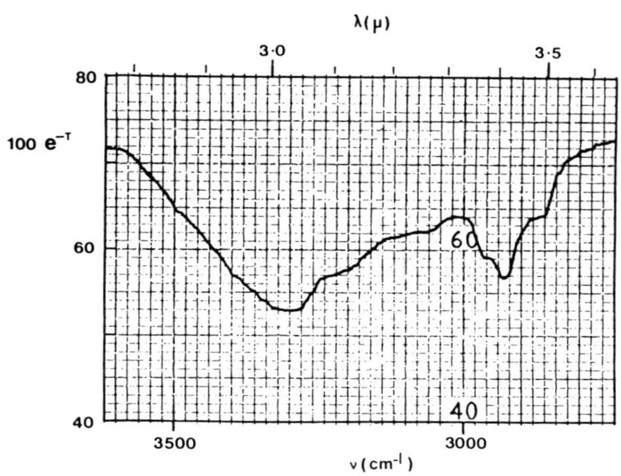

Fig. 9.16 *Enlarged laboratory spectrum giving transmittance values over the 2.6 – 3.6μm waveband for E. coli heated to 350° C.*

wavelength $\lambda(\mu m)$, observed relative fluxes c F_λ, transmittances $e^{-\tau}$ read–off from Fig. 9.16 at the wavelengths in question, and calculated values obtained by a procedure now to be described.

Imagine the radiation from GC–IRS 7 to be collimated to give a beam with intensity distribution $I(\lambda)\,d\lambda$ directed towards the Earth. Because of the scattering and absorption of the radiation which occurs *en route* to the Earth, a terrestrial observer determines the spectrum

$$\exp[-Q_{sca}(\lambda)].\exp[-Q_{abs}(\lambda)].I(\lambda)\,d\lambda\ , \tag{9.5}$$

$Q_{sca}(\lambda)$ and $Q_{abs}(\lambda)$ being the wavelength dependent scattering and absorption integrated along the line of sight. The source of $I(\lambda)$ is inferred from studies of CO absorption at $\lambda \cong 2.4\mu m$ and from near–infrared filter photometry to be an M supergiant with an effective temperature near 3200° K, so that $I(\lambda)\,d\lambda$ is much like the Planck distribution for this temperature. Multiplication by $\exp[-Q_{sca}(\lambda)]$ has the effect over a limited wavelength range of yielding an intensity distribution $I(\lambda)\exp[-Q_{sca}(\lambda)] \cong B(\lambda,T_c)$, where $B(\lambda,T_c)$ is the Planck function for a suitably chosen colour temperature T_c. From the known scattering properties of interstellar grains, T_c can be shown to be likely to lie in the range from 1000° K to 1500° K. In our original work we took $T_c = 1100°$ K, which was consistent with the envelope of the observations of Allen and Wickramasinghe, D.T. Hence (9.5) can be written as

$$AB(\lambda,1100)\exp[-Q_{abs}(\lambda)]\,d\lambda\ , \tag{9.6}$$

where A is a constant depending on the intrinsic emission and distance of the source. If $Q_{abs}(\lambda)$ arises from the absorption values given in Fig. 9.16 then (9.6) takes the form

$$AB(\lambda,1100)\exp[-\alpha\tau(\lambda)]\,d\lambda\ , \tag{9.7}$$

with $\exp[-\tau(\lambda)]$ given by the curve of Fig. 9.16 and α the factor by which the quantity of absorbing material along the astronomical line of sight exceeds the amount used in the laboratory sample for which Fig. 9.16 was obtained.

It is worth noting that $B(\lambda,1100)$ is nearly flat over the wavelength range from $2.8\mu m$ to $3.6\mu m$, varying by about 10 percent, so that the situation is nearly the same as if the interstellar grains were in the laboratory with a flat source function used to obtain their spectrum. This is a favourable situation for using the astronomical observations to infer $\tau(\lambda)$ – i.e. for making the present comparison.

A and α must be specified before explicit numbers can be calculated from (9.7). The constant A disappears when (9.7) is normalised with respect to the scale used for c F_λ (F_λ being the flux), with α remaining as a disposable constant. We chose $\alpha = 1.3$ so as to give the correct depth of the flux curve at the $3.4\mu m$ band centre. All that then remains to decide is the scale factor A, which can evidently be chosen to agree with the observed flux at any one wavelength, but only at one wavelength. The sensitive region for comparing the calculation of (9.7) with the observed fluxes is the range 3.3 to $3.5\mu m$, and the comparison will be all the stronger if we avoid choosing A so as to normalise to one of the data points in this critical range. Explicitly, we have chosen to normalise with respect to the data point at $\lambda = 3.562\mu m$.

The reader now has all the information needed for checking an important result. Fig. 9.16 for transmittance values, the observational data in columns 1 and 2 of Table 9.1, and the simple formula (9.7) with $\alpha = 1.3$ for obtaining the last column of Table 9.7, choosing A with respect to $\lambda = 3.562\mu m$ as standard.

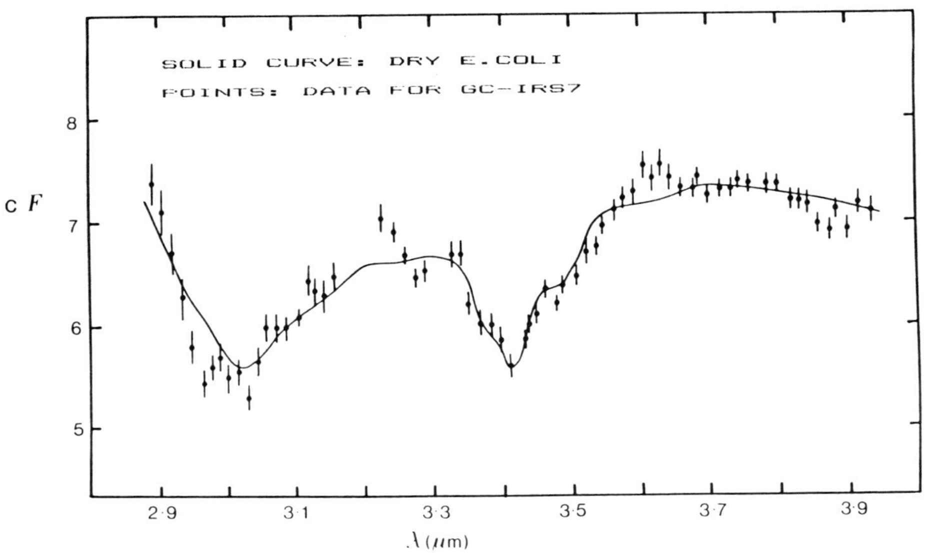

Fig. 9.17 *The agreement between our E. coli model for the parameters given in text (solid curve) and the data for GC–IRS 7 (points with error bars) as supplied by D. T. Wickramasinghe.*

There is no way the reader can really judge the position except by taking the trouble to plot the second and fourth columns of Table 9.1, both against the wavelength values in the first column. First plot the values of AB(λ,1100) exp–$\alpha\tau$, and proceed to draw a firmly–executed curve through the resulting points. Then put in the data points working from left to right. The result should be as in Fig. 9.17. Often enough the next data point kicks away from the curve, as if it had been subject to some arbitrary perturbation. Always as one continues with succeeding points they return eventually to the curve, as if the curve itself were exerting some kind of a restorative force on the points. Rarely do two successive points both head away from the curve. The usual situation is that after a point kicks away from the curve the next point turns around and moves back towards the curve, even if it does not always reach it.

The error bars given by the observers differ at different wavelengths, typical error bars for the data points in various wavelength ranges are shown in Fig. 9.17. It seems significant that where the deviations of points from the curve are largest, the error bars are also largest.

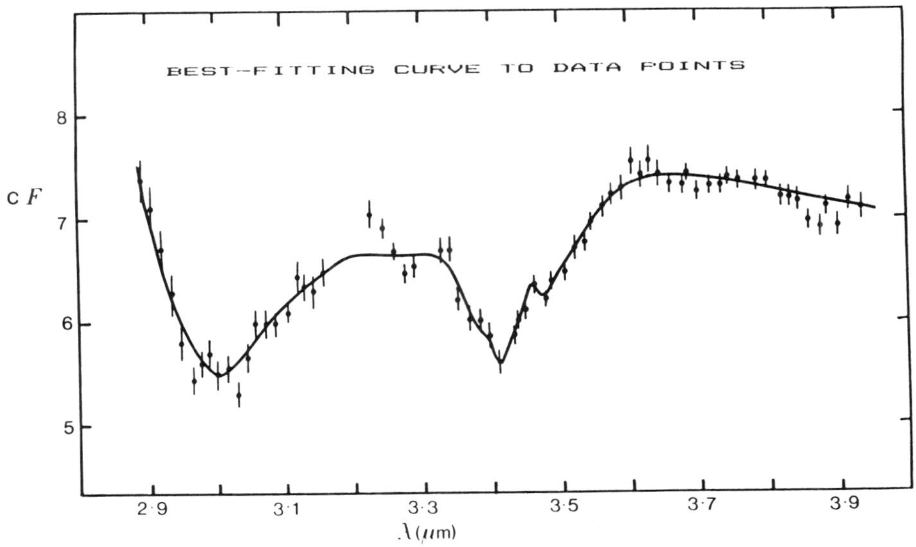

Fig. 9.18 *Points with error bars are the data for GC–IRS 7. Solid curve is the 'best curve ' drawn through these points.*

Assessment of the Goodness of Fit. Doubtless it is possible to find some who will assert that the fit of the points to the curve in Fig. 9.17 is not very good. The curve of Fig. 9.18 is quite certainly in excellent agreement with the data points, for the good reason that it has been drawn deliberately through the data points in order to be so. True, one could have joined up the data points by a jagged curve, but this would have been to overemphasize the accuracy of the data. Subsequent observations of GC–IRS 7 have shown the 1981 data of Allen and Wickramasinghe (*loc. cit*) to be generally correct, but each investigation yields moderate variation in the data, so that the matching of the curve of Fig. 9.18 is as good as it reasonably could be. The generic similarity of this artificial curve to the actual biological curve of Fig. 9.17 is obvious. The question is how much do the two curves differ quantitatively? The artificial curve has been drawn through those particular data points which are most troublesome in Fig. 9.17, the points set out in Table 9.2 with actual transmittance values obtained from Fig. 9.15. (These entries are the same as appeared in Table 9.1.) Below each entry in Table 9.2 an artificial transmittance value is given in brackets, an artificial value which leads by calculation to exact agreement with the data point in question. If the bacterial curve in Fig. 9.15 had had these slightly different transmittance values, then the agreement with the data points would have been essentially perfect.

In Fig. 9.19 we have plotted these 'revised' points over the bacterial curve of Fig. 9.15, a join of which would give transmittance values fitting perfectly the data of Allen and Wickramasinghe (*loc. cit*). The problem for those who seek an abiological explanation of the data is clear from this plot, shown in Fig. 9.19. They would need to obtain what would essentially be the bacterial transmittance curve, but without bacteria! Non–biological models have not been successful in matching these requirements to any degree that might be thought satisfactory.

Non–Biological Explanations. A fundamental problem with such models is that they should possess a disposition of functional groups almost indistinguishable from biology and yet they are required to be manufactured abiotically. Biotic–like material must thus amount to ~ 10^{-26} g cm^{-3} throughout the galactic disk.

TABLE 9.1

$\lambda(\mu m)$	c F_λ	$e^{-\tau(\lambda)}$	AB(λ,1100) exp$-\alpha\tau$ $\alpha = 1\cdot300$	$\lambda(\mu m)$	c F_λ	$e^{-\tau(\lambda)}$	AB(λ,1100) exp$-\alpha\tau$ $\alpha = 1\cdot300$
	Observed		Calculated		Observed		Calculated
2·890	7·4	.620	7·090	3·429	5·9	.590	5·824
2·904	7·1	.605	6·853	3·436	6·05	.605	6·004
2·918	6·7	.596	6·706	3·450	6·15	.623	6·209
2·932	6·3	.580	6·458	3·464	6·35	.638	6·374
2·946	5·8	.570	6·299	3·478	6·2	.640	6·371
2·960	5·45	.558	6·112	3·492	6·35	.641	6·354
2·974	5·60	.549	5·969	3·506	6·45	.661	6·582
2·988	5·75	.541	5·841	3·520	6·7	.688	6·901
3·002	5·5	.532	5·700	3·534	6·75	.698	6·999
3·016	5·56	.531	5·670	3·548	6·95	.708	7·095
3·030	5·3	.530	5·640	3·562	7·1	.711	7·100
3·044	5·66	.536	5·707	3·576	7·25	.716	7·130
3·058	6·0	.543	5·786	3·590	7·3	.719	7·134
3·072	6·0	.553	5·907	3·604	7·55	.727	7·202
3·086	6·0	.566	6·069	3·618	7·45	.730	7·205
3·100	6·1	.570	6·105	3·632	7·55		
3·114	6·45	.574	6·141	3·646	7·45		
3·128	6·35	.579	6·190	3·660	7·3	.745	7·287
3·142	6·3	.583	6·224	3·674	7·3		
3·156	6·5	.594	6·355	3·688	7·45		
3·170		.605	6·486	3·702	7·25	.760	7·365
3·184		.610	6·532	3·716	7·3		
3·198		.615	6·578	3·730	7·3		
3·212		.620	6·623	3·744	7·4		
3·226	7·05	.621	6·612	3·758	7·35	.770	7·337
3·240	6·95	.622	6·600	3·772	7·3		
3·254	6·7	.624	6·602	3·786	7·35		
3·268	6·47	.628	6·631	3·800	7·35		
3·282	6·5	.631	6·646	3·814	7·2		
3·296				3·828	7·2		
3·310		.638	6·687	3·842	7·15		
3·324	6·7	.638	6·660	3·856	6·95		
3·338	6·75	.637	6·619	3·870	6·9		
3·352	6·2	.624	6·416	3·884	7·15		
3·366	6·0	.598	6·045	3·989	6·9		
3·380	6·0	.590	5·915	3·912	7·2	.790	7·150
3·394	5·85	.583	5·798	3·926	7·1		
3·408	5·65	.569	5·593				

TABLE 9.2

λ (μm)	c F_λ (arbitrary scale)	$e^{-\tau(\lambda)}$		$AB(\lambda, 1100)\exp{-\alpha\tau}$ $\alpha = 1{\cdot}3000$
2·890	7·4	.620		7·980
		(.641	⟶	7·40)
3·002	5·5	.532		5·700
		(.5175	⟶	5·50)
3·072	6·0	.553		5·907
		(.560	⟶	6·00)
3·156	6·5	.594		6·355
		(.6045	⟶	6·50)
3·478	6·2	.640		6·371
		(.627	⟶	6·20)
3·534	6·75	.698		6·999
		(.679	⟶	6·75)
3·618	7·45	.731		7·205
		(.749	⟶	7·45)

If one exposes any combination of substances made up mostly of the H, C, N, O elements to an immensely disruptive flux of radiation, a large number of atoms and radicals will be made simultaneously available, a situation that does not occur in interstellar space. The atoms and radicals in such a contrived experiment will reform into more stable molecules, and they will do so through a complex multitude of channels – unlike the simpler reforming of chemical bonds that occurs at low fluxes. Among the multitude of channels there will be some that are unusual, involving only small trickles of recombining particles. In such a multitude of small trickles there could well be a material with a semblance to biomaterial, provided one does not press the comparison too closely.

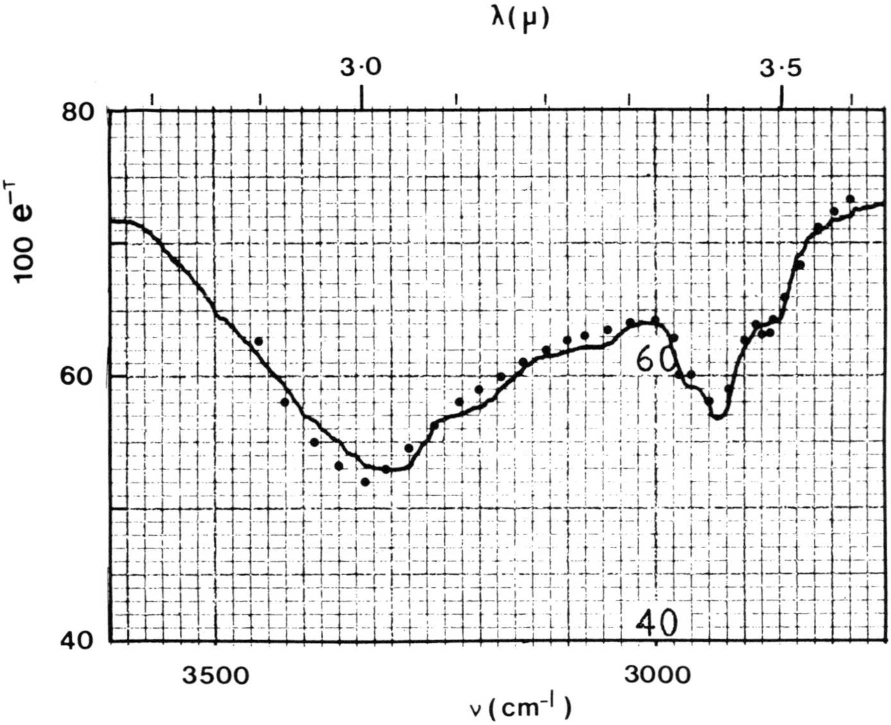

Fig. 9.19 *The laboratory transmittance curve for E. coli (Fig. 9.16) together with points corresponding to the requirements of the best fitting curve shown in Fig. 9.17.*

One of the closest semblances to biomaterial in regard to $\tau(\lambda)$ curves in the $3.2-3.7\mu$m that we have found is for the case of the condensed carbonaceous matter in the Murchison meteorite. A calibrated spectrum kindly supplied by Prof. H.D. Pflug was corrected to take account of a graphitic component, and the residual non–graphite transmittance curve is shown in Fig. 9.20. The two curves in Fig. 9.20 are normalised at the peak of their 3.4μm absorptions by multiplying $\tau(\lambda)$ by a suitable scale factor to allow for the different masses in the spectroscopic samples.

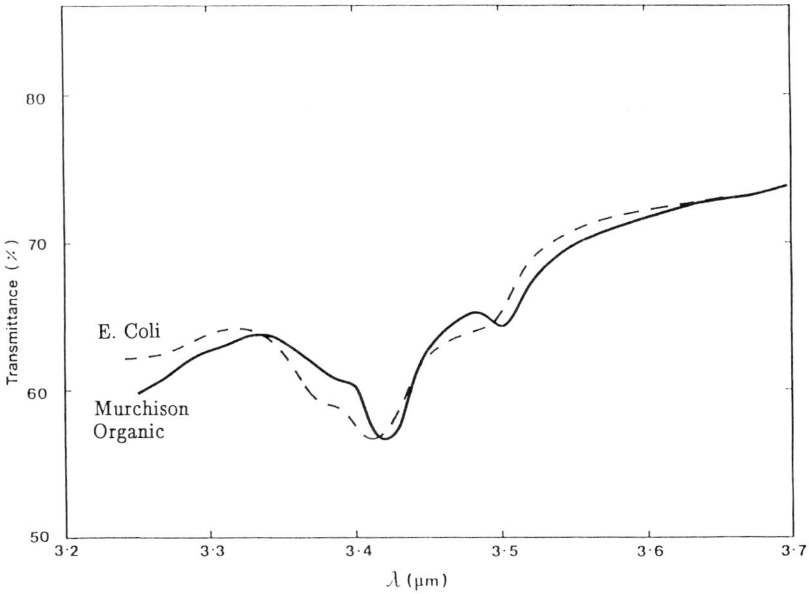

Fig. 9.20 *The absorption spectrum of organic material in the carbonaceous component of the Murchison meteorite after removing a flat graphite spectrum contributing 50% to the measured absorption at 3.289μm. Normalisation as in Fig. 9.16.*

One should also note that Pflug (1982, 1983) has found very many objects in the carbonaceous material of the Murchison meteorite that are morphologically of distinctive biological forms, as for instance the distinctive bacterium *Pedomicrobium*. Hence one can argue that the spectrum of the original material of the meteorite was like biological material for the good reason that it was biological. Even more boldly, one can see Fig. 9.20 as yet another confirmation that life exists outside the Earth.

A proposed abiotic solution that merits close attention is HAC (Hydrogenated Amorphous Carbon). Fig. 9.21(a) shows an uncalibrated transmittance curve for an HAC sample published by Watanabe et al. (1982). Fig. 9.21(b) is a simple superposition of this transmittance curve with a degree of smoothing over the data points of Allen and Wickramasinghe (*op. cit*). This fit proposed by Duley and Williams over the limited 3.3 – 3.6μm band looks promising at first sight, but the situation could be deceptive.

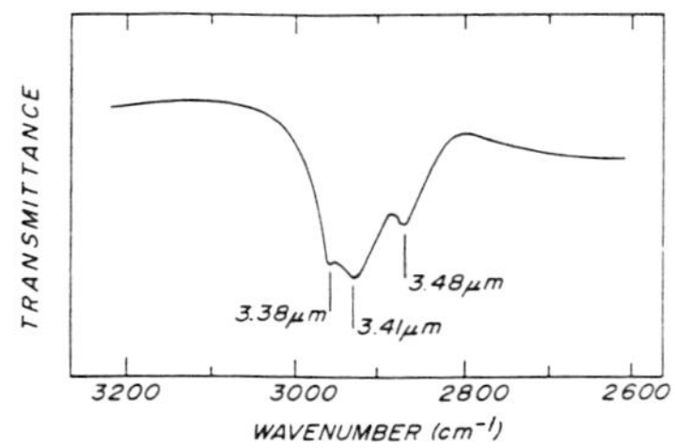

Fig. 9.21a *Absorption spectrum of HAC dust between 2600 and 3200 cm⁻¹ (Watanabe et al., 1982).*

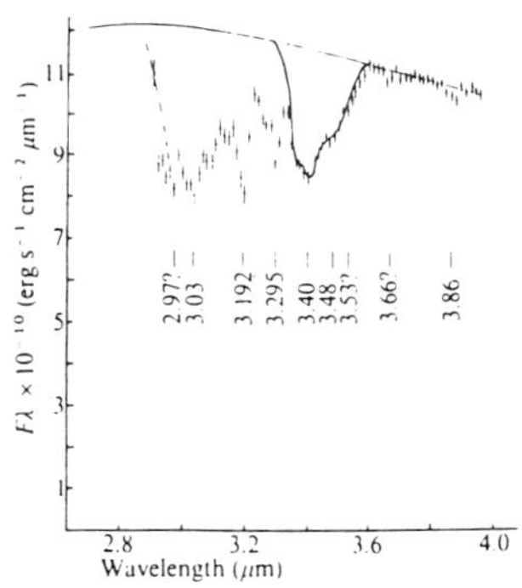

Fig. 9.21b *Duley and Williams' comparsion of HAC spectrum with GC–IRS 7 data (Duley and Williams, 1983).*

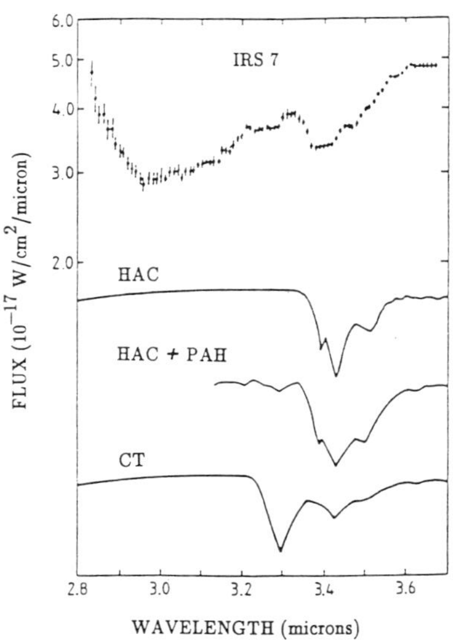

Fig. 9.21c *Ordinate–displaced absorption spectra computed by Colangeli et al, 1989 for some synthetic carbonaceous materials. A spectrum of IRS 7 is given for comparison.*

Unless it happened that the optical depth of the laboratory sample measured by Watanabe et al. was the same as the optical depth of the grains along the path length from the Earth to the galactic centre, this fitting of a laboratory spectrum to the observations is an invalid procedure. The ordinate in Fig. 9.21(a) is marked as 'Transmittance' which means $\exp[-\tau(\lambda)]$, where τ is the optical depth of the laboratory sample as a function of wavelength. When the amount of the sample is changed by a factor a, say, the transmittance curve becomes $\exp[-a\tau(\lambda)]$, which *does not have the same shape* as $\exp[-\tau(\lambda)]$ unless $\tau << 1$. In the case of GC–IRS 7 the observations require $\tau(3.4\mu m)$ of order unity, and this optical depth is not small enough for the shape of the transmittance curve to be regarded as invariant to the degree of accuracy involved in our own fit of the data as in Fig. 9.17. Only when the ordinate scale of a transmittance curve is given numerically can the amount of the sample be changed, and the transmittance curve given by Watanabe et al. does not have a numerical ordinate scale. It cannot therefore be used in the manner in which it was employed by Duley and Williams, except to suggest a *prima facie* case for further exploration. Note here that HAC does not possess any feature at $3\mu m$, so that other (non–ice) OH carrying solid materials are required to explain the observed absorption at this wavelength.

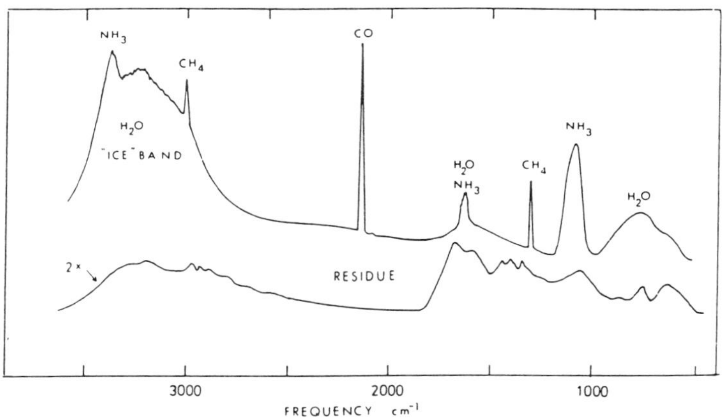

Fig. 9.22 *IR spectra of Greenberg's mixed ices before and after irradiation (From Greenberg, 1983).*

Bussoletti et al. (1978) and Colangeli et al. (1988) have carried out further laboratory investigations on the HAC hypothesis and claim that HAC mixed with polycyclic aromatic hydrocarbons and coal tar could produce a satisfactory $3.4\mu m$ profile. Their spectra for HAC, HAC+PAH and coal tar are shown in Fig. 9.21(c), together with a set of observations relating to GC–IRS 7. Since some of the starting materials in their samples – e.g. cellulose and coal tar – are of biological origin, the improved fit to the biological spectrum may not be too surprising. We observe that their material lacks a $3\mu m$ absorption, however.

A major effort has been directed in recent years to synthesising complex organic polymers by irradiation of mixtures of simple ices (Moore and Donn, 1982;

Greenberg, 1983; Strazulla, 1986; Wdowiak et al., 1988, 1989). The experiments could be looked on as complex solid-state versions of the famous Urey–Miller experiment. The experiments were done with a variety of motives, including the attempt to understand the origin of carbonaceous compounds in meteorites and comets. Initial mixtures of inorganics differ from one experiment to another, and so also does the nature of the ionizing radiation. Moore and Donn used particle

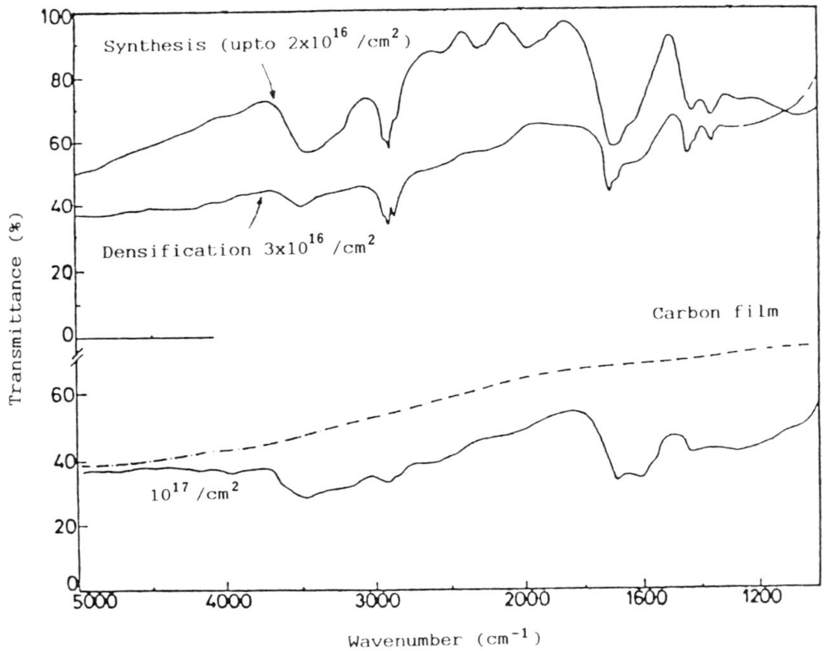

Fig. 9.23 *IR transmission spectra of a residue obtained from CH_4 frozen after 1.5 MeV proton irradiation. The dashed curve is the spectrum of an amorphous carbon film. (From Strazulla, 1986).*

irradiation of a mixture of CH_4, NH_3, H_2O; Greenberg used ultraviolet irradiation of the same materials; Strazulla used MeV proton irradiation of solid methane and water ice. Figs. 9.22 and 9.23 show the spectra of resulting organic residues, from Strazulla and Greenberg respectively. Not surprisingly the final spectra in the two cases are similar in their general appearance. Strazulla's spectra have the advantage that they are shown in a calibrated form so that $\tau(\lambda)$ values can be measured. Furthermore, Strazulla's transmittance data are plotted together with the normalised transmittance values for *E. coli* type material as is shown in Figs. 9.24 and 9.25. By comparing these plots with Fig. 9.20 we see that each of the synthetic curves deviate from the *E. coli* data even more than does the curve for Murchison material. The situation for abiotic organics as an explanation for the 3.4μm feature seems far from satisfactory at the present time. Moreover, the published spectra are of poor quality compared to Fig. 9.16, suggesting that only small quantities of the wanted residues were obtained.

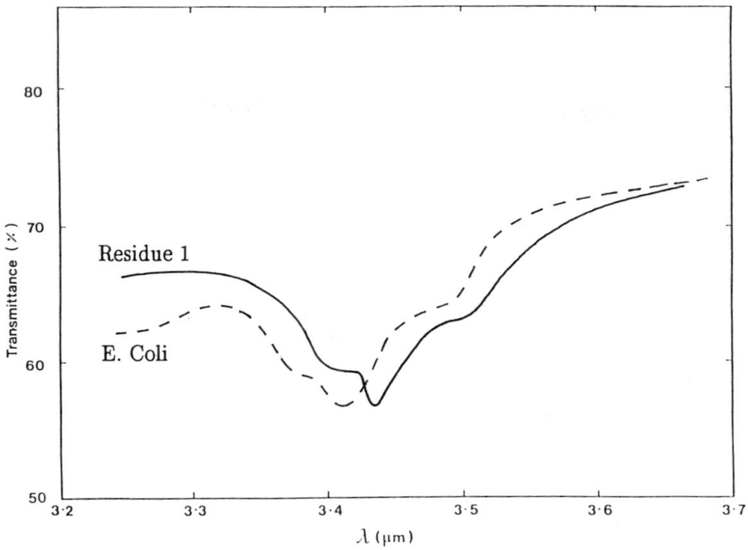

Fig. 9.24 *Comparison between normalised transmittance curves of E. coli and Strazulla's proton irradiated H_2O-CH_4 ices.*

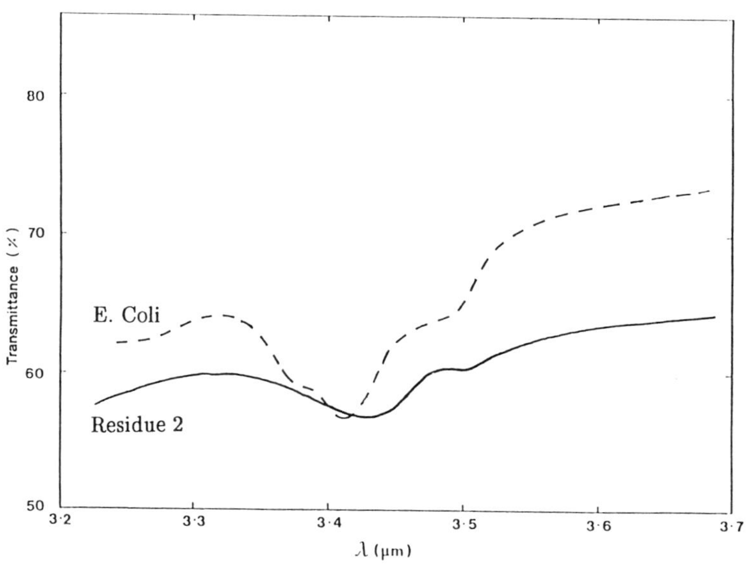

Fig. 9.25 *Same as Fig. 9.24 but with higher dose.*

9.6. HOW MUCH WATER–ICE?

We have seen in Section 9.5 that the absence of a strong 3.1μm ice feature in the spectrum of GC–IRS 7 militates against the presence of a significant amount of ice along the path length to this source. Several infrared sources associated with molecular clouds and stars with anomalous extinction curves (high A_V/E_{B-V} ratios) do show evidence, however, of a broad absorption feature near 3μm which has usually been attributed to water–ice. There is of course every reason to believe that water–ice condensation would occur within molecular clouds where H_2O molecules have indeed been observed. Even thin mantles of water ice on organic grains would produce strong absorption features near $\lambda = 3\mu$m, for the reason that the mass absorption coefficient of water ice is some 50 times that for organic matter typified by bacteria. Fig. 9.26 shows the effect of condensing various quantities of H_2O ice on a dessicated bacterial grain of radius 0.3μm. The calculation is done by taking $k(\lambda) = \alpha\lambda/4\pi$ where α is the average linear absorption coefficient of the mixture and $n(\lambda)$ is computed from the Kramers Krönig relations. Q_{ext} values are computed from the Mie formulae for spheres. The stack of curves is for (bottom to top) mass fractions of H_2O, $\xi = 0$, 0.001, 0.005, 0.01, 0.03, 0.05, .1, .3, .5. We note that even relatively small mass fractions of water lead to a dominance of the 3.1μm feature, causing a developing loss of the feature at 3.4μm.

The points (open circles) in Fig. 9.27 show observations of absorption near 3μm for BN, and also results reported by Whittett et al. (1983) for grains in the Taurus clouds. The data points for the Taurus grains show that the optical depth $\tau(\lambda)$ of the grains is such that

$$\frac{\tau(3.4\,\mu\mathrm{m})}{\tau(3\mu\mathrm{m})} \cong \frac{1}{4.8} \tag{9.8}$$

Attributing the 3.4μm absorption to the C–H stretching mode in organic material and the 3μm absorption to OH in water ice, whether crystalline or amorphous, and noting further that

$$\frac{\tau(3.4\,\mu\mathrm{m})}{\tau(3\mu\mathrm{m})} = \frac{\kappa_{org}(3.4\,\mu\mathrm{m})}{\kappa_{ice}(3\mu\mathrm{m})}\,\frac{\rho_{org}}{\rho_{ice}}, \tag{9.9}$$

where ρ_{org}, ρ_{ice} are the respective mass densities of the two materials, and κ_{org}, κ_{ice} are their mass absorption coefficients,

$$\frac{\rho_{org}}{\rho_{ice}} \cong \frac{1}{4.8} \cdot \frac{\kappa_{ice}(3\mu\mathrm{m})}{\kappa_{org}(3.4\,\mu\mathrm{m})} \tag{9.10}$$

Taking 0.917 as the specific gravity of ice, the absorption maximum of 33,000 cm^{-1} for ice is equivalent to a mass absorption coefficient κ_{ice} of 36,000 cm^2 g^{-1}. On the other hand for organic polymers of a wide range of types including polyformaldehyde and biomolecules, we find from laboratory measurements that

$$\kappa_{org}(3.4\mu\mathrm{m}) < 1200\ \mathrm{cm}^2\ \mathrm{g}^{-1}\,. \tag{9.11}$$

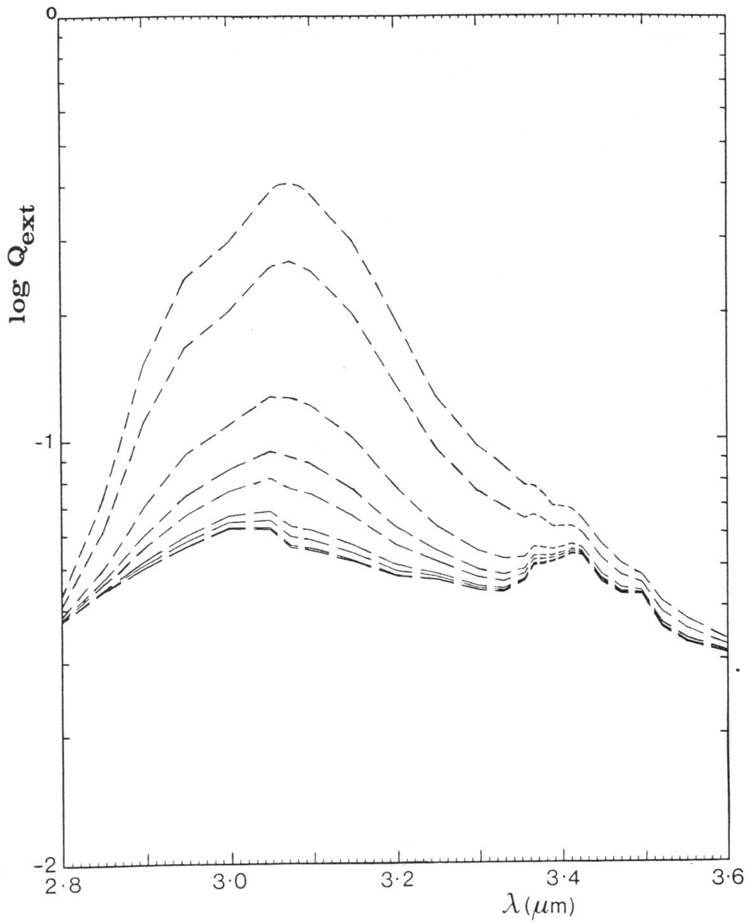

Fig. 9.26 *Calculated $Q_{ext}(\lambda)$ plotted on a log scale for dessicated bacteria of radii 0.3μm with various mass fractions of H_2O–ice. The stack is bounded by the pure dry bacterial curve at the bottom and a 50% by–mass water–ice case at the top. (From the bottom to top are curves for mass fractions of water: 0, .001, .005, .01, .03, .05, .1, .3, .5).*

Hence from (9.10) we have

$$\frac{\rho_{org}}{\rho_{ice}} > \sim 6.25 \,. \tag{9.12}$$

In view of the considerable excess of organic material given by (9.12) it is natural to think of any water–ice there may be as a thin condensation on the organic grains.

There is no hope of explaining the data in terms of water–ice grains alone, as can be seen from the calculations for grains of various sizes given by the curves in Fig. 9.27. These calculations were done from the Mie formulae using the optical constants obtained for amorphous ice by Léger et al. (1983). The data for BN is that of Merrill et al. (1976) (filled circles) and of Whittett et al. (1983) for Taurus (open circles), both the observations and calculations being normalised to give E = 6.4 at the wavelength of maximum absorption.

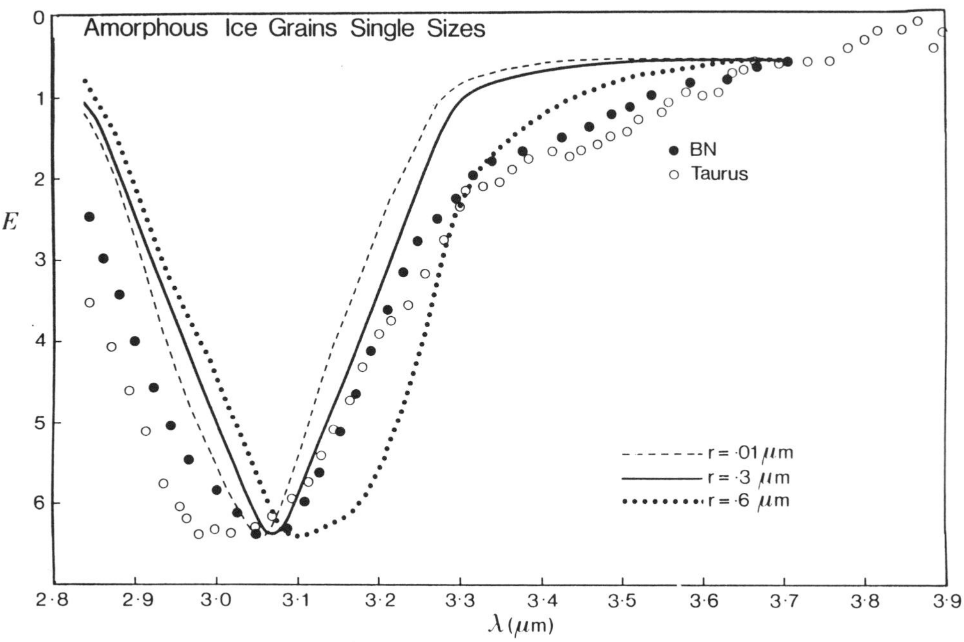

Fig. 9.27 *2.8 – 3.7μm extinction curves for amorphous ice grains, single sizes.*

Fig. 9.28 shows the effect of adding a few percent of ice to freeze–dried bacteria of the type of *E. coli*, and to eukaryotic cells of the type of yeast, the latter being in particularly good agreement with the data for BN. The transmittance data used to obtain Fig. 9.28 was from Fig. 9.14, the calculations being done for a size distribution appropriate to spore–forming bacteria, as discussed in Chapter 8.

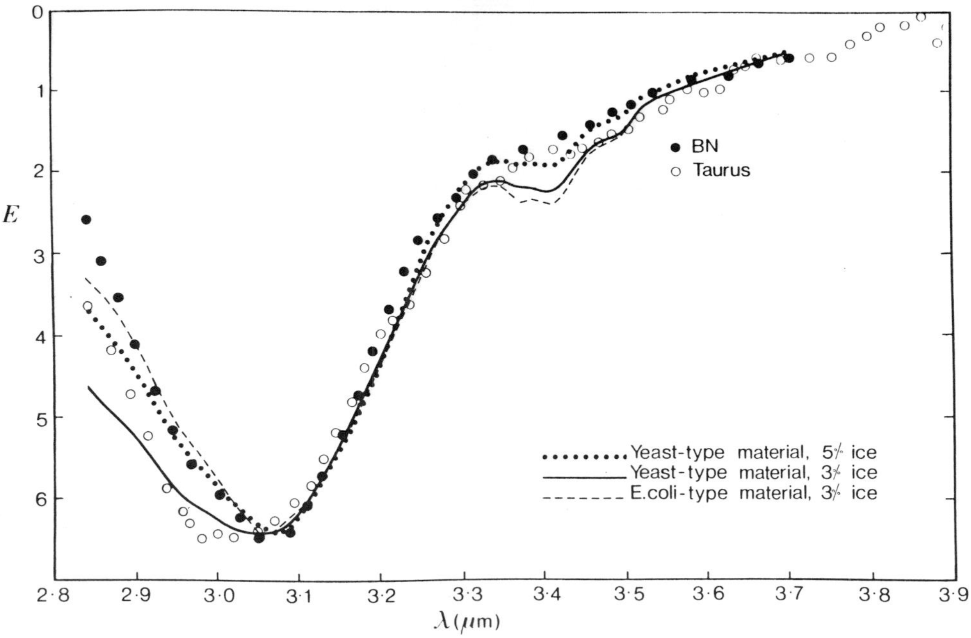

Fig. 9.28 *2.8 – 3.7μm extinction for bacterial size distribution of hollow grains of optical refractive index m = 1.16.*

A consistency check that there cannot be more than ~ 5% water ice in the BN source comes from modelling both the extinction and polarisation data over the 2.9 – 4μm waveband. From the data given by Kobayashi et al. (1980), Capps et al. (1978) and Dyck and Lonsdale (1981), we can calculate P/τ for the observations, and also for infinite cylinder models of ice particles combined with various mass fractions of organic material (typified by data in Fig. 9.14). Taking a grain of radius $a = 0.3\mu m$ we find that the mass fraction of ice must fall well below 10% to obtain a fit to the rather flat P/τ curve defined by the observations. These results are shown clearly in Fig. 9.29.

From a large number of infrared observations now available it appears that astronomical objects where H_2O–ice is a dominant grain species are rare. The highly variable 3.0μm feature is more indicative of the role of an organic component with H_2O–ice occurring only in trace quantity. Unequivocal detections of crystalline water–ice is confined to a few very cold OH/IR circumstellar shells, a striking example being the source IRAS 09371+1212 (Omont et al., 1989). Both the 3.1μm absorption and a 43μm emission feature point to the dominance of crystalline ice in this particular instance.

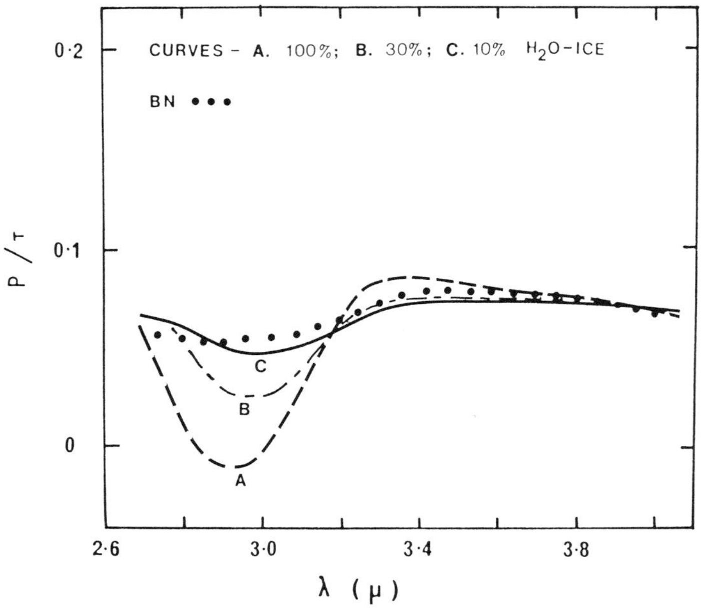

Fig. 9.29 *Polarisation to optical depth ratio of the BN object from 2.6 – 3.9μm compared with calculations for H_2O ice cylinders and for cylinders comprised of organic–ice mixtures (Data from Capps et al, 1978).*

9.7. SOURCES WITH SPECTRA IN THE 2 – 14μM WAVEBAND

With extensions of the wavelength base over which observations are available, the scope for spectroscopic identification expands. Fig. 9.30a,b shows ground–based 2 – 4μm and 8 – 13μm spectra of objects associated with molecular clouds and both oxygen–rich and carbon–rich cool stars (Merrill and Stein, 1976). The oxygen–rich stars show 10μm features in absorption (Fig. 9.30a) and the carbon stars (Fig. 9.30b) show the usual 10.5 – 12.7μm band commonly attributed to SiC. We note that this latter band appears with a more or less constant profile, militating against the SiC explanation which is shape/size dependent. The most curious feature in the carbon star spectra is a broad absorption near $\lambda = 3\mu m$ which from temperature considerations cannot be due to water–ice. In view of the carbon excess associated with these stars, a more likely identification would be that both 3μm and 11μm features are due to organic molecules, the former arising from OH and NH bonds.

The situation in a low density stellar mass flow would inevitably be highly complex. Condensation of molecules and polymers would be expected to produce variable results depending upon the precise set of physical conditions that prevailed. In view of this it is indeed remarkable that we were able to model the spectra of a wide range of sources, as shown in Fig. 9.31 (Hoyle and Wickramasinghe, 1977). The model required absorbing and emitting material with a transmittance curve that was closely similar to that of cellulose as measured in the laboratory, a single material only.

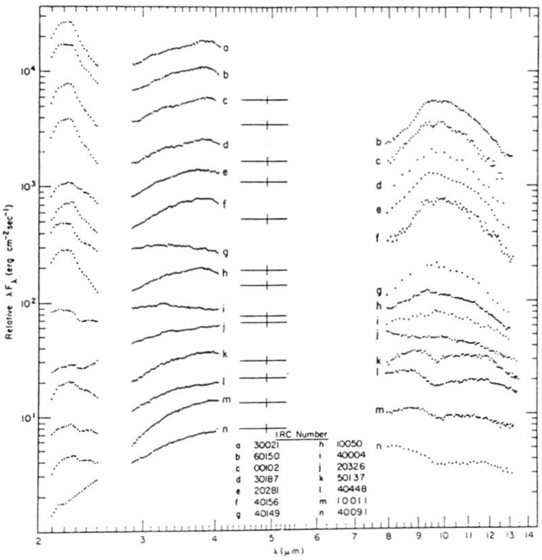

Fig. 9.30a *Ground–based spectrophotometry (in the form λ F_λ) of oxygen–rich CRL sources showing 10–μm absorption. Sources a–d are associated with dense molecular clouds, whereas e and f are cool stars. (From Merrill and Stein, 1976).*

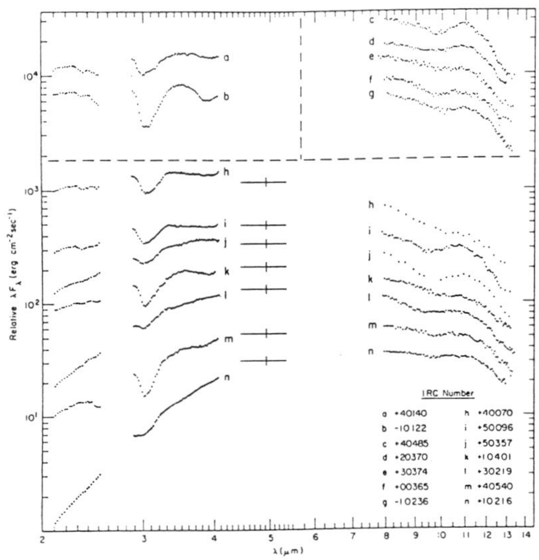

Fig. 9.30b *Ground–based spectra of carbon–rich IRC sources, revealing a 11.5μm emission feature. (From Merrill and Stein, 1976).*

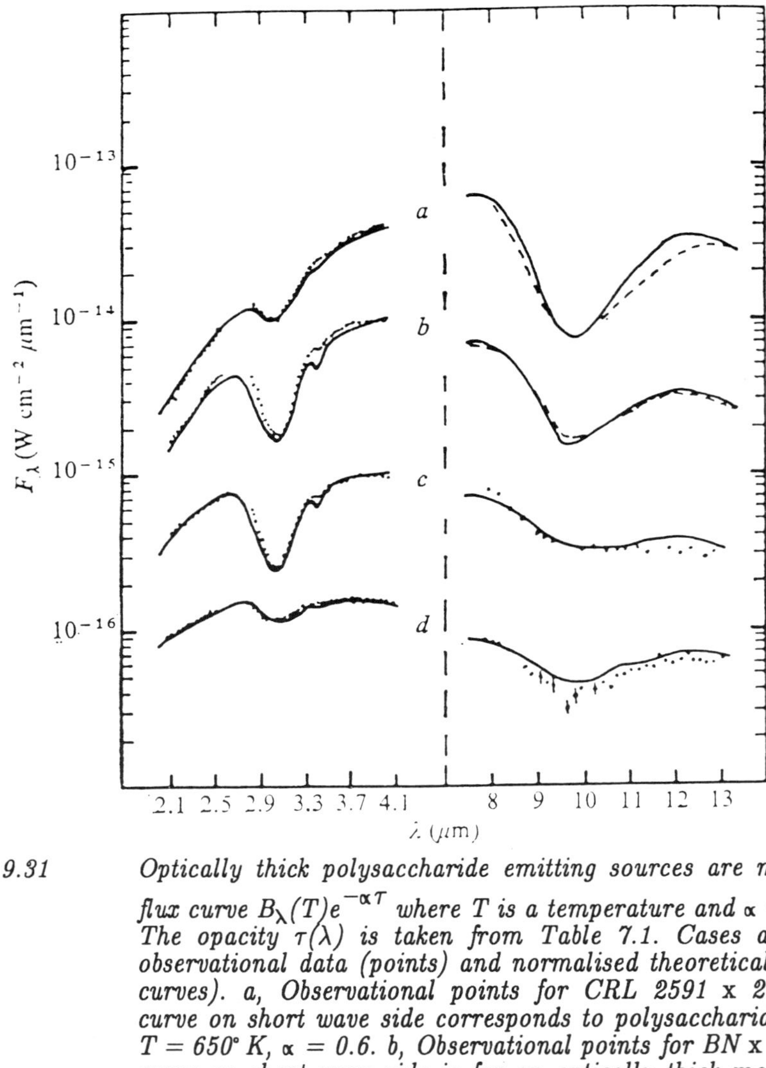

Fig. 9.31 *Optically thick polysaccharide emitting sources are modelled by a flux curve* $B_\lambda(T)e^{-\alpha\tau}$ *where T is a temperature and* α *is a constant. The opacity* $\tau(\lambda)$ *is taken from Table 7.1. Cases a–d represent observational data (points) and normalised theoretical fluxes (solid curves). a, Observational points for CRL 2591 x 20; theoretical curve on short wave side corresponds to polysaccharide model with* $T = 650° K$, $\alpha = 0.6$. *b, Observational points for BN x 4; theoretical curve on short wave side is for an optically thick model with* $T = 650° K$, $\alpha = 1.75$. *c, Observational points for NGC 2264I R; theoretical curve on short wave side is for optically thick model with* $T = 800° K$, $\alpha = 1.7$. *d, Observational points for CRL 490 x 3; theoretical curve on short wave side is for optically thick model with* $T = 850° K$, $\alpha = 0.25$. *The theoretical curves are normalised to match the observations at one wavelength.*

The agreements in Fig. 9.31 suggest the type of mass–flow condensation model discussed in Chapter 8. We cannot from this infer the explicit composition of cellulose, but what it does show is that a water–ice/silicate explanation is not a logically inescapable conclusion to be drawn from spectra such as in Fig. 9.30 or 9.31.

The spectra in Figs. 9.30 and 9.31 lacked data points in the wavelength range 4 – 8μm, for the reason that radiation at these wavelengths is absorbed in the Earth's lower atmosphere. The use of telescopes above the main atmospheric absorbing layer permitted more complete spectra to be obtained covering the entire 2 – 14μm waveband. A large number of such spectra are now available from the use of telescopes such as the 0.9m telescope on the Kuiper Airborne Observatory (KAO). A sample of spectra in the 10μm band and the 3μm band are conspicuously variable.

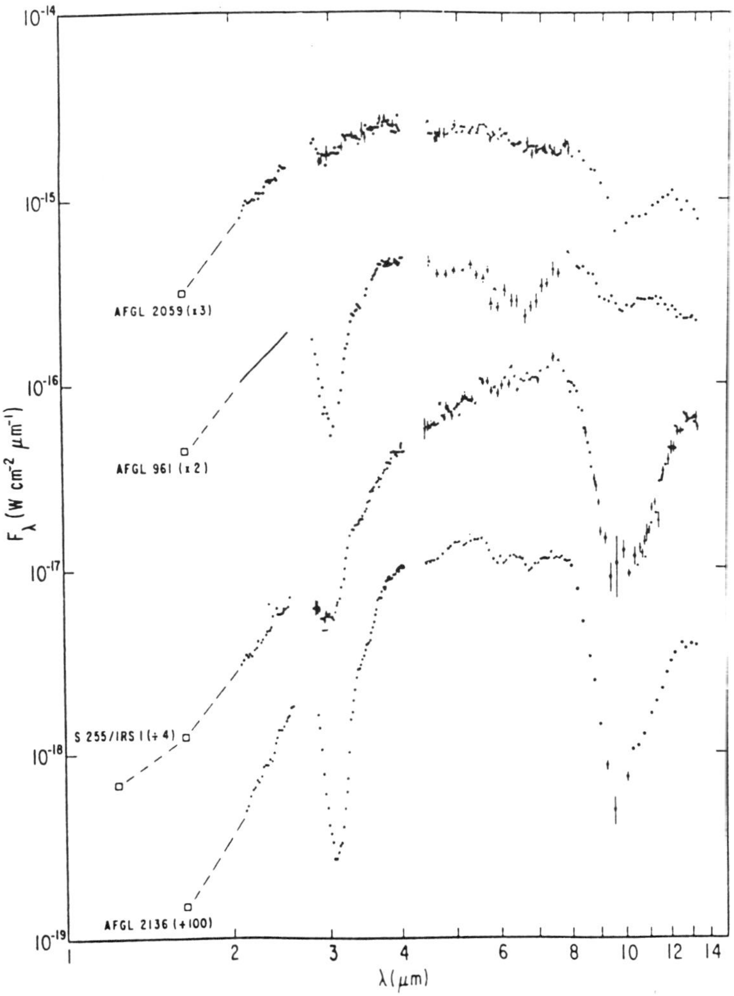

Fig. 9.32 *Spectra of several IR sources from observations using the KAO (Willner et al., 1982).*

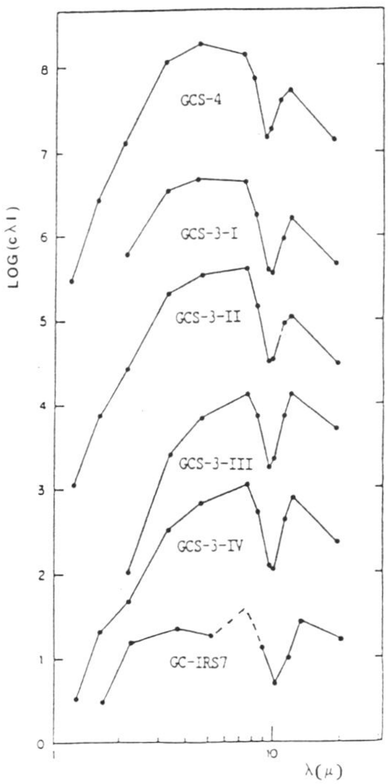

Fig. 9.33 *Infrared spectra of several sources near the galactic centre (Okuda et al., 1989).*

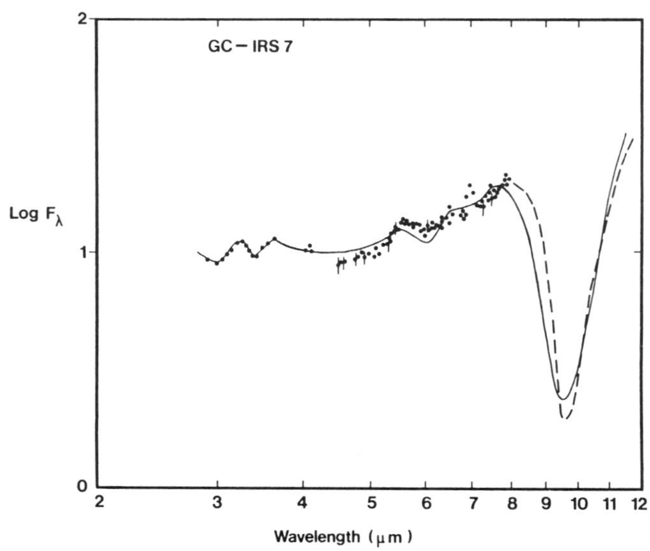

Fig. 9.34 *The observed flux from GC–IRS 7, represented by the dots and the dashed curve (Willner et al., 1979; Allen and Wickramasinghe, D.T., 1981) compared with the calculated flux from the bacteria/silica model discussed in § 9.3 (solid curve).*

Furthermore, we note that the data in Fig. 9.32 reveals a new set of broad, diffuse absorptions in the 4 – 8μm wavelength interval. The characteristic wavelengths of these absorptions are again generally indicative of C–H, C–O, C–O–C, C–C bonds in organic molecules.

For reasons we have already discussed, infrared spectra of sources near the galactic centre have a special significance in that they sample a 10 kpc pathlength to the galactic centre. Fig. 9.33 shows a set of coarse to intermediate resolution infrared spectra for a number of discrete sources distributed within about 1 pc of the source GC–IRS 7 (Okuda et al., 1988). In all cases a 9.7μm extinction of ~ 2.5 mag relative to a neighbouring continuum is seen. This consistency in the value of the central opacity near $\lambda \cong 9\mu$m provides strong confirmation that the extinction arises mostly from grains in the 10 kpc long line of sight rather than in a local environment. The points and dashed line of Fig. 9.34 represent infrared data at higher spectral resolution for the source GC–IRS 7. Over the waveband 2.9 – 3.9μm the observations of Allen and Wickramasinghe (1981) are used together with the data compiled by Willner et al. (1979) over the waveband 3.9μm to 13μm. The two sets of data are matched at $\lambda = 4\mu$m. We note that the envelope of the observations between 2μm and 4μm is closely the same as a Planck curve for T = 1100° K and the slope between $\lambda = 8\mu$m and $\lambda = 12\mu$m corresponds to a Planck curve at about 250° K. Accordingly we adopt an underlying source for GC–IRS 7 to correspond to a superposition of two black bodies, the cooler contributing to the bulk of the radiation at $\lambda = 8\mu$m and the hotter to the bulk of the radiation at $\lambda = 3.3\mu$m. Such a situation may be realised in the case of a hot star surrounded by an extended dust cloud with a radial temperature gradient.

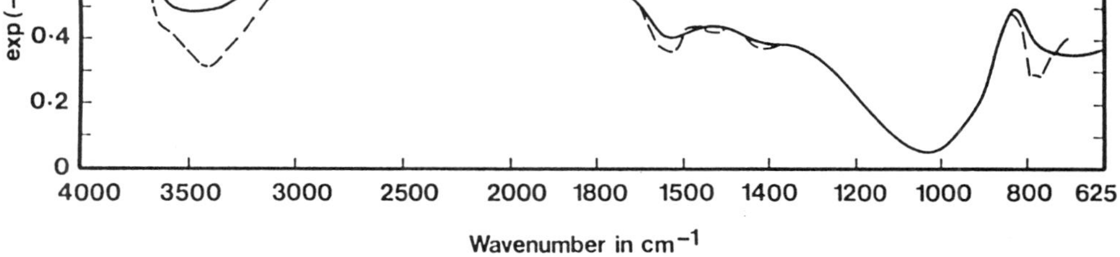

Fig. 9.35 *Transmittance data for dried–out bacteria–diatom silica mixture (solid curve); data for 'damp' sample (dashed curve). Solid curve over the 2.5 – 4μm waveband coincides with spectrum of disc heated in a cell flushed with* N_2 *to 180° C.*

In Section 9.3 we showed that over the 8 – 35μm waveband the spectrum of the Trapezium nebula corresponded closely to the flux predicted from a mixture of diatom silica and a mixed bacterial sample. The dashed curve of Fig. 9.35 is the transmittance curve for such a best–fitting organo–siliceous mixture. On account of the strongly hygroscopic properties of diatom silica the absorption at ~ 3μm is still dominated by a trace quantity of water which is absorbed. Using the laboratory spectrum of liquid water to correct for this absorption and the ratio of optical depths at 2.9μm and 3.4μm corresponding to fully desiccated organisms (Fig. 9.15) we obtain the solid curve of Fig. 9.35 as the transmittance appropriate for fully dried out diatoms. We have confirmed that this procedure yields a spectrum that almost exactly corresponds to measurements of a disc containing the mixture heated to 180° C in a cell flushed with N_2 at high pressure. We have, however, chosen to use the solid curve in Fig. 9.35 (Table 9.3) in our calculation because we think that it represents an asymptotic situation with zero water content. To compute the flux for GC–IRS 7 we use the formula

$$F_\lambda \propto [B_\lambda(1100K) + \xi B_\lambda(240K)]\, e^{-\alpha\tau} \tag{9.14}$$

with ξ chosen so that the cooler black body contributes 96% of the flux at $\lambda = 8\mu m$ and the hotter black body 99% of the flux at $\lambda = 3.3\mu m$. The value of α is taken to be $\alpha = 1.14$. The calculations are normalised to 1.00 at 4.5μm to correspond with the observations. The resulting flux values are also tabulated in Table 9.3 and plotted as the solid curve in Fig. 9.34.

TABLE 9.3

Optical depth of bacteria–silica mixture and the calculated flux for GC–IRS 7

$\lambda(\mu)$	τ	F_λ	$\lambda(\mu)$	τ	F_λ
2·9	0·67	0·95	6·2	0·89	1·19
3	0·69	0·91	6·5	0·82	1·51
3·1	0·58	1·01	7	0·97	1·61
3·2	0·49	1·10	7·5	0·97	1·96
3·3	0·46	1·11	8	1·20	1·77
3·4	0·62	0·90	8·5	1·71	1·12
3·45	0·51	1·01	9	2·53	0·49
3·5	0·48	1·03	9·5	3·22	0·24
3·6	0·37	1·13	10	3·00	0·33
3·7	0·36	1·11	10·5	2·12	0·92
4·0	0·36	1·04	10·75	1·77	1·40
4·5	0·37	1·00	11	1·56	1·80
5·0	0·43	1·06	11·5	0·97	3·57
5·5	0·49	1·26	12·5	0·92	3·78
6·0	0·87	1·09	13·0	1·02	3·34

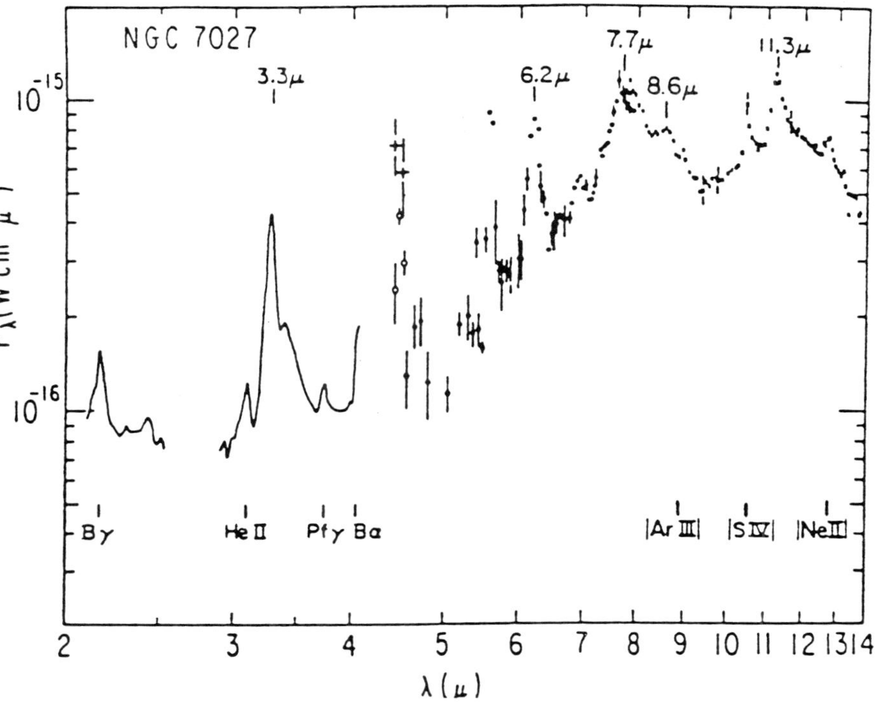

Fig. 9.36 *2 – 14μm spectrum of NGC 7027 showing the unidentified infrared emission features together with atomic transitions. (From Russell et al., 1977b).*

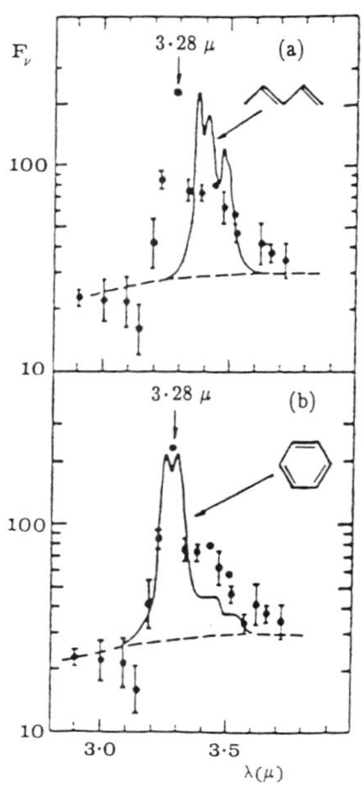

Fig. 9.37 *Spectra of C_6H_6 and C_5H_{12} compared data for NGC 2023 (Léger and Puget, 1984).*

Over the limited wavelength range $\lambda = 2.9 - 3.9\mu m$ the correspondence between the astronomical data of Allen and Wickramasinghe (1981) and our earlier model calculation (Fig. 9.17) is precisely reproduced. In addition, we find excellent agreement over the $4 - 8\mu m$ spectral region and over the $8 - 12\mu m$ waveband. This close agreement points, in our view, to submicron grains with average properties very similar to the organic/silica mixture we have described in Section 9.2. The mixture it will be recalled was in the usual cosmic proportions.

9.8. EVIDENCE FOR PAH

Diffuse infrared emission features at the wavelengths 3.3, 3.4, 6.2, 7.6, 8.7 and $11.3\mu m$ were known to exist in several types of astronomical object from about 1975 (see Wilner, 1984). The confirmation of the existence of these diffuse emission bands in the mid-infrared waveband $4 - 8\mu m$ was one of the major triumphs of the Kuiper Airborne Observatory. Fig. 9.36 shows the spectrum of the planetary nebula NGC 7027 in which most of the diffuse IR emission bands were noticed clearly for the first time (Russell et al., 1977). Amongst the classes of object in which some or all of these bands have been observed are planetary nebulae, reflection nebulae, HII regions and active galaxies. The first attempt to understand a physical mechanism for the excitation of these bands was due to Sellgren (1984), who proposed that very small grains, comprised of less than ~ 100 atoms, are transiently heated to temperatures of ~ 1000° K when they absorb UV photons (see discussion in Chapter 6). Such heated grains then emit infrared radiation which bear the spectroscopic signatures of constituent molecules.

Léger and Puget (1984) proposed that diffuse IR emissions may be due to polycyclic aromatic molecules, a grain model which was in fact first dicussed by Donn (1968). They considered polycyclic aromatic hydrocarbons – particularly structures in a compact graphitic sequence of which benzene is the smallest. Fig. 9.37 shows their calculated fits to the $3.28\mu m$ band profile in NGC 2023 for both benzene (an aromatic molecule) and C_5H_{12} an aliphatic chain. A combination of the two seems necessary. Fig.9.38 shows the emission bands of a PAH sequence weighted with Planck functions at different temperatures (Léger et al., 1989). The dashed curve is a representation of Sellgren et al's. (1985)observations of NGC 2023 to which a $6.2\mu m$ band as observed in M82 (Willner et al., 1977) was added. The general correspondences with the data are good for some bands, but not so good for others. The worst case appears to be at $11.3\mu m$ where a principal band in the laboratory spectra (e.g. for coronene) is significantly displaced. What is clear, however, is that aromatic molecules of some types are involved in causing infrared band emissions as are observed in such objects as M82 an NGC 2023.

Until 1988, PAH emission bands seemed confined to localised sources in our galaxy and to external galaxies. The remarkable discovery of Giard et al. (1988) was the detection of the $3.3\mu m$ aromatic emission feature in the diffuse galactic radiation which clearly arose from small heated grains. Spectrometric data in the $2.8 - 3.6\mu m$ waveband obtained from a balloon-based instrument are reproduced in Fig. 9.39, along with data from other workers and including IRAS observations at 12, 25, 60 and $100\mu m$. The long-wave part of this flux curve is clearly due to 15° K classical-sized interstellar grains, whereas at wavelengths less than $5\mu m$ PAH-type particles must dominate to produce both the underlying continuum as well as the spectral features.

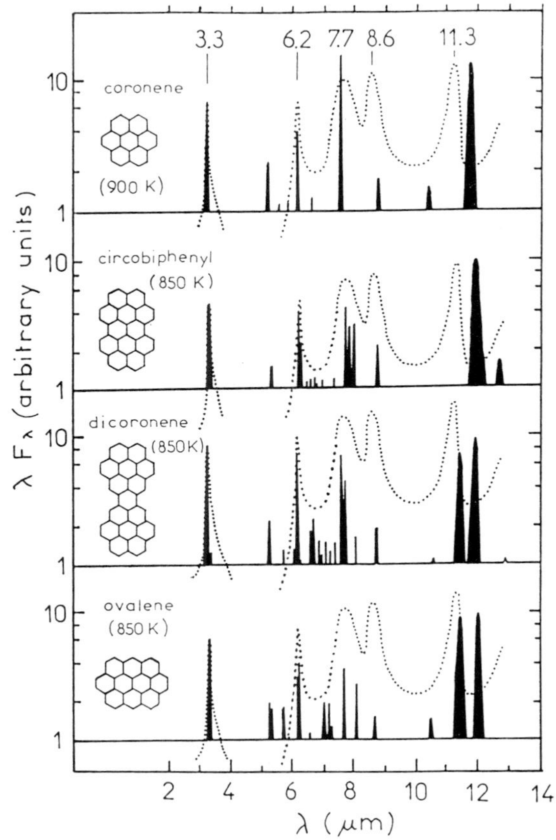

Fig. 9.38 *Emission spectra of several compact PAH's at different temperatures calculated from absorption spectra measured in the laboratory by A. Léger. The dotted line is Léger's representation of the data for NGC 2023 from the observations of Sellgren et al. (1985) (Courtesy A. Léger).*

3.3μm Emission and 2200 Å Absorption. Infrared emission throughout the 3 – 12μm waveband must arise from an absorption of ultraviolet starlight that is subsequently degraded into the infrared. The total fraction of carbon tied up in the form of PAH molecules around sources may be typically in the region of ~ 3% (Léger and Puget, 1984), and the same mass fraction of total galactic carbon seems to be needed for explaining the data of Fig. 9.39.

Organic molecules in very small grains in the quantities involved here must contribute appreciably to the extinction of starlight at ultraviolet wavelengths. Aromatic molecules invariably possess strong absorptions in the ultraviolet with band strengths amounting to ~ 500,000 $cm^2\ g^{-1}$ in typical cases and with the peak UV absorption wavelength varying in the range ~ 1800 – 2600 Å. A density of ~ 10^{-27} $g\ cm^{-3}$ in the form of aromatic molecules would thus produce an absorption

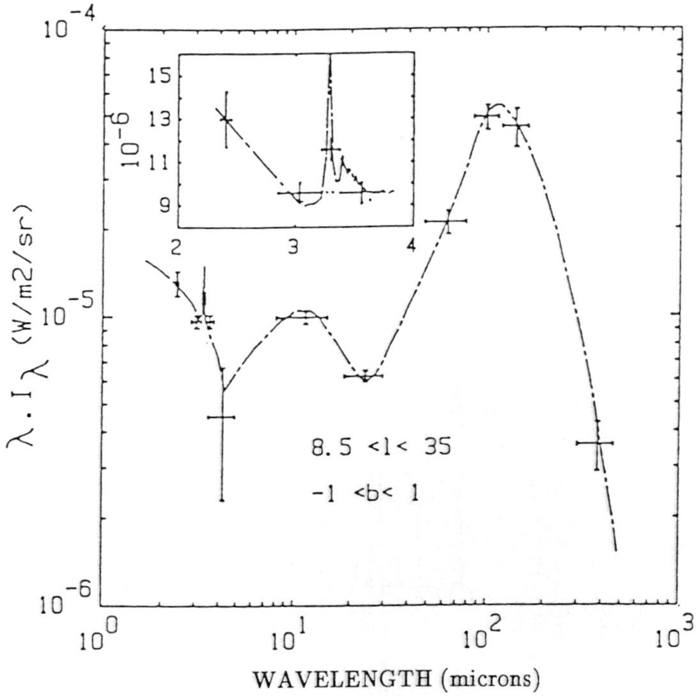

Fig. 9.39 *Giard et al. (1988) data of the averaged $\lambda.I_\lambda$ values for the inner galaxy (8.5° < ℓ < 35°, −1° < b < + 1°) in the IR and submillimeter range. The insert shows a detail of the AROME measurements. A typical observed profile of the 3.3μm aromatic feature in Fig. 9.37 has been scaled to match this data.*

of ~ 2 mag/kpc at the centre of the UV band, which is close to the value appropriate to the well–known λ 2175 Å interstellar absorption feature. It would seem natural, therefore, to connect the 3.28μm diffuse galactic emission with the λ 2175 Å interstellar absorption. A candidate molecule for the 3.28μm emission that conspicuously fails to produce an extinction band at 2175 Å could be deemed unsatisfactory for this reason (Hoyle and Wickramasinghe, 1989).

We have already seen that a compact graphitic–PAH typified by coronene gives a reasonable degree of correspondence in the positions and relative strengths of emissions at 3.28, 6.2, 7.7 and 8.6μm. However, the sequence of molecules from naphtalene, tetracene, perylene, coronene does not meet the condition of the previous paragraph for reasons clearly evident in Fig. 9.40. We see here the UV spectra for members of this sequence starting from naphtalene. Naphtalene has an absorption peak at ~ 2200 Å, which is indeed satisfactory, but further members have peaks that are systematically shifted to longer wavelengths, tending eventually to a peak near 2600 Å, which is appropriate to the absorption coefficient of bulk graphite. The higher order compact PAH's, which are expected to behave like bulk graphite, are therefore less than satisfactory models for the interstellar 3.28μm emission band.

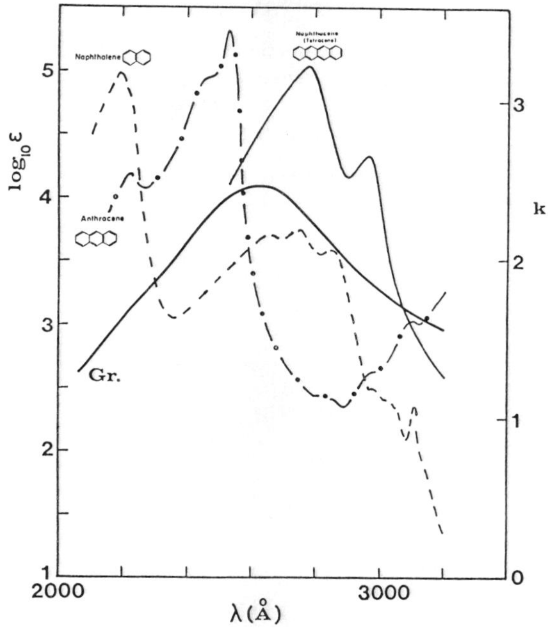

Fig. 9.40 *The absorption spectra of naphtalene, anthecene, tetracene and graphite. The ordinate scale for molar extinction coefficient (cm^{-1}) (gm mol/litre)$^{-1}$ refer to the molecules; the absorption coefficient k refers to graphite (Gr.).*

We pointed out some years ago that the distribution of ultraviolet chromophores in a wide range of naturally occurring molecules peaks strongly at a wavelength close to ~ 2200 Å (Hoyle and Wickramasinghe, 1977, 1979). In particular we noted that several bicyclic aromatic structures involving N–substitutions in a benzene ring had spectra similar to the λ 2175 Å interstellar feature (Hoyle and Wickramasinghe, 1977). Quinoline and its derivatives are amongst the substances we considered which have UV peaks at wavelengths ranging from 2180 Å to 2300 Å, as can be seen from Table 9.4 (Scott, 1964). The infrared spectra of two members of this group are displayed in Fig. 9.41 (Stadler IR Spectra, 1978). Using the transmittance data from this figure, we now compute the profile of the 3.28μm emission in NGC 2023 according to the equation

$$F \propto \tau B_\lambda(T) , \tag{9.15}$$

where τ is the optical depth of the laboratory sample. The two curves in Fig. 9.42

TABLE 9.4

Principal absorption wavelength and molar extinction coefficient of quinolines and isoquinolines (Scott, A.I., Interpretation of the Ultraviolet Spectra of Natural Products – Pergamon Press, 1964).

Compound	λ(nm), (ε)
QUINOLINES	
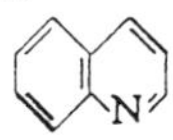	226 (35,500)
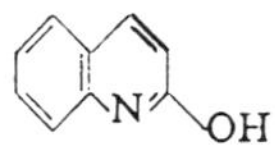	224 (26,700)
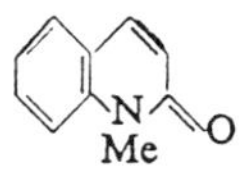	228 (34,100)
ISOQUINOLINES	
	218 (79,000)
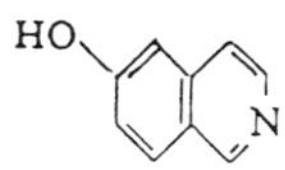	229 (43,600) 225 (25,600)
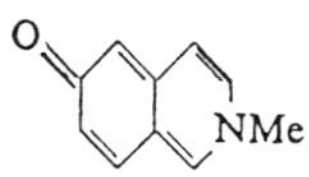	230 (34,400)

are for a temperature T = 400° K and for (a) quinoline (or isoquinoline) and (b) a mixture of quinoline and 7–methylquinoline in a proportion such that at 3.28μm the ratio of opacities is 2 : 1. The latter case gives a better agreement in the 3.4μm wing of the band, and it is also significantly better in this regard than coronene and other compact PAH's belonging to the graphitic sequence. If the diffuse galactic IR emission includes the same bands at 6.2, 7.7, 8.6 and 11.3μm as in M82 and NGC 2023 (Selgren et al., 1985; Willner et al., 1977), then 7–methylquinoline (in the form of very small particles) can offer an explanation for these feature at 11.3μm which the compact graphitic sequence (including coronene) does not. We do not wish to suggest that clusters of quinoline or its derivatives are precisely identifiable as being responsible for the astronomical features we have discussed. These molecules, however, seem to be representative of the inevitably complex molecular mix that must exist in interstellar space.

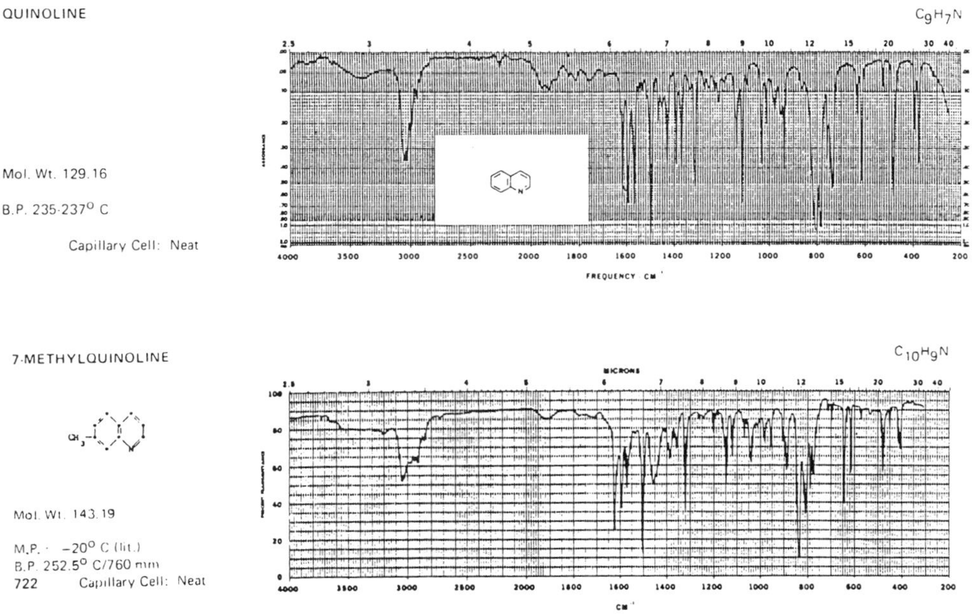

Fig. 9.41 *Infrared spectra of quinoline and 7–methylisoquinoline from the Stadler Handbook (1978).*

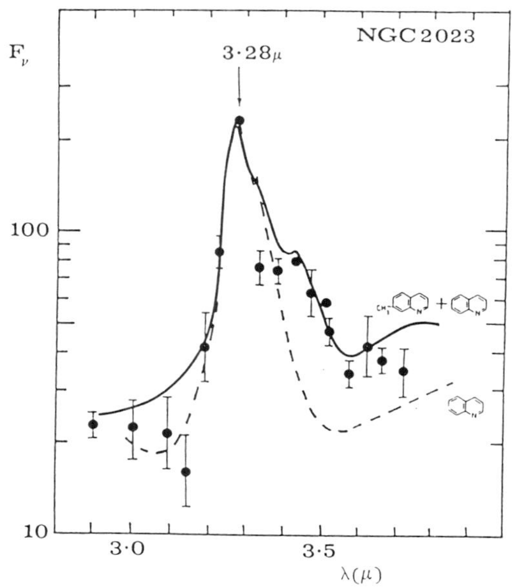

Fig. 9.42 *The flux from the planetary nebula NGC 2023 in the 2.9 – 3.7μm waveband compared with the prediction for emission by quinoline and a mixture of quinoline and 7–methylisoquinoline. The proportions in the mixture are chosen so that the components contribute to the 3.28μm absorption in the ratio 2 : 1.*

The Biological Model. We now consider the possible relationship between the biological grain model discussed in Chapters 7 and 8 and the diffuse interstellar bands. If the grains with typical diameters of $1\mu m$ which cause the visual extinction of starlight are predominantly biological in character their degradation products would also be expected to occur under interstellar conditions. Organic particles with total numbers of atoms ranging from several hundred to a few tens might be thought to result from the break up of larger grains. The smallest particles in this range could include individual aromatic molecules which might be expected to have relatively short dissociative lifetimes. The radioastronomical organic molecules, which are mostly linear structures, have abundances relative to hydrogen in the relatively narrow range $10^{-8} - 10^{-11}$, more or less independently of complexity or molecular weight (Mann and Williams, 1980). This feature is consistent with the view that such molecules are degradation products of more complex structures. The molecules observed so far by radioastronomical techniques probably represent only the tip of an iceberg, with an extensive population of molecular species remaining unobserved for purely practical reasons. Simple aromatic molecules such as benzene have proved difficult to detect, partly because of the low oscillator strengths of the relevant transitions between rotational levels.

The formation of aromatic molecules on a vast scale in the galaxy, as is implied by the data we have dicussed, poses a major difficulty for inorganic theories of astrochemistry. Aromatic molecules, however, are commonplace in biology and such molecules are expected to arise quite naturally within fragments of biological grains as they become disrupted in conditions of intense ultraviolet or particle irradiation. It is possible that the ratio of the numbers of condensed aromatic to aliphatic molecules might increase substantially above an initial value in the parent grains under certain irradiation conditions, due to the greater stability of ring structures. The observation that the $3.3\mu m$ aromatic CH stretching invariably dominates in emission spectra whilst the $3.4\mu m$ aliphatic stretching dominates in absorption may have an interesting relevance to this question. In this connection we also note that the insoluble organic matter in carbonaceous chondrites is comprised mainly of a polymer with an aromatic skeleton (Hayes, 1967).

TABLE 9.5

Categories of molecules considered

α–β unsaturated acids and esters
α–β unsaturated lactones
Indole chromophores
Pyridines
Quinolines, isoquinolines, and acridines
Pyrimidines
Purines
α –oxo and γ –lactones and derivatives
Hydroxyanthraquinones

We now consider a sample of naturally occurring aromatic molecules listed under various catagories in several tables given by Scott (1964). Table 9.5 lists the classes of compounds considered.

For 115 aromatic molecules under the above categories that were listed explicitly by Scott (1964) we examined infrared spectra in the Stadler Atlas (1978) and in Butterworth's Index Cards of Spectra (1966, 1970, 1972). We first constructed a histogram of the distribution of the wavelength of maximum CH absorption in the 3.2 to 3.5μm waveband, which is shown in Fig. 9.43. The average band profile defined by the curve in this figure peaks at $\lambda = 3.28\mu$m, in good agreement with the behaviour of an individual heterocyclic ring structure such as quinoline that was discussed earlier.

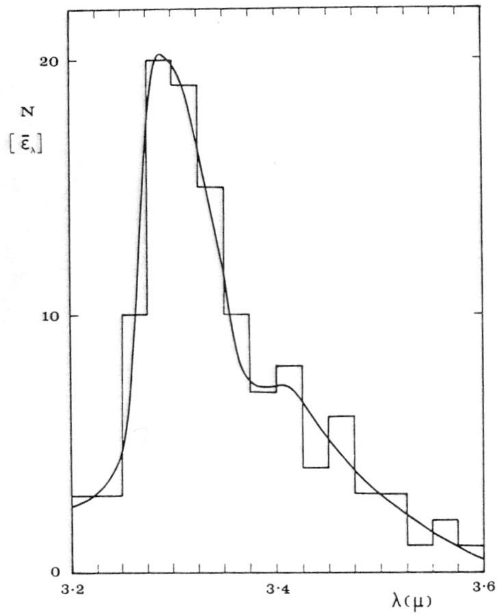

Fig. 9.43 *The distribution of the wavelength of the maximum of the CH band in 115 aromatic compounds in categories listed in Table 9.5 and set out in several tables of Scott (1964). See also Hoyle and Wickramasinghe (1979). The curve represents a mean emissivity profile of the mixture in the waveband 3.2 – 3.6μm.*

We next proceed to estimate the average emissivity of the ensemble (including 115 molecules) listed in Table 9.5 over the wavelength range 2.5 – 12.5μm. Each of the laboratory ϵ_λ curves from the standard atlases were electronically digitised and scaled so that $\epsilon_\lambda = 1$ at the wavelength of maximum absorption in the 2.5 – 4μm waveband. The set of normalised ϵ_λ functions were then averaged with equal weightings to yield a normalised $\bar{\varepsilon}_\lambda$ curve for the entire ensemble. We envisage that such molecules will be contained within fragments of organic grains comprised typically of between 50 and 300 atoms and attaining temperature spikes in the general range ~ 200 to 1200° K.

If the constituent molecules of such an aromatic ensemble are distributed uniformly over all the grain fragments irrespective of size, the radiation flux arising from the entire ensemble will be given by

$$F_\lambda \propto \bar{\varepsilon}_\lambda \{\Sigma B_\lambda(T)\, S(T)\}\,, \tag{9.16}$$

where $\bar{\varepsilon}_\lambda$ is the average emissivity and S(T) is the total surface area of fragments with temperature T at any instant and $B_\lambda(T)$ is the Planck function.

We next make the simplifying assumption that the distribution of temperatures in equation (9.16) average under the summation sign to give an approximately flat wavelength dependence of the function within curly brackets. In this case log F_λ for

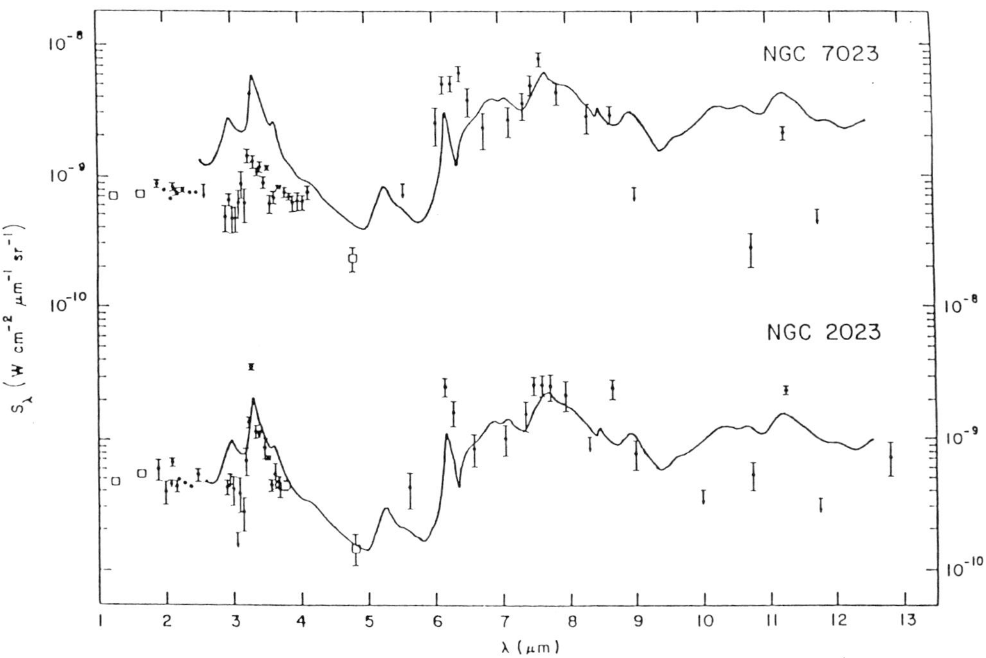

Fig. 9.44 *The spectra of NGC 7023 and NGC 2023 from Sellgren et al. (1985) compared to a model involving a mixture of 115 aromatic molecules.*

an astronomical source would differ from log $\bar{\varepsilon}_\lambda$ only by a constant and a direct comparison between the observational data and the laboratory infrared spectra for the substances in question becomes possible. Fig. 9.44 shows the result of such a comparison, with the theoretical curve displaced by an arbitrary amount on the ordinate scale. The observational points are from Sellgren et al. (1985). Table 9.6 lists the wavelengths of the principal absorption peaks in the astronomical data.

TABLE 9.6

Wavelengths of Principal Absorption Peaks (μm)

NGC 7027, NGC 2023:		3·3	3·4		6·2	7·7	8·6	11·3
Mixture of Aromatics:	2·9	3·3	3·4	5·25	6·2	7·7	8·8	11·3

The agreement of the positions of the peaks and the relative strengths of the bands are broadly satisfactory. Departures from the observations seen over the short and long wavelength ends in NGC 7023 are not serious and could easily be removed by truncating the temperature distribution at a suitable extremal value. Minor modifications of the integrated spectrum including a suppression of unwanted weaker features (e.g. the 2.9μm band) might be possible by including the effects of thermal modification of the aromatic ensemble.

For very small grains comprised of biological aromatic molecules to be a viable model for the diffuse bands we require their ultraviolet spectrum to peak at $\lambda \approx 2175$ Å for the reason we have already discussed. For the set of aromatic molecules considered earlier, we plot in Fig. 9.45 the distribution of main ultraviolet peaks in the 1900 – 2600 Å wavelength interval. The solid curve is an average absorption curve computed on the assumption that each molecule in the ensemble has an absorption profile similar to the quinazoline isomers (see Fig. 8.7). This profile to a good approximation may be represented by

$$\tau(\nu) = \text{const}\,\frac{\gamma^2}{(\nu-\nu_0)^2 + \gamma^2}$$

$$\gamma = 0.2\mu\text{m}^{-1} \text{ for } \nu < \nu_0 \qquad (9.17)$$

$$\gamma = 0.35\mu\text{m}^{-1} \text{ for } \nu > \nu_0\,,$$

where $\nu_0 = 4.6\mu\text{m}^{-1}$. We note that this curve does indeed peak at $\lambda \simeq 2200$ Å. With an estimated mean mass extinction coefficient of $\sim$ 500,000 cm^2 g^{-1} at $\lambda = 2200$ Å for our organic mixture, we find that $\sim$ 6% of the interstellar C is required to be in such a form so as to provide an interstellar absorption amounting to $\sim$ 2 mag/kpc at the band centre. Thus the mass requirements deduced from both ultraviolet and infrared data would appear to be satisfactorily consistent in the biological model.

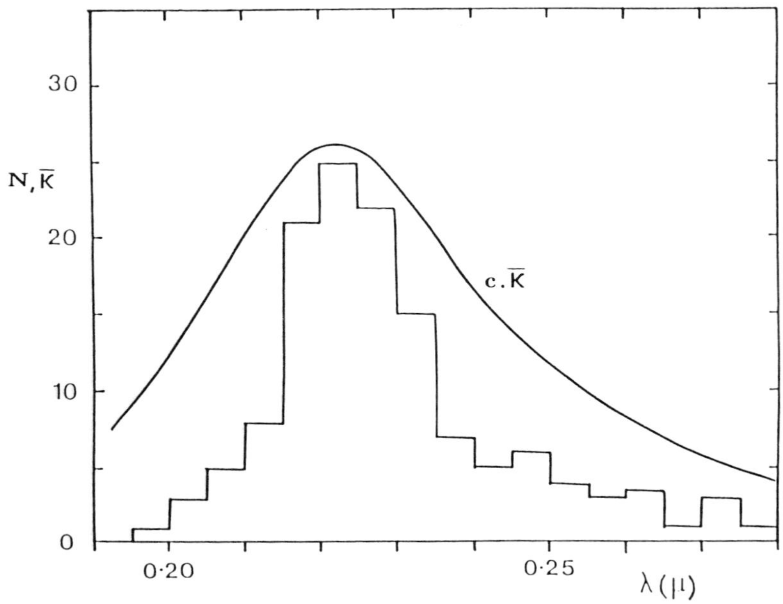

Fig. 9.45 *Histogram shows the distribution of principal ultraviolet absorption peaks for a set of naturally occurring aromatic molecules of types listed in Table 9.5. The curve is the calculated mean absorption profile for the ensemble.*

9.9. AROMATIC MOLECULES AND THE DIFFUSE OPTICAL BANDS

In Chapter 3 we noted that over 20 diffuse interstellar absorption features occur at well–defined wavelengths in stellar spectra over the visual waveband extending from ~ 4400 Å to ~ 7000 Å. These bands generally correlate in their relative strengths one with another and also with other indices relating to extinction such as E(B–V). The indications are that the causative agent is in some way connected with grains but is not necessarily within them. The diffuse bands are characterised by a spread of widths and strengths, the strongest and broadest being a 30 Å wide feature centred on λ 4430 Å. This band is the most extensively studied feature and is found to occur in stellar spectra of nearby external galaxies as well. Comparable in strength is one centred on λ 6180 Å with a bandwidth of ~ 25 – 40 Å. The majority of the diffuse bands are, however, much weaker and have widths in the range 2 – 4 Å.

Attempts to explain these features on the basis of inorganic models have been singularly lacking in success (see for instance, Wilson, 1964). Even for the narrowest of the diffuse bands the observed widths are too large for atomic lines or lines that can be attributed to small molecules. Whilst we ourselves had discussed the possible

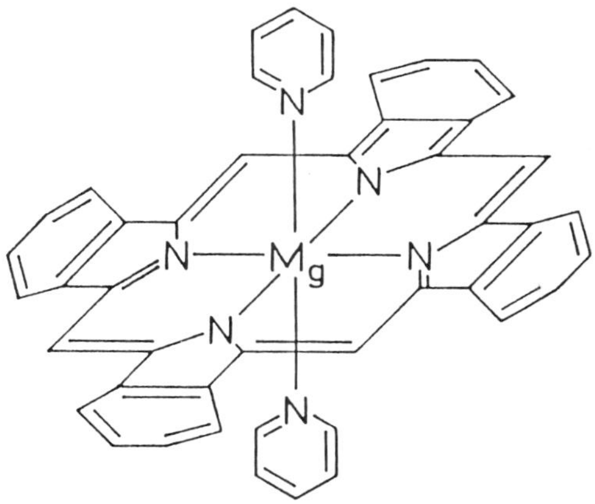

Fig. 9.46 *Structure of dispyridylmagnesium tetrabenzoporphine.*

role of lattice interactions in broadening absorption lines arising from impurity atoms and molecules, an appropriate universally occurring host–impurity system continued to elude us (Hoyle and Wickramasinghe, 1967).

Against such a backdrop of failure F.M. Johnson's (1967) attempt to identify the diffuse band carrier with an aromatic macromolecule is worthy of praise (see also Johnson, 1972; Johnson et al, 1973). The molecule proposed was in fact a porphyrin, dispyridyl magnesium tetrabenzoporhine ($MgC_{46}H_{30}N_6$) which belongs to the class of heterocyclic aromatic molecules. The structure of this molecule is depicted in Fig. 9.46. It is worth noting that Johnson's suggestion predated the radioastronomical discoveries of complex organic molecules by a few years and the now extensive discussions of polyaromatic hydrocarbon molecules (PAH's) by over two decades. It is therefore no wonder that Johnson's ingenious ideas of the 1960's scarcely received any attention and now seems to be largely forgotten. We shall point out here that the porphyrin model for the diffuse bands can now be placed in proper context and has indeed a good deal to commend it.

Firstly, let us note that porphyrins have a unique importance in living systems and therefore have a profound relevance to the biological grain model we have already discussed. Chlorophyll (a porphyrin) plays a crucial role in photosynthesis and cytochrome (also a porphyrin) is vital for respiration. Porphyrins (including non–magnesium metalloporphyrins) have predominantly planar configurations and are endowed with a high degree of thermal stability. The particular molecule considered by Johnson is reported to have a "sublimation" temperature approaching 900° K, but the dissocation temperature in vacuum could be even higher. If such heterocyclic aromatic molecules arise from the break–down of larger biological grains they are likely to have a long–term persistence under interstellar conditions.

Even a cursory glance at published spectral data of chlorophylls, metalloporphyrins and related pigments shows the possibility of the strongest visual absorptions occurring at wavelengths that overlap with λ 4430 Å and some of the other diffuse interstellar features (Hoyle and Wickramasinghe, 1979; Stern and

Timmons, 1970). Although the apparent clustering of measured absorption wavelengths close to 4430 Å was an encouraging aspect at the outset the laboratory techniques that are normally employed had introduced several limitations to modelling. In most instances spectroscopic data is secured at room temperatures for samples dispersed in a variety of solvents. Such procedures in general introduce significant wavelength shifts (upto 10 Å) as well as a merging of individual lines into relatively broad bands.

F.M. Johnson's pioneering contributions followed essentially from the development of an experimental technique that overcame some of these difficulties, a technique originally used by Shpol'skii (1962) for obtaining detailed spectra of organic molecules. The sample is suspended in an appropriately chosen paraffin matrix and cooled to 77° K. The technique effectively yields high resolution spectra which are comprised of sharp lines rather than broad bands and also permits an analysis of vibrational levels of the absorbing molecule

TABLE 9.7

Laboratory Spectra of $MgC_{46}H_{30}N_6$ Compared with Diffuse Interstellar Bands (Adapted from Johnson *et al*, 1973) [f ≡ strong flourescence lines; a ≡ strong absorptions]

Laboratory data for MgTBP		Astronomical data	
Central wavelength (Å)	Range (Å)	Central wavelength (Å)	Range (Å)
6663 (f)	1–2	6661	1–2
6614 (f)	1–2	6614	1–2
6289 (f)	1–2	6284	4
6284 (f)	1–2		
6174 (a)	14	6175	~ 20
4428 (a)	40	4428	~ 30

Table 9.7 lists the principal wavelengths for the strong bands that are found in both absorption and fluorescence spectra for $MgC_{46}H_{30}N_6$. It is most remarkable that two of the strongest diffuse interstellar bands and 3–4 others are immediately matched by this model to a high degree of precision. Johnson also obtained infrared spectra of the same molecule in KBr pellets to construct a detailed energy level diagram based on experimental data of the electronic (0,0) states corresponding to wavelengths 6175 Å (state A) and 4428 Å (state B).

Fig. 9.47 shows Johnson's energy level configurations in which he utilises vibrational frequencies γ = 231, 280, 721, 753, 1051, 1110 cm^{-1} together with their harmonics and combinations. Johnson has shown that all the 25 or so diffuse interstellar features could arise from transitions involving various vibrationally excited levels. The question immediately arises as to why most of the diffuse interstellar bands originate from vibrationally excited levels. To resolve this matter Johnson invoked an interstellar chemistry that led to the formation of molecules

Fig. 9.47

Johnson's energy level diagram for disphyridylmagnesium tetrabenzoporphine compared with the wavelengths of the diffuse interstellar bands.

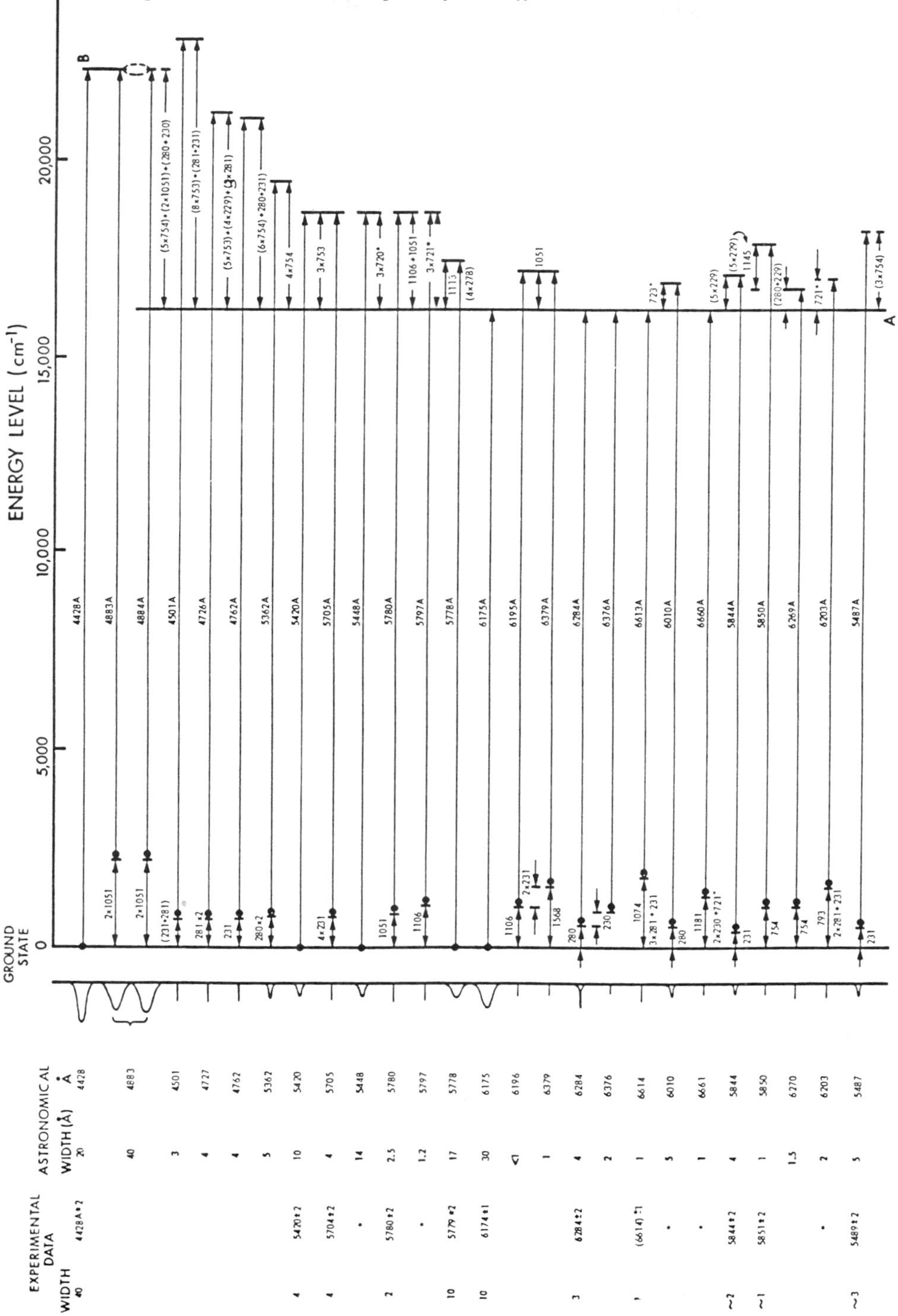

preferentially in excited states. It is more likely, in our view, that spike–heating of aromatic clusters by UV photons, as discussed earlier (Chapter 5 and section 9.8), would lead to a distribution of internal temperatures that in turn serves to populate vibrational states. Optical photons encountering such vibrationally excited molecules could readily lead to the absorption spectra as calculated by Johnson. On the basis of such a model we expect the transitions corresponding to vibrationally excited levels to be comparatively weaker as well as narrower, as indeed they are found to be.

References

Aitken, D.K., Roche, P.F., Spenser, P.M. and Jones, B., 1979, *Astron. Astrophys.*, **76**, 60.

Aitken, D.K., 1981, in C. G. Wynn–Williams and D. P. Cruikshank (eds.) *Infrared Astronomy*, D. Reidel).

Aitken, D.K., Roche, P.F., Bailey, J.A., Briggs. G.P., Hough, J.H. and Thomas, J.A., 1986, *Mon. Not. Roy. Astr. Soc.*, **218**, 363.

Allen, D.A. and Wickramasinghe, D.T., 1981, *Nature*, **294**, 239.

Bussoletti, E., et al., 1978, *Astr. Ap. Suppl. Ser.*, **70**, 257.

Butterworths Documentation of Molecular Spectroscopy: 1966, 1970, 1972, Butterworth Scientific Publications.

Cameron, A.G.W., 1970, *Space Sci. Rev.*, **15**, 121.

Capps, R.W., Gillett, F.C. and Knacke, R.F., 1978, *Astrophys. J.*, **226**, 863.

Cohen, M., 1980, *Mon. Not. Roy. Astr. Soc.*, **191**, 499.

Colangeli, L., Schwehm, G., Bussoletti, S., Fonti, S., Blanco, A. and Orofino, V., 1989, *ESA–SP*, **290**, 49.

Danielson, R.E., Woolf, N.J. and Gaustad, J.E., 1965, *Astrophys. J.*, **141**, 116.

Day, K.L., 1979, *Astrophys. J.*, **234**, 158.

Donn, B., 1968, *Astrophys. J. Lett.*, **152**, L129.

Dorschner, J., Friedemann, C. and Gurtler, J., 1977, *Astron. Nachr.*, **298**, 279.

Duley, W.W. and Williams, D.A., 1979, *Nature*, **277**, 40.

Duley, W.W. and Williams, D.A., 1983, *Mon. Not. Roy. Astr. Soc.*, **205**, 67.

Dyek, H.M. and Lonsdale, C.J., in C. G. Wynn–Williams and D. P. Cruikshank (eds.) *Infrared Astronomy*, D. Reidel.

Forrest, W.J., Gillett, F.C. and Stein, W.A., 1975a, *Astrophys. J.*, **195**, 423.

Forrest, W.J., Gillett, F.C. and Stein, W.A., 1975b, *Astrophys. J.*, **192**, 351.

Forrest, W.J., Houck, J.R. and Reed, R.A., 1976, *Astrophys. J.*, **208**, L133.

Giard, M., Pajot, F., Lamarre, J.M., Serra, G., Caux, E., Gispert, R., Leger, A. and Rouan, D., 1988, *Astron. and Astrophys. Let.*, **201**, L1.

Gilra, D.P., 1971, *Nature*, **299**, 237.

Hayes, J.M., 1967, *Geochim. Cosmochim. Acta*, **31**, 1395.

Hoyle, F. and Wickramasinghe, N.C., 1967, *Nature*, **214**, 969.

Hoyle, F. and Wickramasinghe, N.C., 1969, *Astrophys. Space Sci.*, **66**, 77.

Hoyle, F. and Wickramasinghe, N.C., 1969, *Nature*, **223**, 459.

Hoyle, F. and Wickramasinghe, N.C., 1977, *Nature*, **268**, 610.

Hoyle, F. and Wickramasinghe, N.C., 1977, *Nature*, **270**, 323

Hoyle, F. and Wickramasinghe, N.C., 1979, *Astrophys. Sp. Sci.*, **65**, 241.

Hoyle, F. and Wickramasinghe, N.C. and Al Mufti, S., 1982, *Astrophys. Sp. Sci.*, **86**, 63.

Hoyle, F. and Wickramasinghe, N.C., 1977, *Nature*, **268**, 610.

Hoyle, F. and Wickramasinghe, N.C., 1989, *ESA SP.*, **290**, 67.

Hoyle, F. and Wickramasinghe, N.C., Olavesen, A.H., Al–Mufti, S. and Wickramasinghe, D.T., 1982, *Astrophys. Sp. Sci.*, **83**, 405.

Hoyle, F. and Wickramasinghe, N.C., 1980, *Astrophys. Sp. Sci.*, **69**, 511.

Johnson, F.M., 1967, *Astron. J.*, **72**(3), April 1967.

Johnson, F.M., 1972, *Ann. N.Y. Acad. Sci.*, **187**, 186.

Johnson, F.M., Bailey, D.T. and Wegner, P.A., 1973, in J.M. Greenberg and H.C. van de Hulst (eds.) *Interstellar Dust and Related Topics*, D. Reidel.

Khare, B.N. and Sagan, C., 1973, *Icarus*, **20**, 311.

Knacke, R.F., Cudaback, D.D. and Gaustad, J.E., 1969, *Astrophys. J.*, **158**, 151.

Knacke, R.F., 1977, *Nature*, **269**, 132.

Knacke, R.F. and Kratschmer, W., 1980, *Astron. Astrophys.*, **92**, 281.

Kobayashi, U., Kawara, K., Sato, S. and Okuda, H., 1980, *Publ. Astron. Soc.* Japan, **32**, 295.

Léger, A., Gauthier, S., Defourneau, D. and Rouan, D., 1983, *Astron. Astrophys.*, **117**, 164.

Léger, A., d'Hendecourt, L. and Defourneau, D., 1989, *Astron. Astrophys.*, in press.

Léger, A. and Puget, J.L., 1984, *Astron. Astrophys.*, **137**, L5.

Little–Marenin, I.R., 1986, *Astrophys. J.*, **307**, L15.

Mann, A.P.C. and Williams, D.A., 1980, *Nature*, **283**, 721.

Merrill, K.M., Russell, R.W. and Soifer, B.T. (1976), *Astrophys. J.*, **207**, 763.

Moore, M.H. and Donn, B., 1982, *Astrophys. J.*, **257**, L47.

Okuda, H., Shibai, H., Nakagawa, T., Matsuhara, H., Kobayashi, Y., Kaiful, N., Nagata, T., Gatley, I. and Geballe, T., 1988, preprint.

Omont, A., Moseley, S.H., Forveille, T., Glaccum, W., Harvey, P.M. and Likkel, L., 1989, Proc. 22nd Eslab Symp. *ESA.SP.*, **290**, 379.

Pflug, H.D., 1982, Private communication.

Pflug, H.D., 1983, in C. Wickramasinghe (ed.) *Fundamental Studies and the Future of Science*, , U.C.C. Press.

Russell, R.W., Soifer, B.T. and Willner, S.P., 1977, *Ap. J. Lett.*, **217**, L149.

Russell, R.W., Soifer, B.T.and Forrest, W.J., 1975, *Astrophys. J. Lett.*, **198**, L41.

Scott, A.I., 1964, *Interpretation of the Ultraviolet Spectra of Natural Products*, Pergamon Press, Oxford.

Sellgren, K., 1984, *Astrophys. J.*, **277**, 623.

Sellgren, K., Alamandola, L.J., Bregman, J.D., Werner, M. and Wooden, D., 1985, *Astrophys. J.*, **299**, 416.

Stadler Handbook of Infrared Spectra, 1978 (Stadler Research Laboratories, USA).

Stern, E.S. and Timmons, C.J., 1970, *Electronic Absorption Spectroscopy in Organic Chemistry*, Edward Arnold, Ltd., London.

Strazulla, G., 1986, *Light on Dark Matter* (ed. F. P. Israel), D. Reidel.

Tarafdar, S.P. and Wickramasinghe, N.C., 1975, *Astrophys. Sp. Sci.*, **35**, L41.

Watanabe, I., Hasegana, S. and Kurata, Y., 1982, *Japanese Journal of Applied Physics*, **21**, 856.

Wdowiak, T.J., Flickinger, G.C., and Cronin, J.R., 1988, *Ap. J. (Letters)*, **328**, L75–L79.

Wdowiak, T.J., Donn, B., Nuth, J.A., Chappelle, E. and Moore, M. 1989, *Ap. J.*, **336**, 838–842.

Wickramasinghe, D.T. and Allen, D.A., 1980, *Nature*, **287**, 518.

Wickramasinghe, N.C., 1974, *Nature*, **252**, 462.

Wickramasinghe, N.C., Hoyle, F. and Majeed, Q., 1989, *Astrophys. Sp. Sci.*, **158**, 335.

Whittet, D.C.B., 1979, *Nature*, **281**, 708.

Whittet, D.C.B., Bode, M.F., Longmore, A.J., Baines, D.W.T. and Evans, A., 1983, *Nature*, **303**, 218.

Willner, S.P., 1984, in M. F. Kessler and J. P. Phillips (eds.) *Galactic and Extragalactic IR spectroscopy*, D. Reidel.

Willner, S.P., Russell, R.W., Puetter, R.C. Soifer, B.T. and Harvey, P.M., 1979, *Astrophys. J.*, **229**, L65.

Willner, S.P., Soifer, B.T., Russell, R.W., Joyce, R.R. and Gillett, F.C., 1977, *Astrophys. J. Lett.*, **217**, L121.

Willner, S.P. et al., 1982, *Astrophys. J.*, **253**, 174.

Woolf, N.J., 1973, in J. M. Greenberg and H. C. van de Hulst (eds.) *Interstellar Dust and Related Topics*, D. Reidel.

10. Dust in External Galaxies

10.1. INTRODUCTION

Interstellar dust is not a phenomenon in any way peculiar to our own galaxy. Photographs of external galaxies show striking evidence for dust, particularly in spiral and irregular systems (Sandage, 1961). Dust lanes often serve to delineate spiral arms, young stars and HII regions that are present in these galaxies. One of the most dramatic examples of extragalactic dust is to be seen in NGC 4594 (the Sombrero Hat, Fig. 10.1) where the galaxy is divided through its central plane by an opaque dust layer. Scarrott et al. (1987) dicovered linear polarisations of a few percent perpendicular to the dust layer near the extremities of the disk, and attributed this to scattering by grains. A galaxy such as ours viewed edge on would look like this, with a dust layer some 150 pc or so thick along its central plane. Fig. 10.2 shows the dust lanes in M51, which is a spiral galaxy viewed almost face–on. The integrated light from such galaxies shows linear polarisation, indicating the presence of aligned grains. The galaxy NGC 891 shown in Fig. 10.3a is interesting in having protruberences in the dust layer extending normal to its plane, to angular distances of about 30 sec of arc, corresponding to heights of ~ 100 pc. It can be argued that such filaments are evidence of dust being lifted by radiation pressure against the gravitational potential energy of the galaxy. There is other evidence of entire galaxies being shrouded in dust, showing that dust generated within the galaxy is somehow expelled. A polarised halo is found to exist around the exploding galaxy M82 (Fig. 10.3b), together with conspicuous dust lanes across it, indicating that the entire galaxy is shrouded in a cloud of gas and dust. This is also confirmed by observations of a $10\mu m$ absorption feature to which we shall refer again below. Several extragalactic objects also show evidence of the interstellar λ 2175 Å feature, and galaxies such as the LMC and SMC, in which an extinction law has been determined, show similarities to the galactic extinction law ($\propto 1/\lambda$) at visible wavelengths. All these observations point to the existence of grains with properties very similar to interstellar grains on an extragalactic, cosmological scale.

10.2. THE MAGELLANIC CLOUDS: LMC AND SMC

These two irregular galaxies, literally at our cosmic doorstep, offer the best scope for investigating the properties of interstellar matter in external galaxies. The Large Magellanic Cloud (LMC) is at a distance of ~ 50 kpc while the Small Magellanic Cloud (SMC) is at a distance of ~ 65 kpc. Individual stars, groups of stars an HII regions have been studied in these objects and both galaxies have been mapped in CO as well as H_2 (Young et al., 1989; Cohen et al., 1988). Fig. 10.4 shows the 30 Doradus Complex, a giant HII region some 250 pc across in the LMC, containing a total mass of some 5×10^6 M⊙ of gas, dust and young stars. Studies of HII regions in both the LMC and SMC show that abundance ratios of metals (including C, N, O) relative to H are considerably less than galactic values (Dufour, 1984). One explanation is that these elements are more efficiently locked up in grains. Spectrophotometry and polarimetry of individual stars have yielded P/E_{B-V} values generally very close to galactic values, implying similarly efficient alignment mechanisms. The electric vectors of polarization are systematically aligned across much of the Clouds, and rather remarkably shows a tendency to point to each other, possible implying the existence of a magnetic field linking the two galaxies.

Fig. 10.1 ***Sombrero Hat Galaxy NGC 4594.***

Fig. 10.2 *M51 – a spiral galaxy seen face–on.*

Fig. 10.3a *The galaxy NGC 891.*

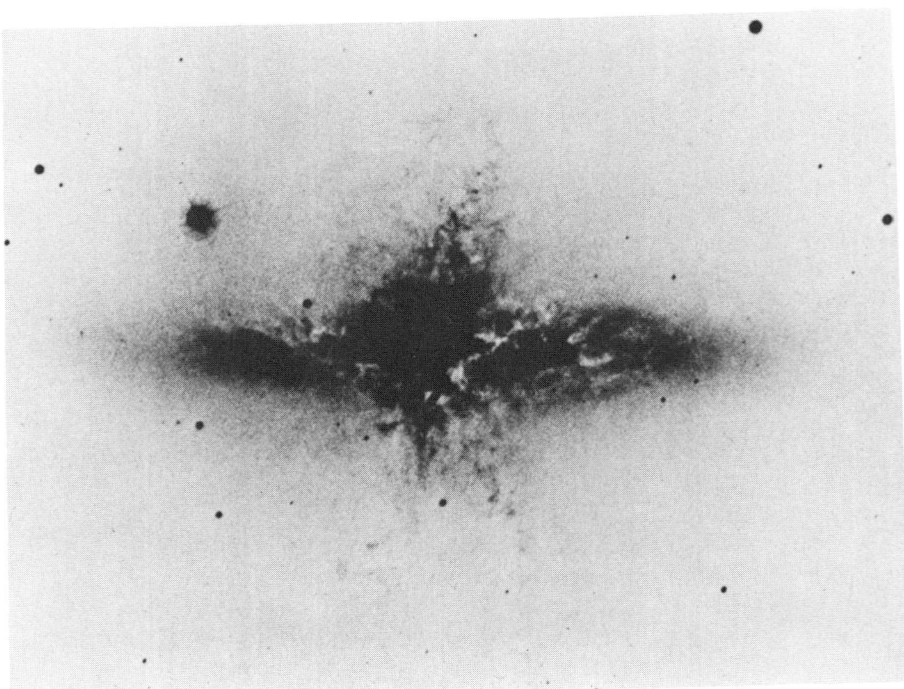

Fig. 10.3b *Exploding galaxy M82.*

Fig. 10.4 30 Doradus Complex – A giant HII region in the LMC.

From UBV photometry of the brightest stars in the Magellanic Clouds, Feast et al. (1960) found evidence to indicate an extinction law closely similar to the $1/\lambda$ galactic extinction law.

From infrared observations of early–type supergiants in both the SMC and LMC, Morgan and Nandy (1982), Kornreef 1982) and Nandy et al. (1984) discovered that the infrared extinction curve from 5500 Å to 2.2μm was essentially the same in the two Clouds and was identical to the extinction law in the galaxy. Extensions of the extinction curve into the ultraviolet were also made by Nandy and his colleagues (Nandy et al., 1982; Howarth, 1983; Morgan and Nandy, 1982). The extinction curve for these two galaxies, compiled from their published data, are plotted in Fig. 10.5a, the normalisation being to $\Delta m = 1.8$ mag at $\lambda^{-1} = 1.8\mu m^{-1}$. Also shown in the Figure is the extinction curve for the Galaxy similarly normalised. The data indicate variability in the density of the carrier of the 2175 Å feature. The LMC has less 2175 Å carriers than the Galaxy, and the SMC is probably completely lacking in them. If, as we suggest, the 3.3μm emission is generally related to the strength of the 2175 Å interstellar absorption, both being due to very small particles comprised of aromatic molecules, we should not expect to find strong 3.3μm emissions in either of these galaxies. Likewise the mid–infrared diffuse band emissions should also be reduced.

Roche et al. (1987) obtained 8 – 13μm spectra of two HII regions, one in each of the Clouds. They found a 8 – 12μm emission feature in the nebula N44A of LMC

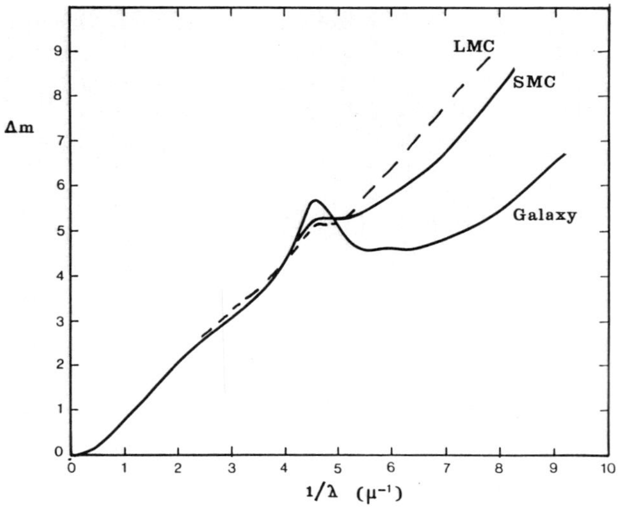

Fig. 10.5a *Normalised extinction curves for the LMC (Howarth (1983)) and the SMC from data of Nandy et al. (1982) compared with the extinction curve of the galaxy. Normalisation is to $\Delta m = 1.8$ at $\lambda^{-1} = 1.8\mu m^{-1}$.*

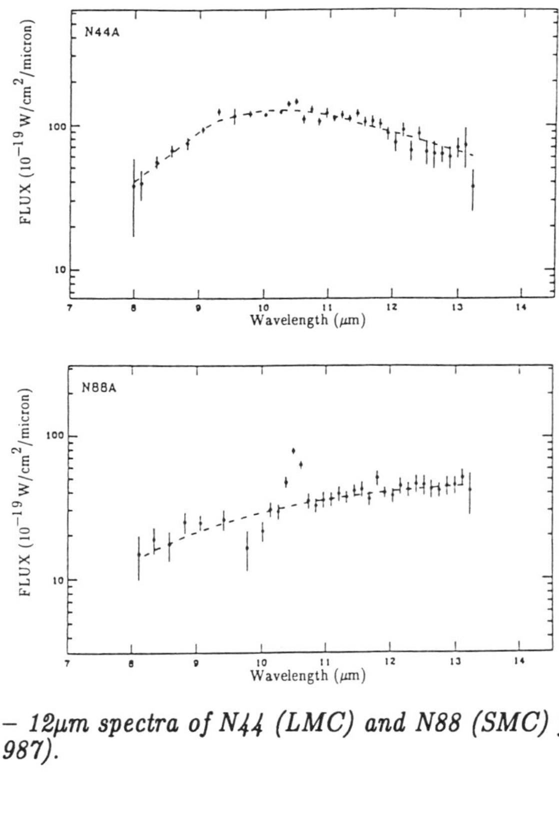

Fig. 10.5b *8 – 12μm spectra of N44 (LMC) and N88 (SMC) from Roche et al. (1987).*

but no such feature in the source N88A of the SMC (see Fig. 10.5b). The former fits well to a Trapezium emissivity model, whereas the latter is consistent with emission from carbon (or graphite) dust. This positive detection of Trapezium–type emission, which we saw in Chapter 9, is probably due to both organics and silica, is an interesting indication of the ubiquity of these materials.

Whether or not the absence of a 10μm feature in N88A is representative of the entire SMC is still unknown. Since not all HII regions in our own galaxy behave like the Trapezium nebula, a wide–ranging conclusion for the SMC would seem unwarranted at the present time

Another observation relevant to grain composition is that the fraction of carbon stars among the cool giants is higher in both the LMC and SMC than it is in the Galaxy (Blanco et al., 1978). If these stars produce 0.02μm graphite grains, the weakness of the 2175 Å feature would be a mystery. To resolve this problem it appears to be necessary to argue that graphite grain injection from carbon stars provides only a small contribution to the 2175 Å feature in the Galaxy.

10.3. M82 AND OTHER GALAXIES

The exploding star–burst galaxy M82 (classified as irregular II) is one of the most extensively studied of extragalactic objects. Apart from the visual evidence of dust in Fig. 10.3, the existence of an extended halo of dust was indicated by the polarimetric work of Elvius (1963) and Bingham et al. (1976), which suggested polarizations of ~ 30% arising from scattering by dust in a halo around the galaxy. Lynds and Sandage (1962) suggested that the polarization was due to synchrotron radiation in the filaments but this mechanism appears to be inconsistent with later

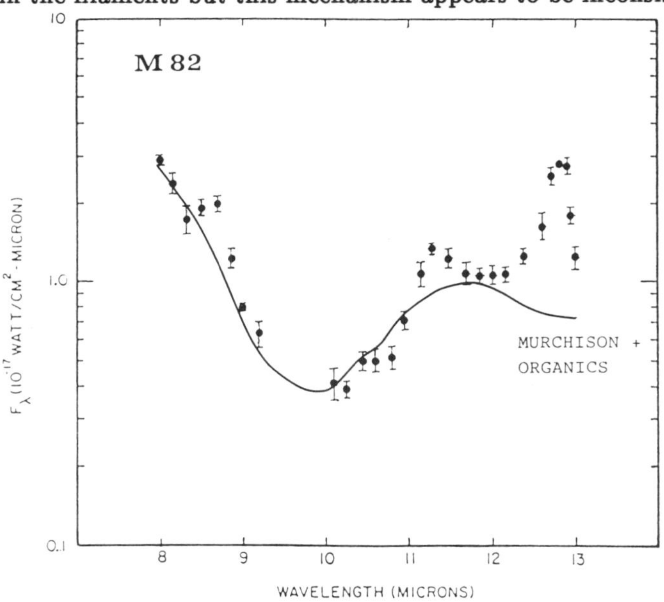

Fig. 10.6 *8 – 10μm spectrum of nuclear region of M82 (Gillett et al, 1975; Willner et al, 1977).*

observations (Visvanathan, 1974), leaving scattering by dust as the likely explanation. The dust in and around M82 could possibly have been formed in the explosion of a massive object in the central regions of the galaxy.

The presence in M82 of cool dust similar to that in our galaxy is demonstrated by the occurrence of a deep 10μm absorption feature shown by the points in Fig. 10.6 (Gillett et al., 1975; Willner et al., 1977; Jones and Rodriguez–Espinosa, 1984; Le Van, P.D. and Price, S.D., 1987). The emission feature at 11.25μm is attributed to aromatic molecules. The curve is for a model involving a mixture of crushed Murchison material, biomaterial and organics, the details of which were described by Wickramasinghe et al. (1989). There is also clear–cut evidence for the presence of the 3.3, 6.2, 7.7, 8.6, and 11.25μm diffuse emission bands from hot dust in M82, as shown in Fig. 10.7 (Gillett et al., 1975; Willner et al., 1977). These are attributed to PAH molecules, or to very small grains comprised of mixtures of aromatic molecules.

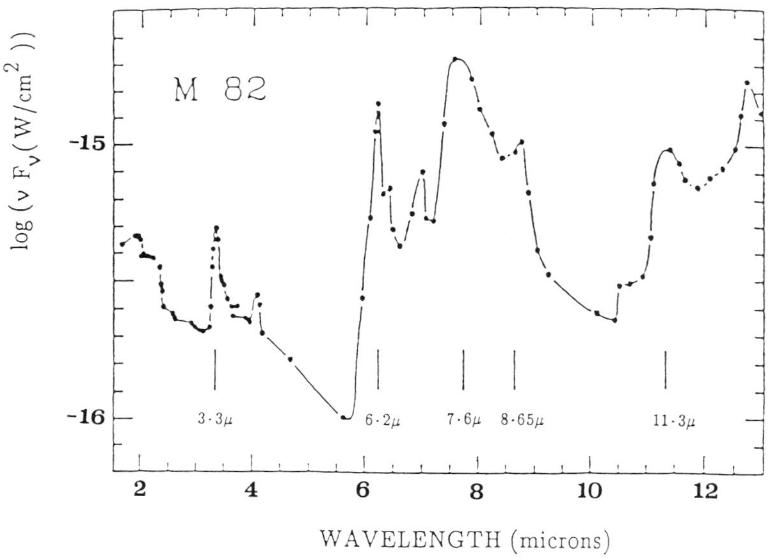

Fig. 10.7 *3 – 12μm emission from M82 (Gillett et al, 1975; Willner et al, 1977).*

Some or all of the diffuse infrared features found in M82 are observed in a number of other active galactic nuclei, Seyfert galaxies and quasars. The 10μm feature has also been observed in many extragalactic objects, including a newly–discovered class of infrared–lumious, optically–faint galaxies (Aitken et al., 1981; Roche, 1986). A particularly striking example of a deep 10μm feature in an object in this latter class is to be seen in the data for NGC 4418 (see Fig. 10.8).

The IRAS Satellite has offered many new insights concerning the distribution of dust in the universe. Before the advent of IRAS the number of known

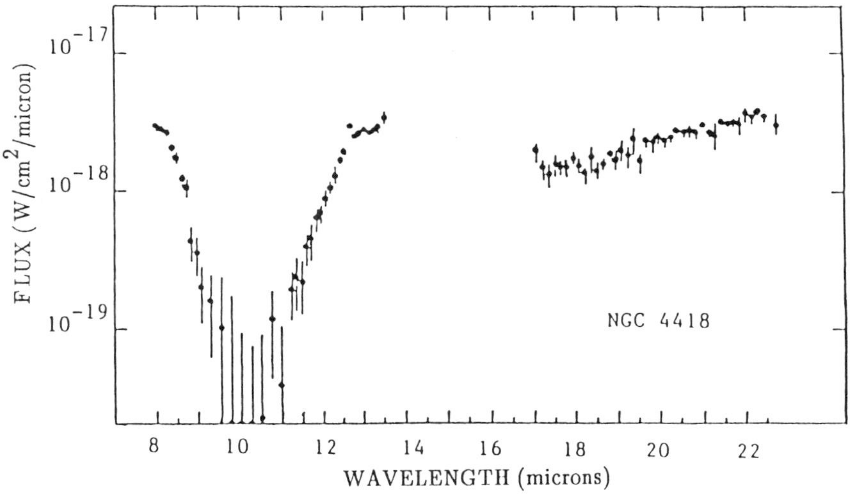

Fig. 10.8 *10 and 20μ absorptions in NGC 4418 (Aitken et al, 1981; Roche, 1986).*

infrared–bright galaxies was about 100, whereas we now have a catalogue of 20,000 such galaxies. The infrared luminosity in all cases is dominated by dust emission which includes both the effects of small grains transiently heated by UV photons, and the effect of thermal radiation from larger cool grains. The primary energy input into the dust could be variable, however. In galaxies with luminosities $< 10^{11}$ L⊙ the energy for heating the dust can be derived from OB associations and normal stars, whereas in more luminous objects the energy may well be derived from intense sources of non–thermal radiation. In many instances a significant fraction of the total energy output is in the infrared.

Figs. 10.9a and b show the infrared energy distributions for our own galaxy and for several extragalaxtic sources. Most spectra are seen to peak at $\lambda \approx 100\mu m$, corresponding to thermal emission from cool grains. But a secondary peak near $2 - 4\mu m$ is present in many galaxies, showing the effect of smaller hotter grains.

A curious result discovered from the IRAS surveys was the exceedingly tight correlation that exists between the far infrared luminosities and the radio luminosities of galaxies (Helou et al., 1986). This correlation, extending over 3 orders of magnitude of luminosity, is shown in Fig. 10.10. The correlation has, in our view, not been satisfactorily accounted for at the present time. One explanation might be based on radio emission from small spinning grains. Grains of $\sim 10^{-7}$ cm needed in our earlier model for this process (see Chapter 5) can now be identified with small aromatic grains.

Another explanation is that supernovae, which are the source of cosmic ray particles causing synchrotron emission, also produce long enough (or large enough) grains to serve as a thermalising agent, yielding high emissivity in the far infrared.

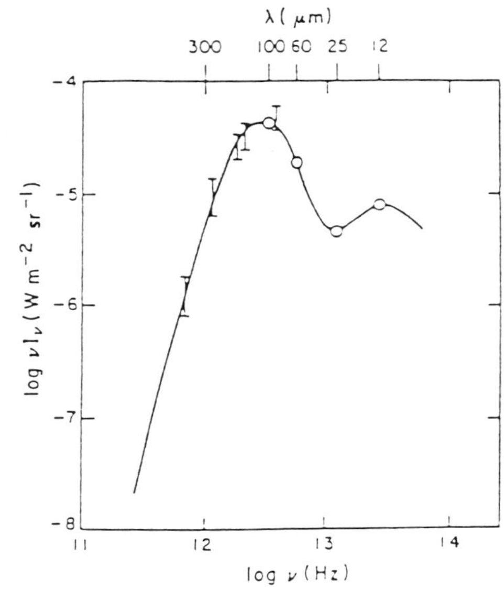

Fig. 10.9a *IR flux from our galaxy (Adapted from Beichman, 1987).*

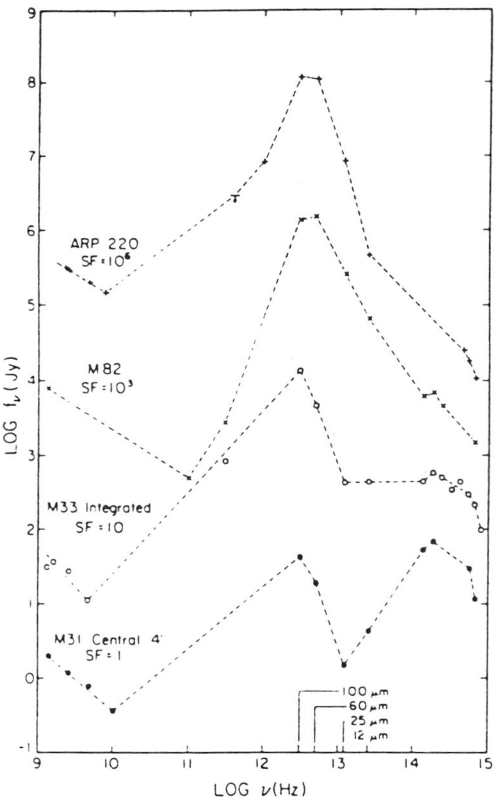

Fig. 10.9b *IR flux from several external galaxies (from Soifer et al, 1987).*

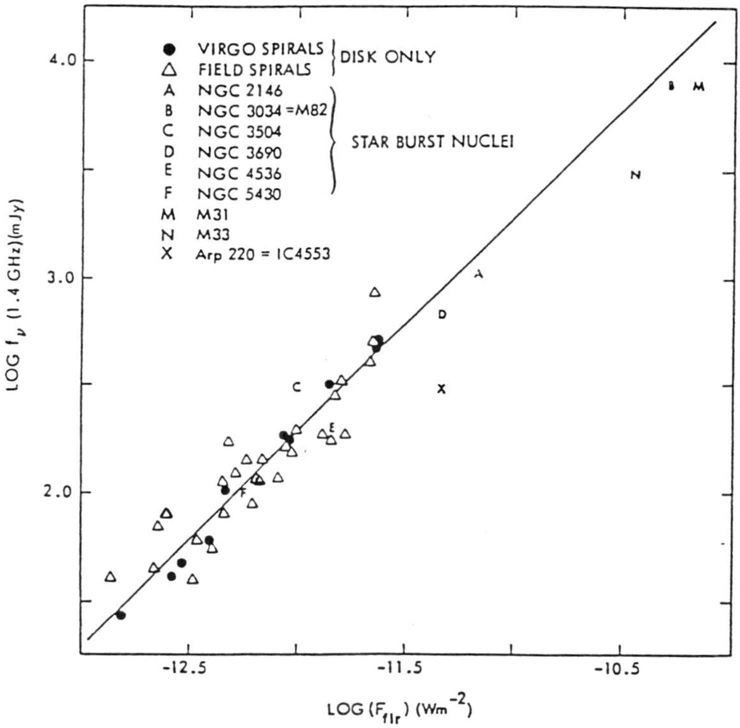

Fig. 10.10 *Correlation between far infrared and radio fluxes of galaxies (Helou et al., 1986).*

The supernova rate in a galaxy must further be correlated with the optical output from O–B associations if the effect seen in Fig. 10.10 is to be explained in this alternative way. Whether such a tight correlation of supernova rate with stellar luminosity can be justified is a matter that is as yet unresolved.

10.4. PARTICLES OF HIGH INFRARED EMISSIVITY

In view of the high luminosities of many galaxies in the far infrared it is remarkable that grain models are largely concerned with particles having only very low emissivities in this region of the spectrum. There is an element of contradiction in this situation, which can be overcome only it seems if thin threadlike metallic particles are also accorded a role.

The slow cooling of metallic vapours is found experimentally to lead to the condensation of so–called whiskers, particles that are threadlike in form with diameters characteristically of only ~ 100 Å but with lengths of about 1 mm, giving an enormous ratio of ~ 10^5 for length–to–diameter (Sears, 1957; Gomer, 1957, 1958; Dittmar and Neumann, 1958; Nabarro and Jackson, 1958). The rationale behind this phenomenon is as follows. With the onset of supersaturation of vapour pressure as cooling proceeds, condensation of small nuclei occurs, growing in the number of metal atoms per nucleus as supersaturation increases. Because of the dynamic nature of these early condensations, with atoms both joining them from the vapour and leaving them by evaporation, the situation is initially fluid. The fluid state is not that of lowest energy, however. By undergoing a cooperative transition to a crystalline state considerable energy is released at some stage, a stage which eventually defines the diameter of the ultimate particle.

The energy released within the first crystal causes dislocations, among which some will be of the helical form known as screw dislocations. These have the special property that further condensation occurs through an extension of the helix, leading to a linearly growing particle, rather than to a more or less spherically growing one. In effect, the sticking coefficient α in the case of a helically–growing particle is greater than that for a spherically growing one. Denoting the latter by β we have

$$\frac{da}{dt} = \frac{\beta}{4s} n \bar{v} m \tag{10.1}$$

for a spherical form of radius a, where n is the vapour density of atoms of mass m and $\bar{v}$ is their mean velocity, while

$$\frac{dl}{dt} \simeq \frac{\alpha}{s} n \bar{v} m \frac{1}{d} \tag{10.2}$$

for a linearly–growing particle of fixed diameter d. Not only is $\alpha > \beta$, but (10.2) gives an exponentially–growing length 1, the mass density in both (10.1) and (10.2) being s. In contrast, (10.1) gives only a linearly–growing radius, the mass increase for a spherical particle becomes much less than for helical particles as time goes on.

Those early nuclei that happen to develop a screw dislocation therefore mop up vapour at preferentially high rates, causing most of the vapour to condense into long cylindrical threads. The exponential growth in length according to (10.2) depends, however, on the sticking coefficient α being independent of 1. For this to be so, atoms adhering to the cylindrical surface of a thread must be able to move in a random–walk process to one or other of the ends in a time less than that for escape back into the vapour phase, the escape being from the adsorption potential at the cylindrical surface, not from the much stronger binding at the ends of the particle. Because the time for random walk to the ends increases as 1^2, there is a stage at which atoms evaporate before reaching either end, a stage reached abruptly when 1 has increased to about 1 mm. Thus α is set by the relation of the random–walk time to the evaporation time, the details being such as to lead to $1/d \simeq 10^5$.

Provided the electromagnetic wavelength is not too long, and provided the optical constants of the metal in question are known as a function of wavelength, Mie–type calculations for infinite cylinders can be used to obtain the mass absorption coefficients of whiskers.

Optical constants are available for the metals of main interest, chiefly iron, at wavelengths shorter than 10μm, and of course at long wavelengths greater than ~ 1 cm. The important parameter in the calculations is the conductivity σ. At the shorter wavelengths $\sigma \sim 10^{16}$ s^{-1}, while for long wavelengths and low temperatures $\sigma \sim 10^{18}$ s^{-1}, or even greater than this for pure metal. Unfortunately there does not seem to be laboratory data determining the interpolation between ~ $10^{18} - 10^{19}$ s^{-1}

TABLE 10.1

Mass absorption values ($cm^2\ g^{-1}$) for infinite cylinders for wavelengths λ and conductivities σ

$\lambda(\mu)$ \ $\sigma(s^{-1})$	1·0E16	1·0E17	1·0E18	1·0E19
0·5	7·58E4	3·76E4	1·29E4	4·17E3
1·0	1·10E5	7·27E4	2·04E4	6·44E3
10	1·71E5	9·10E5	3·06E5	6·92E4
100	1·78E5	1·70E6	7·18E6	1·64E6
1000	1·79E5	1·78E6	1·69E7	5·59E7
10000	1·79E5	1·79E6	1·78E7	1·68E8

at 1 cm and $\sim 10^{16}\ s^{-1}$ at 10μm. Somewhere between 10μm and 1 mm the conductivity rises steeply but we do not know exactly where. This difficulty can be obviated in a fair measure by drawing up a table of calculated values for various σ and various wavelengths λ.

The mass absorption values at the shorter wavelengths where $\sigma \sim 10^{16}\ s^{-1}$ is seen to be $\sim 10^5\ cm^2\ g^{-1}$. Remarkably, even if σ were appreciably higher than $10^{16}\ s^{-1}$ the mass absorption would still be $\sim 10^5\ cm^2\ g^{-1}$ or less at these wavelengths. There is no way in which a conducting particle can give more than this at visual and near infrared wavelengths. At long wavelengths, on the other hand, the mass absorption values, and hence the emissivities, increase more or less proportionally to σ, attaining enormous values in excess of $10^7\ cm^2\ g^{-1}$ at $\sigma \sim 10^{18}\ s^{-1}$ or more. At microwavelengths the absorption values and emissivities are of the order of a hundred times greater than at wavelengths less than $\sim$ 10μm.

Mie calculations for infinite cylinders eventually fail for sufficiently long wavelengths, even for whiskers with a length to diameter ratio as large as 10^5. The period ν^{-1}, where ν is the frequency, eventually becomes so long that electric charge accumulations towards the ends of a whisker build up sufficiently to cut back the flow of current within the particle, lowering the ohmic loss rate and hence reducing the absorptivity of the particles. The requirement for the validity of the above Mie calculations was found by Hoyle and Wickramasinghe (1988) to be

$$\nu > \sim \sigma \ln \frac{1}{d} \cdot \left[\frac{d}{l}\right]^2 , \qquad (10.3)$$

which for $\sigma = 10^{18}$ s^{-1}, d/l = 10^{-5} gives $\nu > \sim 10^9$ Hz, i.e., $\lambda < \sim 30$ cm. The cut–off is abrupt, implying a fall–off in the mass absorption coefficient from high values $\sim 10^7$ cm^2 g^{-1} at centimetre wavelengths to lower values in the range from ~ 10 cm down to 30 cm, and essentially zero absorptivity at still longer wavelengths.

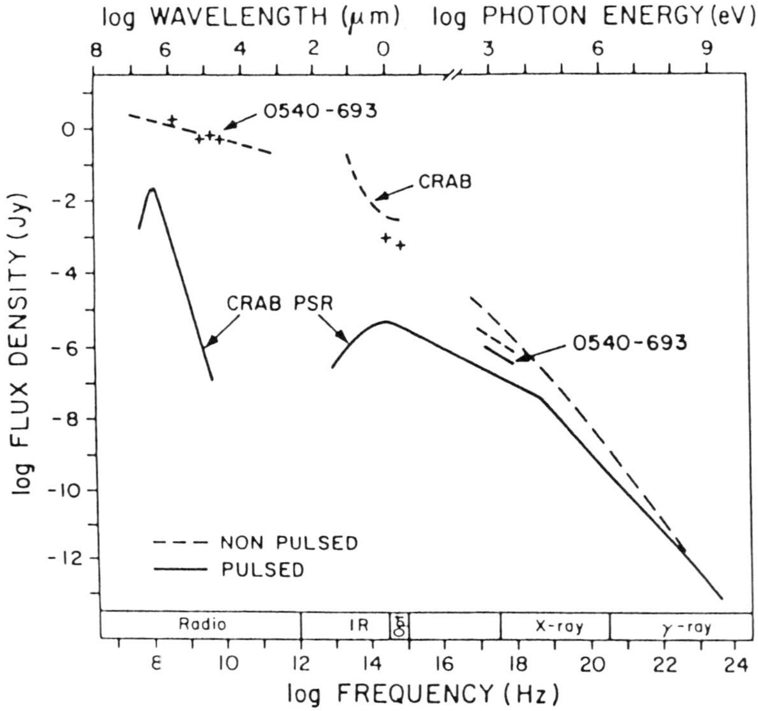

Fig. 10.11 *The spectrum of the Crab Nebula and SNR–0540–693. Solid lines show pulsed spectra, dashed lines represent steady emission. Crosses refer to data for SNR–0540–693. (Adapted from Seward et al., 1985).*

Figure 10.11 shows the emission spectrum of the Crab pulsar PSR 0531+21 (Seward et al, 1985). Unless the gap in the spectrum from ~ 10^9 to ~ 3.10^{13} Hz is an inherent property of the pulsar itself, which we would regard as unlikely, it must be attributed to absorption by the supernova ejecta still surrounding the pulsar but not surrounding the outer regions of the Crab. The gap accords well to what would be expected from metallic whiskers condensed within the ejecta.

10.5. THE EJECTION OF IRON WHISKERS FROM GALAXIES

Owing to their very high mass absorption coefficient in the far infrared, taken with the high infrared luminosities of many galaxies, there must be a marked tendency for iron whiskers to experience strong radiation pressure leading to their expulsion into extragalactic space. According to Hoyle and Wickramasinghe (1988) the escape speed is

$$\sim \left[\frac{\pi}{4} s\, d\, \frac{\text{Mass absorption coefficient x anisotropy of radiation field}}{\text{Density of interstellar gas}}\right]^{1/2} \tag{10.4}$$

For $s = 7.86$ for iron, $d = 10^{-6}$ cm, a mass absorption of 10^7 cm^2 g^{-1}, an anisotropic radiation field of $\sim 10^{-13}$ erg cm^{-3} directed away from the plane of the galaxy, and an interstellar gas density of 10^{-24} g cm^{-3}, the expulsion speed is $\sim$ 25 km s^{-1}, sufficient for the particle to be expelled from a disk of interstellar material of width 100 pc in only $\sim 10^{14}$ s. Once in the general interstellar medium, whiskers are thus expelled from the medium in only a few million years. Even in a cloud of radius 1 pc and a high density of 10^{-18} g cm^{-3}, the retention time is increased to little more than this, since in the neighbourhoods of such clouds the anisotropy of the local far infrared field is likely to be correspondingly increased. The conclusion is that whiskers cannot be retained for very long inside their parent galaxies.

Because of the decrease in the density of the ambient gas as a whisker comes clear of the disk of interstellar material there is an increase in its outward speed, essentially as the inverse square root of the ambient density. Taking the latter to be $\sim 10^{-28}$ g cm^{-3} as the confines of a galaxy are reached, the expulsion speed rises to $\sim$ 2500 km s^{-1}, or to even more for galaxies of high infrared emission. In a Hubble time of $\sim 3.10^{17}$ s whiskers would reach a distance of $\sim$ 25 mpc or more from their parent galaxies. Thus the distribution of iron whiskers in extragalactic space is averaged on the scale of superclusters of galaxies.

The very high emissivity of whiskers in the far infrared and microwave regions of the spectrum requires them to be cool, at no more than a few degrees Kelvin, when foreign atoms and molecules in the gas through which they are passing can adhere and become condensed on their surfaces. Some sputtering would occur as whiskers picked–up speed, after escaping from the galactic disk of interstellar material, but it can be argued that such sputtering would largely be confined to the removal of foreign materials, with the inner iron atoms of the particle remaining in a condensed state, i.e. with the integrity of the whisker unimpaired on account of its protective coating

10.6. THE MICROWAVE BACKGROUND

The mass absorption values of Table 10.1 show that metallic whiskers have a remarkable ability to absorb and re–emit radiation in the far–infrared and microwave regions of the spectrum. They act as exceedingly efficient thermalisers of such fields. Had their properties been appreciated in advance of the discovery in 1965 of the cosmic microwave background (Penzias and Wilson, 1965) it would have seemed natural to think that they were somehow involved with the genesis of the background. The immediate question would have been the energy density to be expected of astrophysically–generated radiation fields, rather than an immediate plunge into the supposition that the background was a relic of the manner of origin of the entire universe, a supposition which by its very nature cannot be observationally confirmed, and which by the strictest pragmatic standards of science must be regarded as a nil hypothesis, i.e. a hypothesis in which all is put in at the beginning.

It would surely have been noticed that the known average density of galactic material, determined from observation at 3.10^{-31} g cm^{-3}, implied an average helium density of 7.5×10^{-32} g cm^{-3}, and that such a helium density implied an energy production close to 4.5×10^{-13} erg cm^{-3}. (The transformation of hydrogen to helium gives close to 6×10^{18} erg g^{-1}). Such an energy production per unit volume, converted to a thermalised radiation field, yields a temperature of $(4.5 \times 10^{-13}/a)^{1/4} = 2.78°$K. Here a is the radiation constant, 7.56×10^{-15} erg cm^{-3} deg^{-4}. The closeness of this estimate to current estimates, would undoubtedly have seemed impressive, especially as the relic hypothesis gave no predicted temperature at all. In big–bang cosmology, as it has developed over the past 25 years, the temperature is a further hypothesis, adding one nil supposition to another.

The general large–scale smoothness of the background on the sky presented problems for big–bang cosmology, resolved a decade ago by the introduction of inflationary models, according to which the universe has expanded over a much greater range of scale than was permitted by earlier models, except in the case of the steady–state theory. The latter involved a similar spacetime geometry to the inflationary models, which also took over the same explanation of the flatness of the universe as had been offered in the steady–state theory. The picture became a smooth featureless background maintaining a thermalising interchange with matter so long as the matter remained sufficiently ionised. A stage was reached, however, inflation having ceased for some reason that remains not too clear, at which the temperature fell to a point where ionisation could not be sustained. Thereafter, radiation became decoupled from matter in an atomic sense. But not in a gravitational sense. Subsequent changes in the disposition of matter, especially in the formation of galaxies, would produce slight non–uniformities in the radiation field, non–uniformities that were calcutated and predicted to be present on angular scales of several arc minutes in the effective temperature of the background radiation on the sky. Unfortunately for the theory, these predicted fluctuations were not found. It therefore became necessary to modify the big–bang theory so that its predictions of fine–scale angular fluctuations were reduced. This process has by now gone through several iterations, the present situation being that forms of the theory have been devised which just manage to predict fluctuations a little below the limits set by observation. Such an iteration process cannot proceed indefinitely, and the present consensus view is that the avoidance of a confrontation of the theory with observation has now been pressed about as far as it can go.

This is not a satisfactory state of affairs, and despite the reluctance of cosmologists to think outside the patterns to which they have become accustomed, it may be worthwhile to consider the possible relevance of iron whiskers to the observed smoothness and black–body distribution of the background. The supernova rate per galaxy may be taken as 1 per 30 years. With a production of 0.1 M⊙ of iron whiskers per supernova, the amount produced in 10^{10} years would be $\sim 3.10^{7}$ M⊙, about 1 part in 3000 of the typical mass of a galaxy. Since the smoothed–out cosmological density of all galactic material is $\sim 3.10^{-31}$ g cm^{-3}, the contribution to the smoothed–out density from iron whiskers would be $\sim 10^{-34}$ g cm^{-3}, giving 10^{-6} g cm^{-3} for the content of a unit column of cosmological length, $\sim 10^{28}$ cm. Thus for a mass absorption coefficient in the range 10^{7} to 10^{8} cm^{3} g^{-1} for iron whiskers in extragalactic space, microwave radiation would experience 10 to 100 absorptions and re–emissions on a cosmological timescale, ample to ensure thermalisation. The corresponding visual opacity for a mass absorption coefficient of $\sim 10^{5}$ cm^{2} g^{-1}

remains low, however, about 0.1 on a cosmological scale. Extragalactic space also becomes effectively transparent at wavelengths longer than 10 cm, for reasons concerning whisker lengths discussed above.

Once a radiation field has become thermalised, its smoothness becomes independent of irregularities in the distribution of the thermalising agent. Smoothness depends then only on uniformity in the energy density of the field itself. Because radiation travels $\sim 10^{28}$ cm in a cosmological time scale, radiation diffuses to a considerable distance – despite being absorbed and re–emitted a number of times – from its region of origin. For as many as 100 absorptions and re–emissions the diffusion distance is as large as 300 mpc, again greatly smoothing irregularities in the origin of radiation on the scale of clusters of galaxies.

To sum up the situation: The present data on the fine–scale smoothness of the background makes it hard to suppose that the last thermalisation of the radiation occurred at an epoch before the condensation of galaxies. A commonsense interpretation of the data suggests that the microwave radiation has been thermalised at an epoch subsequent to the formation of galaxies. To understand how this might have been possible, the properties of iron whiskers appear important, and may indeed be essential.

References

Aitken, D.K., Roche, P.F. and Phillips, M.M., 1981, *Mon. Not. Roy. Astr. Soc.*, **196**, 101p

Beichman, .A., 1987, *Ann. Rev. Astron. Ap.*, **25**, 521

Bingham, R,G., McMullan, D., Pallister, W.S., White, C., Axon, D.J. and Scarrott, S.M., 1976, *Nature*, **259**, 463.

Blanco, B.M., Blanco, V.M. and McCarthy, M.P., 1978, *Nature*, **271**, 638.

Cohen, R.S., Dame, T.E., Garay, G., Montani, J., Rubio, M. and Thaddeus, P., 1988, *Astrophys. J.*, **331**, L95.

Dittmar, W. and Neumann, K., 1958 in R. H. Daramas, B. W. Roberts and D. Turnbull (eds) *Growth and Perfection in Crystals*, J. Wiley, New York.

Doufour, R.J., 1984, in van den Berg, S. and de Boer (eds.) *Structure and Evolution of the Magellanic Clouds*, K.S., D. Reidel.

Elvius, A., 1963, *Lowell Obs. Bull.*, **5**, 281.

Feast, M.W., Thackeray, A.D. and Wesselink, A.J., 1960, *Mon. Not. Roy. Astr. Soc.*, **121**, 25.

Gillett, F.C., Kleinmann, D.E., Wright, E.L. and Capp, R.W., 1975, *Astrophys. J. Lett.*, **198**, L65.

Gillett, F.C., Kleinmann, D.E., Wright, E.L. and Capp, R.W., 1975, *Astrophys. J. Lett.*, **198**, L68.

Gomer, R., 1957, *J. Chem. Phys.*, **26**, 1333.

Gomer, R., 1958, *J. Chem. Phys.*, **28**, 437.

Howarth, I.D., 1983, *Mon. Not. Roy. Astr. Soc.*, **203**, 301.

Hoyle, F. and Wickramasinghe, N.C., 1988 *Astrophys. Sp. Sci.*, **147**, 245.

Hoyle, F. and Wickramasinghe, N.C., 1988, *Astrophys. Sp. Sci.*, **147**, 245.

Jones, B. and Rodrigues–Espinosa, J.M., 1984, *Astrophys. J.*, **285**, 580.

Koorneef, J., 1982, *Astr. Astrophys.*, **107**, 247.

LeVan, P.D. and Price, S.D., 1987, *Astrophys. J.*, **312**, 592.

Morgan, D.H. and Nandy, K., 1982, *Mon. Not. Roy. Astr. Soc.*, **199**, 979.

Nandy, K., McLachlan, A., Thompson, G.I., Morgan, D.H., Willis, A.J., Wilson, R., Gondhalekar, P.M. and Houziaux, L., 1982, *Mon. Not. Roy. Astr. Soc.*, **201**, 1P.

Nandy, K., Morgan, D.H. and Houziaux, L., 1984, *Mon. Not. Roy. Astr. Soc.*, **211**, 895.

Naborro, F.R.N. and Jackson, P.J., 1958, in R. H. Daramus, B. W. Roberts and D. Turnbull (eds) *Growth and Perfaction in Crystals*, J. Wiley, New York.

Roche, P.F., 1986, in F. P. Israel (ed.) *Light on Dark Matter*, D. Reidel).

Sandage, A., 1961, *'The Hubble Atlas of Galaxies'* (Carnegie Inst. Washington).

Sears, G., 1957, *Ann. New York Acad. Sci.*, **65**, 388.

Seward, F.D., Harnden, F.R. and Elsner, R.F., 1985, in M. C. Kafatos and R, B. C. Henry (eds) *The Crab Nebula and Related Supernova Remnants*, Cambridge University Press, Cambridge.

Scarrott, S.M., Ward–Thompson, D.W. and Warren–Smith, R.F., 1987, in R. Beck and R. Grave (eds.) *Interstellar Magnetic Fields*,Springer, NY.

Sorfer, B.T., Houck, J.R. and Neugebauer, G., 1987, *Ann. Rev. Astron. Ap.*, **25**, 187.

Visvanathan, N., 1974, *Astrophys. J.*, **194**, 319.

Wickramasinghe, N.C., Hoyle, F., Majeed, Q., Al–Mufty, S. and Wallis, M.R., 1989, *ESA SP–290*, 73.

Willner, S.P., Soifer, B.T., Russell, R.W., Joyce, R.R. and Gillett, F.C., 1977, *Astrophys. J. Lett.*, **217**, L121.

Young, J.S., Xie, S., Kenney, J.D. and Rice, W.L., 1989, *Astrophys. J. Suppl.*, in press.

INDEX